全国高等卫生职业教育创新型人才培养“十三五”规划教材

供医学美容技术等专业使用

面部护理技术

主　编　熊　蕊　陈丽君

副主编　张　颖　王　艳　高　琼

编　者　（以姓氏笔画为序）

王　艳　湖北中医药高等专科学校

阮夏君　北京留指间玉健康管理有限公司

李文娟　湖北职业技术学院

杨海腾　湖北职业技术学院

张　颖　湖北职业技术学院

陈丽君　皖北卫生职业学院

徐　静　湖北职业技术学院

高　琼　孝感市中心医院

韩　慧　皖北卫生职业学院

熊　蕊　湖北职业技术学院

编写秘书

张　颖　湖北职业技术学院

華中科技大學出版社

http://www.hustp.com

中国·武汉

内容简介

本书是全国高等卫生职业教育创新型人才培养"十三五"规划教材。

本书的编写以"创新型"人才培养为指导思想，以尽量满足高职医学美容技术专业的教学需求和医学美容机构、美容企业工作岗位对医学美容技术专业人才知识、能力、素质的要求为宗旨，以实现高素质技术技能型医学美容技术专业人才培养为目标。全书共七个项目，包括概述、面部护理服务流程、面部皮肤护理、面部常见损美性皮肤护理、眼部护理、唇部护理和面部护理常用美容仪器等。

本书适合高职医学美容技术、美容美体艺术、人物形象设计等专业使用。

图书在版编目(CIP)数据

面部护理技术/熊蕊，陈丽君主编．—武汉：华中科技大学出版社，2017.1(2023.8 重印)
全国高等卫生职业教育创新型人才培养"十三五"规划教材．医学美容技术专业
ISBN 978-7-5680-1265-2

Ⅰ.①面… Ⅱ.①熊… ②陈… Ⅲ.①面-皮肤-护理-高等职业教育-教材 Ⅳ.①TS974.1

中国版本图书馆 CIP 数据核字(2017)第 000360 号

面部护理技术 熊 蕊 陈丽君 主编
Mianbu Huli Jishu

策划编辑：居 颖
责任编辑：熊 彦
封面设计：原色设计
责任校对：李 琴
责任监印：周治超
出版发行：华中科技大学出版社(中国·武汉) 电话：(027)81321913
武汉市东湖新技术开发区华工科技园 邮编：430223
录 排：华中科技大学惠友文印中心
印 刷：武汉科源印刷设计有限公司
开 本：787mm×1092mm 1/16
印 张：7.5
字 数：202 千字
版 次：2023 年 8 月第 1 版第 7 次印刷
定 价：38.00 元

全国高等卫生职业教育创新型人才培养“十三五”规划教材（医学美容技术专业）

编委会

前言

QIANYAN

为了贯彻落实《高等职业教育创新发展行动计划(2015—2018年)》,推动高职医学美容技术专业教育教学改革,培养高素质技术技能型医学美容技术人才,在总结近几年高职医学美容技术专业面部护理技术教学经验的基础上,结合美容行业职业标准,依据各医疗美容机构、美容会所岗位能力要求,分析典型工作任务,确定教学内容及各项目并编写了《面部护理技术》教材。本书是湖北省高等学校省级教学研究项目"高职医学美容技术专业顶岗实习标准的研究与实践"(项目编号:2014466)和校级课题"医学美容技术专业学分制改革的研究与实践"成果之一。

本书遵循"三基五性"的基本原则,即基本理论、基本知识、基本技能和科学性、先进性、实用性、针对性和启发性,突出学生综合职业能力的培养。教材编写具有以下特点:一是以美容行业职业标准及岗位能力要求为依据,以工作过程为导向,立足高职医学美容技术专业人才培养目标,将教学内容整合序化为七个项目:概述、面部护理服务流程、面部皮肤护理、面部常见损美性皮肤护理、眼部护理、唇部护理和面部护理常用美容仪器。在内容的安排上,以项目为主导,以技能培养为主线,理论联系实际,淡化了教材内容的纯理论性,兼顾了基础性。二是在编写体例上,针对高职高专学生基础薄弱、思维活跃等特点,注重激发学生的学习兴趣,每个项目都展示了学习目标,以便于学生目标清晰地学习并抓住学习要点;以"导学案例"为引导,提出相应的思考问题,设计教学活动,使学生通过对案例的分析来获得知识与技能,培养其分析问题和解决问题的能力;穿插知识拓展,激发学生的学习兴趣。每一个项目后面都附有能力检测题和重点内容提示。三是强调职业针对性。结合项目操作流程,以顾客为中心,模拟工作情境。在教材的编写中,充分考虑工作情境对教学过程、教学效果的影响,利用仪器、设备及案例营造具有真实工作情境(职业环境)特点的教学环境。最后,教材内容及文字简明,安排合理,详略得当,重点突出,图文并茂,充分体现了教材的实用性。

本书是高职高专医学美容技术专业及相关专业的教学用书,也可作为社会人员的培训和自学教材,还可以作为中、高级美容师职业技能鉴定辅导教材。

在教材的编写过程中得到了各位编者及相关用人单位的大力支持,在此表示衷心的感谢!由于医学美容技术专业的特殊性,加上编者水平有限,书中难免会有不足和疏漏之处,恳请广大读者谅解并予以指正。

熊蕊　陈丽君

目录

MULU

项目一　概　　述

学习目标

（1）掌握皮肤的解剖结构和生理功能及头面部按摩常用穴位。

（2）熟悉面部护理的概念和作用。

（3）了解面部护理的起源与发展。

（4）培养良好的职业美感和美容师职业素养。

项目描述

本项目主要介绍面部护理的起源与发展、概念及作用，常用头面部经络和腧穴以及皮肤的解剖结构和生理功能，使学生对面部护理技术有一个粗浅的认识，为后续面部护理知识和技能的学习打下基础。

一、面部护理的起源与发展

据史料记载，早在上古三代时期，就有"禹选粉""纣烧铅锡作粉""周文王敷粉以饰面"等美容护肤记录。两汉时期，《毛诗注疏》中说，兰，香草也，汉宫中种之可著粉，说明当时人们已能够从植物中提取、制作美容用的粉。盛唐时期，美容开始向养颜和调整皮肤生理机能方面发展。大医学家孙思邈编撰的《备急千金要方》和《千金翼方》中，除收集了各种美容保健及治疗的内服、外用方 200 余首，还专辟"面药"和"妇人面药"两篇，收集美容秘方 130 首。有很多可以治疗面部疾病和美化面容、皮肤、毛发、肢体的方剂，还提供了养生驻颜的其他方法。相传武则天曾炼益母草泽面，皮肤细嫩滋润，到了 80 多岁仍保持美丽的容貌。她的女儿太平公主曾用桃花粉与乌鸡血调和敷面，其面色红润，皮肤光滑。

宋代，人们同样注重皮肤的养护，并沿袭和发展了唐代的美容秘方，美容技术不断提高，制出了专门的珍珠膏。到了元代，一些北方游牧民族的妇女盛行"黄妆"，即在冬季用一种黄粉涂面，直到春暖花开才洗去。这种粉是将一种药用植物的茎碾成粉末，涂了这种粉可以抵御寒风和沙砾的侵袭，开春后才洗去，皮肤会显得细白柔嫩。

明代，人们用珍珠粉擦脸，使皮肤滋润。名医李时珍将医学与养生紧密结合，编撰出巨著《本草纲目》，书中记载了数百种既是药物又是食物，既营养肌肤又美化容颜的验方，仅"面"一篇中就列载 164 种之多。在所有这些美容养颜方法中，有外用的、也有内服的，药用原理主要是根据皮肤反映出来的现象或从内部调整，或从外部加以润泽或保护，既科学又很少有副作用。

清代，宫廷的美容方法集历代之大成，比较注重饮食营养，形成了一套系列化的养颜健体的独特方法。慈禧太后在美容上大下工夫，脸抹鸡蛋清，身洒西桂汁，口服珍珠粉，沐浴用人乳。

新中国成立后，随着医学、生物学、化学、物理学、营养学和遗传学的发展，使人们能从科学的角度掌握皮肤的生理及病理的内外因果关系。改革开放以后，美容行业更是欣欣向荣，科学技术的日新月异，美容仪器的使用，能够很好地完成手工难以完成的细致工作。人们对养颜护肤、健美护肤的强烈要求，也促进了美容化妆品工业的迅速发展，品种齐全、功能全面的美容护肤用品、化妆用品应运而生。这些都为更好地进行皮肤护理创造了良好的条件。

近年来，随着社会经济的发展及科技的进步，使美容业得到发展，许多新的化妆品及保养品纷纷上市，皮肤的保养更趋向于多元化。根据美容产品和专业美容护理项目的逐步延伸，面部护理采用了各种方法和技艺进行养颜驻容、祛斑除皱、增白润肤和延缓皮肤衰老。各种美容技术迅速普及，从传统的皮肤磨削、液氮冷冻、倒模面膜、化学祛斑等，到现代激光美容、间充质注射、化妆品护肤，以芳香疗法、各种现代化美容仪器的使用以及高效原生态化妆品的研制开发，美容市场日渐繁荣，取得了令人瞩目的成就。

二、面部护理的概念与作用

（一）面部护理的概念

面部护理，又称为脸部保养，是运用科学的方法，专业的美容技艺、美容仪器及相应的美容护肤品维护和改善面部皮肤，使其在结构、形态和功能上保持良好的健康状态，延缓其衰老的进程。面部皮肤护理可分为以下两类。

1. 预防性皮肤护理 预防性皮肤护理是利用清洁、去角质、按摩等护理方法来维护皮肤的健康状态。

2. 改善性皮肤护理 改善性皮肤护理是针对一些常见皮肤问题，如色斑、痤疮、老化、敏感等，利用相关的美容仪器、疗效性护肤品对其进行特殊的保养和处理，达到改善皮肤状况的效果。

（二）面部护理的作用

与预防疾病的健康观一样，预防面部皮肤的问题要比治疗容易得多。要拥有好的皮肤，保养和护理工作非常关键。在全身皮肤中，面部皮肤因环境因素而受到的损害最大，容易出现敏感、晒伤、痤疮、老化等皮肤问题。正确的皮肤护理有助于改善皮肤表面的缺水状态，可保持毛孔通畅，淡化色斑，加速皮肤的新陈代谢，减少微细皱纹，有助于延缓皮肤衰老，保持皮肤的健康状态。面部皮肤护理可以起到以下 5 个方面的作用。

1. 清洁皮肤 定期到美容院或自行做适当的面部深层清洁，能有效地清除老化的角质，有助于保持毛孔畅通，加快新陈代谢，减少痤疮的形成。

2. 预防衰老 正确的面部皮肤护理有助于抵御不良环境因素的影响，减缓皮肤老化，从而保持皮肤健康、年轻。

3. 改善皮肤 通过洁肤、按摩、润肤、防护等一系列正确的面部皮肤护理操作，可加强皮肤的保护、再生和自我修复功能，有助于改善皮肤的不良状况，如晦暗、色斑、肤质粗糙等，从而保持皮肤健康、美丽。

4. 减轻压力 面部皮肤护理过程中，正确的按摩手法、舒适的环境、轻松的音乐，美容师与顾客贴心的沟通与交流，有助于其神经和肌肉的放松，舒缓压力。

5. 增强自信 面部护理在改善皮肤不良状况的同时，不仅能促进面部皮肤健康美丽，更能增添被护理者的自信心。

三、面部护理相关医学基础知识

（一）头面部的经络和腧穴

经络是运行全身气血，联络脏腑，沟通内外，贯穿上下的通路，是人体功能的调控系统，是经脉和络脉的总称。经络系统由经脉和络脉两大系统组成。经脉是经络系统中纵行的主干，多循行于人体的深层，可分为十二正经和奇经八脉两大类；络脉是经脉的分支。经络具有联络脏腑器官，沟通表里上下，通行运气，濡养脏腑组织，感应传导和调节机体平衡的作用。

腧穴又称穴位，是分布在经络循行线上脏腑经络之气血输注于体表的部位，是内脏在体表的反应点，是美容进行穴位施术的刺激部位。腧穴通过经络与脏腑密切相连，脏腑的生理、病理变化可以反应到腧穴，而腧穴的感应又通过经络传于脏腑，因此，腧穴、经络、脏腑之间就形成了既相互联系，又相互影响的整体关系。

1. 头面部的经络 分布于头面部的经络主要包括手阳明大肠经、手太阳小肠经、手少阳三焦经、足阳明胃经、足太阳膀胱经、足少阳胆经，均是十二经脉中的阳经。奇经八脉的督脉、任脉、冲脉、带脉、阴跷脉、阳跷脉也直接与头面部联系。因督脉和任脉有专穴，故与十二经脉并称“十四经”。奇经八脉对加强人体各经脉之间的联系、调节其气血的盛溢有极其重要的作用。通过对经络的刺激，以疏通经络、调节气血、调理脏腑、濡养皮肤，使面容润泽。

2. 头面部的腧穴

(1) 攒竹：眉头凹陷处，即目内眦角直上，当眶上切迹处(图 1-1)。应用：视力保健及眼周皮肤美容的要穴之一。通过经穴按摩、刮痧，可养目，除去眼部细纹等。

(2) 睛明：目内眦上方 0.1 寸，眶骨内缘凹陷处(图 1-1)。应用：视力保健以及眼周皮肤美容的要穴之一。通过经穴按摩可养目，除去眼部细纹等。

(3) 承泣：目正视，瞳孔直下，眼球与眶下缘之间(图 1-2)。应用：面部美容养生常用要穴之一。通过经穴按摩、面部刮痧，可改善眼周发黑、眼周眼袋、眼睑水肿、眼睑跳动等。

(4) 四白：瞳孔直下，眶下孔凹陷处(图 1-2)。应用：面部养生美容常用要穴之一。通过经穴按摩、面部刮痧，可改善眼周发黑、眼周眼袋、眼睑水肿、眼睑跳动、面赤痛痒、头面疼痛、眩晕等。

(5) 巨髎：瞳孔直下，平鼻翼下缘处，鼻唇沟外侧(图 1-2)。应用：面部养生美容常用要穴之一。通过经穴按摩、面部刮痧，可改善黑眼圈、眼袋、眼部细纹、皮肤松弛、面色无华、痤疮、黄褐斑、眼睑跳动、齿痛、唇颊痛等。

(6) 地仓：口角外侧旁开 0.4 寸，上直对瞳孔(图 1-2)。面部养生美容常用要穴之一。通过经穴按摩、面部刮痧，可减轻口周皱纹、口唇皲裂、齿痛、流涎等。

(7) 迎香：鼻翼外缘中点，旁开 0.5 寸，当鼻唇沟中(图 1-3)。面部养生美容常用要穴之一。通过经穴按摩达到面部养生美容的目的。可用于治疗鼻塞、面痒、痤疮、口周皮炎等。

(8) 神庭：前发际正中直上 0.5 寸(图 1-4)。头面部养生美容常用要穴之一。通过经穴按摩、刮痧达到头部养生美容的目的。可用于治疗脱发、早白、面部水肿、瘙痒、皮屑多、头痛、眩晕、失眠等。

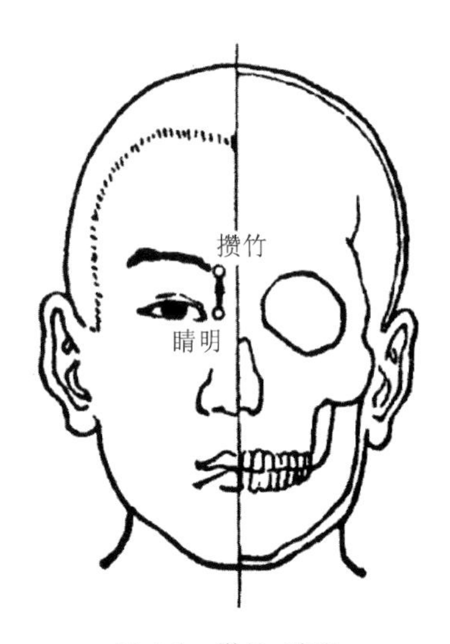

图 1-1 攒竹、睛明

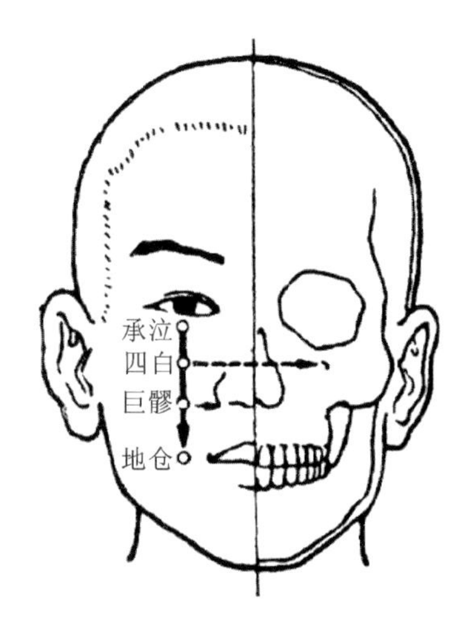

图 1-2 承泣、四白、巨髎、地仓

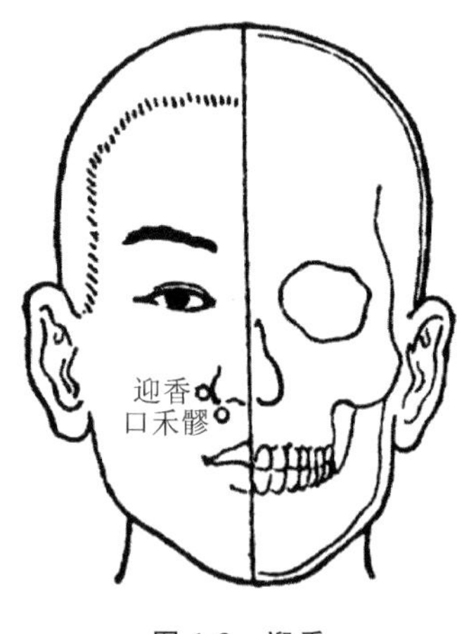

图 1-3 迎香

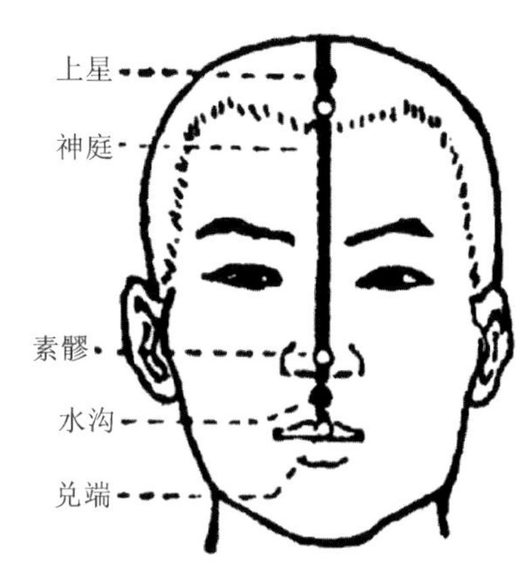

图 1-4 神庭、水沟

(9) 水沟:人中沟上 1/3 与中 1/3 交点处(图 1-4)。面部美容常用穴位,也为急救穴位之一。可用于治疗昏迷、癫痫、口角㖞斜、口周皱纹等。

(10) 印堂:位于面部两眉头连线中点(图 1-5)。面部美容常用要穴,通过经穴按摩、刮痧达到面部养生美容的目的。可用于治疗头痛、眩晕、眉弓痛、目痛、失眠、抽搐等。

(11) 鱼腰:位于眉毛中间,目正视,直对黑眼珠正中(图 1-5)。眼部美容保健穴位。通过经穴按摩达到眼部保健、美容的目的等。

(12) 球后:当眶下缘外 1/4 与内 3/4 交界处(图 1-5)。眼部美容常用经穴。通过经穴按摩达到眼部养生美容的目的。可用于治疗目疾等。

(13) 鼻通:又名上迎香,在鼻孔两侧,鼻唇沟上方(图 1-5)。面部养生美容穴位之一。可灸、经穴按摩、面部刮痧。用于治疗鼻炎、头痛、鼻塞等。

(14) 承浆:当颏唇沟的正中凹陷处(图 1-6)。面部养生美容穴位之一。通过经穴按摩、刮痧达到面部养生美容的目的。可用于治疗口眼㖞斜、口干、口臭、齿痛、流涎、癫狂等。

(15) 太阳:眉梢与目外眦之间,向后约一横指凹陷处,即眼角延长线上方(图 1-7)。面部养生美容要穴。通过经穴按摩、刮痧可改善鱼尾纹、眼袋、黑眼圈、眼睑浮肿。可用于治疗偏正头痛、面痛、目疾、口眼㖞斜、牙痛等。

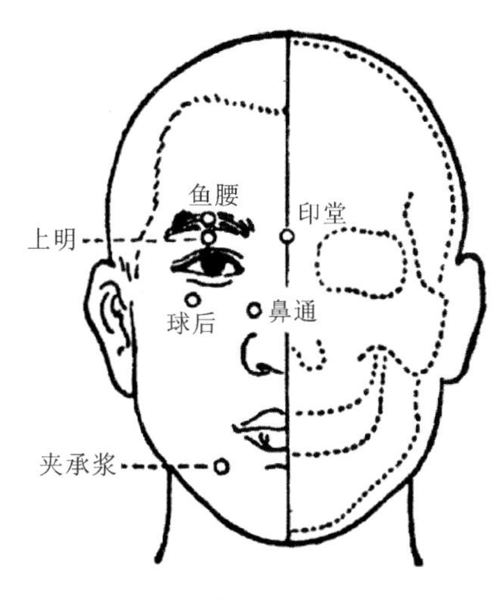

图 1-5 印堂、鱼腰、球后、鼻通

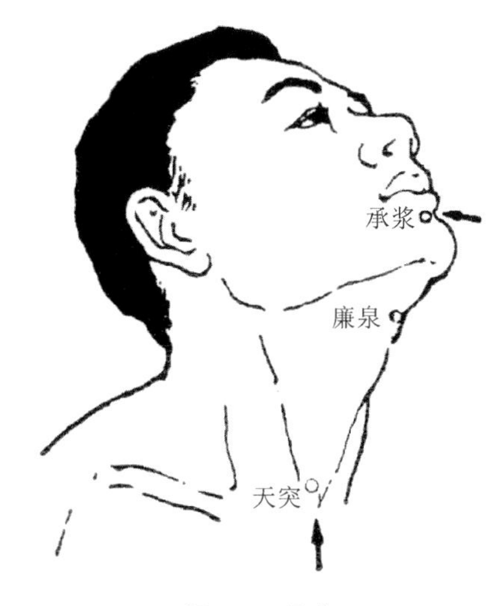

图 1-6 承浆

(16) 安眠：在项部，翳风穴与风池穴连线的中点（图 1-7）。颈部养生美容要穴。通过经穴按摩、刮痧，可减轻失眠、头痛症状，达到镇静、安眠的目的等。

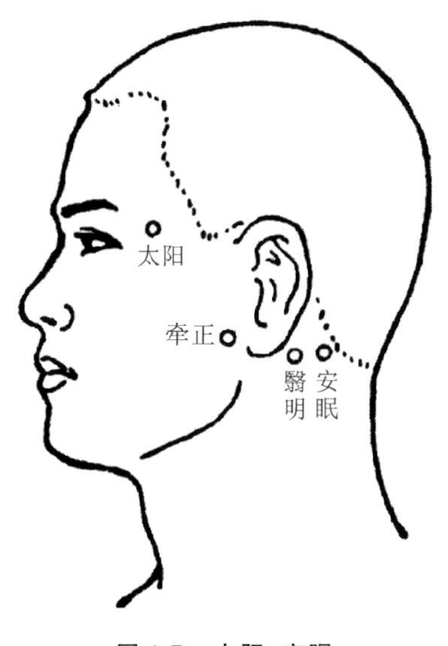

图 1-7 太阳、安眠

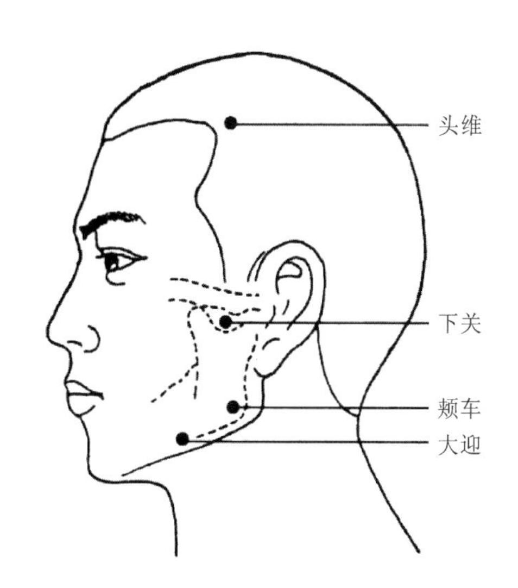

图 1-8 头维、下关、颊车、大迎

(17) 头维：在头部侧面，当额角发际线上 0.5 寸，头正中线旁开 4.5 寸处（图 1-8）。头面部养生美容常用穴位。通过经穴按摩，可减轻头痛、目眩、脱发、失眠、神经衰弱等症状。

(18) 下关：耳前方约一横指，当颧弓与下颌切迹所围成的凹陷处，合口有孔，张口闭（图 1-8）。面部养生美容常用穴位。通过经穴按摩、刮痧，可改善面部色素沉着、颜面皱纹，用于治疗齿痛、面痛、下颌关节痛、耳鸣、耳聋等。

(19) 颊车：下颌角前上方一横指（中指），咀嚼时咬肌隆起处，按之凹陷处（图 1-8）。面部养生美容常用穴位。通过经穴按摩、面部刮痧，可减少面颊皱纹，用于治疗下颌关节功能紊乱、齿痛、面肌抽搐等。

(20) 大迎：下颌角前方，咬肌附着部的前缘，面动脉搏动处（图 1-8）。面部养生美容常用穴位。可灸、经穴按摩、面部刮痧。用于治疗齿痛、颊肿、面肿、面肌瞤动等。

(21) 丝竹空:眉毛外端凹陷中。面部美容保健常用穴位(图 1-9)。通过经穴按摩、面部刮痧,可改善眼周鱼尾纹、眼袋、色斑、头痛目眩、目赤肿痛、眼睑跳动等。

(22) 角孙:耳尖直上入发际处(图 1-9)。可灸、经穴按摩。用于治疗脱发、白发、面部皱纹、颜面色斑、耳鸣、目翳、齿痛、项强等。

(23) 耳门:耳屏上切迹的前方,下颌骨髁突后缘,张口有凹陷处(图 1-9)。通过经穴按摩,可改善面部皱纹、颜面色斑,用于治疗耳鸣、耳聋、齿痛等。

(24) 翳风:耳垂后方,当乳突与下颌角之间的凹陷中(图 1-9)。头面部养生美容要穴。可灸、经穴按摩。用于治疗耳鸣、耳聋、面瘫、齿痛、腮腺炎、扁桃体炎等。

(25) 颧髎:目外眦直下,颧骨下缘凹陷处。面部养生美容常用穴位(图 1-10)。可经穴按摩。用于治疗眼睑跳动、口角㖞斜、齿痛唇肿、面肌痉挛等。

(26) 听宫:耳屏前,下颌骨髁状突的后方,张口时呈凹陷处(图 1-10)。通过经穴按摩养颜聪耳,用于治疗耳鸣、耳聋、齿痛、下颌关节肿痛等。

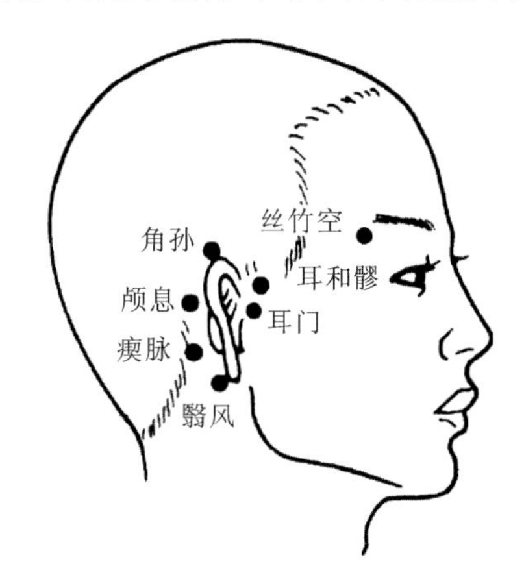

图 1-9 丝竹空、角孙、耳门、翳风

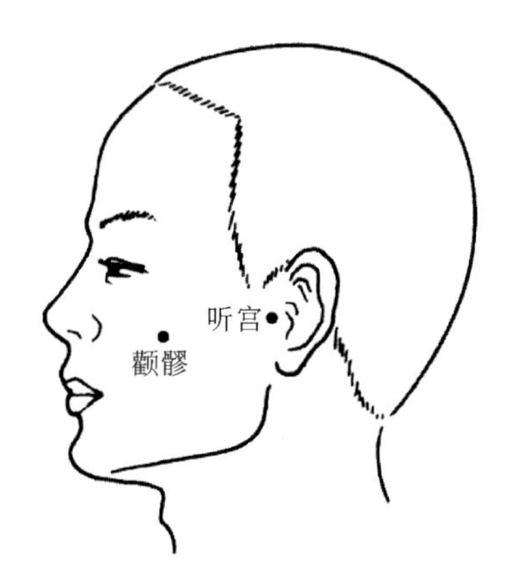

图 1-10 颧髎、听宫

(27) 上关:在耳前,下关直上,颧弓的上缘凹陷处(图 1-11)。耳部养生美容常用经穴。可灸、经穴按摩、面部刮痧。用于治疗偏正头痛、耳聋、耳鸣等。

(28) 瞳子髎:目外眦旁,当眶外缘处(图 1-11)。眼部美容养生常用经穴,通过经穴按摩、面部刮痧,可养眼除皱,去掉眼角皱纹,改善眼圈发黑;用于治疗头目胀痛等。

(29) 听会:当耳屏切迹的前方,下颌骨髁状突的后缘,张口有凹陷处(图 1-11)。耳部美容养生常用经穴,通过经穴按摩、面部刮痧,可益聪利耳。用于治疗耳鸣、耳聋、齿痛、头痛、面痛等。

(30) 阳白:目正视,瞳孔直上,眉上 1 寸处(图 1-12)。额部养生美容常用要穴。可灸、经穴按摩、面部刮痧。用于改善眼额皱纹、前额头痛、目眩等。

(31) 百会:头正中线上,前发际线正中直上 5 寸,或两耳尖连线的中点处(图 1-13)。头部养生美容要穴。经穴按摩、灸法能调护脏腑下垂。可用于治疗脱发、早白、头痛目眩、中风失语、脱肛、久泄、健忘、失眠等。

(32) 风府:位于颈部,后发际线正中直上 1 寸,枕外隆凸直下,两侧斜方肌之间凹陷处(图 1-13)。头、颈、肩部美容养生要穴,宜经穴按摩。可用于治疗感冒、头痛项强、脱发早衰、风疹瘙痒、目眩头晕等。

(33) 风池:在枕骨之下,胸锁乳突肌与斜方肌上端之间的凹陷处,与风府穴相平(图

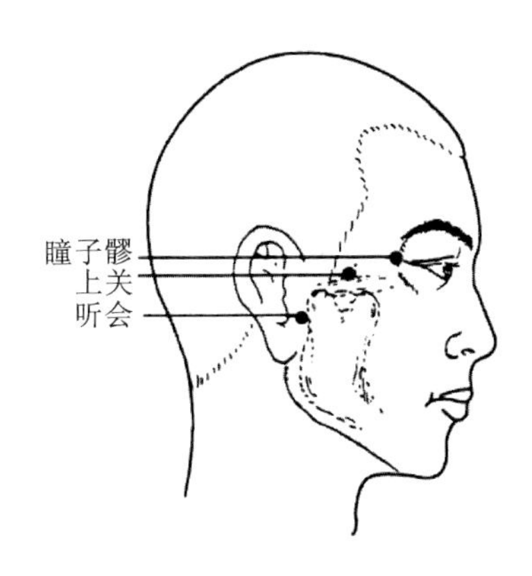

图 1-11 上关、瞳子髎、听会

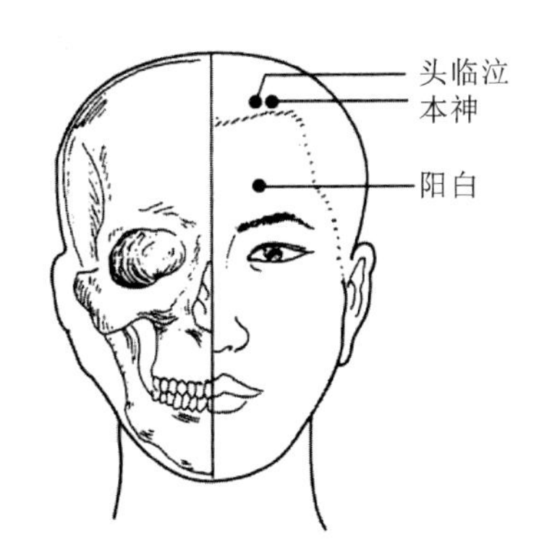

图 1-12 阳白

1-14)。可灸、经穴按摩。头颈部养生美容常用穴位。用于治疗感冒、头项强痛、发热等。

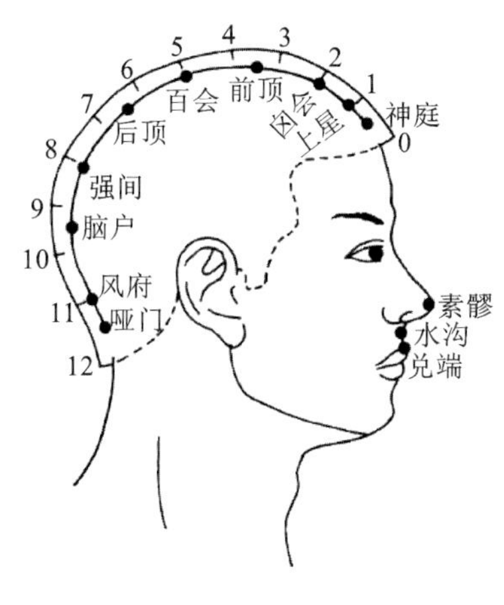

图 1-13 百会、风府

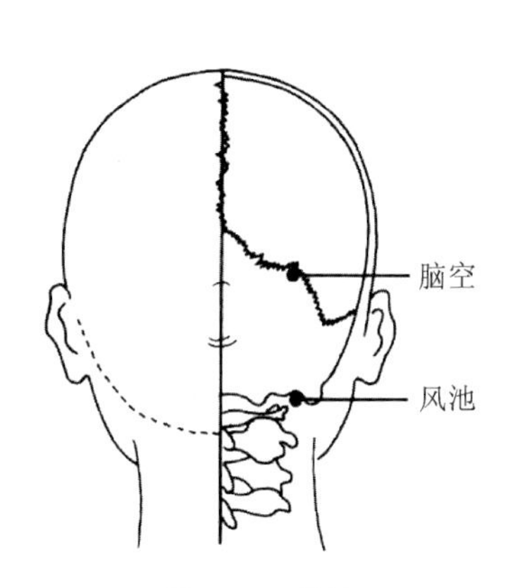

图 1-14 风池

(34) 大椎:后正中线上,第七颈椎棘突下凹陷中(图 1-15)。头面颈部养生美容常用穴位,可灸、经穴按摩、拔罐。用于治疗颜面痤疮、黄褐斑、皮肤过敏、头痛项痛、肩背痛、腰脊强

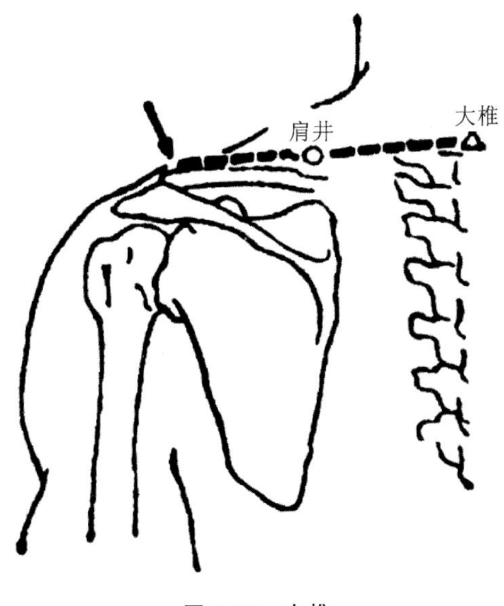

图 1-15 大椎

痛、肩颈部肌肉痉挛、感冒、发热、咳嗽等。

（二）皮肤的解剖结构

皮肤是一个多功能的器官，覆盖于体表，在口唇、肛门、尿道外口和阴道口等处与各器官的黏膜相移行，移行处称为皮肤黏膜连接。皮肤表面有许多纤细的条状凹陷称为皮沟，沟与沟之间呈平行的隆起称为皮嵴。皮沟和皮嵴相间组成皮纹，以手的掌侧和足底的皮纹尤为明显。每个人的指纹都具有独特性，不与他人相同，这是由遗传因素决定的。成年男性皮肤的面积约 1.6 m^2，女性约 1.4 m^2。皮肤的重量为体重的 14%～16%，厚度一般为 0.5～4 mm。其中眼睑的皮肤最薄，仅 0.5 mm；手掌、足底的皮肤最厚，2～4 mm。

皮肤具有移动性和延展性，这对许多美容手术提供了一个极为方便、有利的条件，如面部除皱或减皱术、瘢痕切除后的修补术、皮瓣移植术等等。皮肤的移动性与皮肤受皮下组织固定的程度密切相关，固定得多，移动范围就小，反之则大。皮肤的延展性也有一定的限度，超过限度皮肤就变形。皮肤由浅层的表皮和深层的真皮构成，借皮下组织与深部组织相连。

1. 表皮 表皮是人体最外层的皮肤，由复层扁平上皮细胞组成，其厚度因部位而异，一般厚度为 0.07～0.12 mm。表皮内无血管，但有许多细小的神经末梢，能感知外界刺激，产生触觉、痛觉、压力觉、温觉、冷觉等感觉。表皮由两类细胞构成：一类是角质形成细胞，构成表皮的主体；另一类为非角质形成细胞，数量少，散在角质形成细胞之间，包括黑素细胞、朗格汉斯细胞和梅克尔细胞。根据角质形成细胞分化的程度和层次，表皮由深层到浅层可分为基底层、棘层、颗粒层、透明层、角质层（图 1-16）。

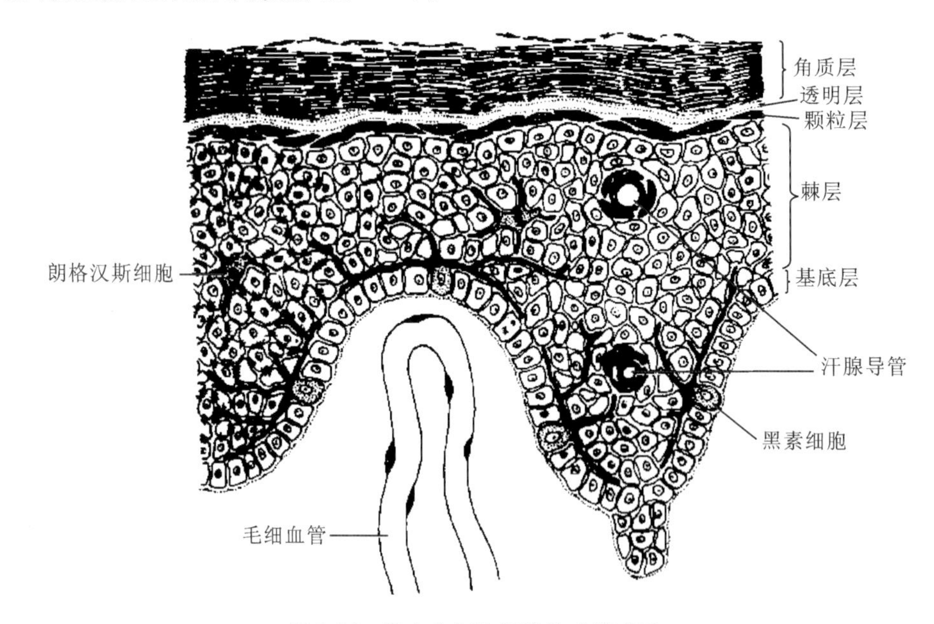

图 1-16 表皮分层和细胞构成模式图

（1）基底层：又称生发层，为表皮的最深层，借基膜与真皮的乳头层相接。由一层矮柱状或立方状细胞组成，细胞核呈椭圆形，色深，位置偏下。基底层细胞具有旺盛的有丝分裂能力，一般情况下，每天约有 10%的细胞增殖产生新的角质形成细胞，并向上有顺序移动，到颗粒层的最上层约 14 天，再移到角质层表面脱落又需约 14 天，此共 28 天时间称为表皮通过时间。基底层有感觉、增殖和抵挡紫外线对人体伤害的功能。

（2）棘层：位于基底层的上面，由 4～10 层多角形细胞组成，细胞表面有许多呈棘状的突起，故称为棘细胞。棘细胞越远离基底层，分化越好，细胞趋向扁平。胞核圆、色深而大，胞体亦大。此层细胞之间具有亲水性，利于与周围进行物质交换。此层细胞的主要功能是增强表皮的黏合能力，以适应皮肤的伸张、牵引等机械作用。该层的深部细胞也有增生分裂能力。因此，棘层有增强皮肤的黏合能力、抗牵拉、感受刺激、提供营养和防御功能。

(3) 颗粒层：位于棘层上面，由 2～5 层较厚的梭形细胞组成。胞核扁圆。最大特点是在胞质中出现了大量形态不规则的嗜碱性透明角质颗粒，常以胞吐方式排入细胞间隙，形成多层膜状结构，成为阻止物质透过表皮的主要屏障。所以，颗粒层有屏障和抵挡紫外线对人体伤害的功能。

(4) 透明层：位于颗粒层和角质层之间，由 2～3 层薄扁平细胞组成，排列呈波浪带状，此层已失去细胞结构，呈均质透明状，折光能力强。此层富含磷脂，有防止水分和电解质通过的屏障作用，故又常称屏障带。因此，透明层有屏障和抵挡紫外线对人体伤害的功能。

(5) 角质层：为表皮的最浅层，由 5～20 层薄扁平的死亡角化细胞重叠堆积而成，较坚韧，有抵御酸、碱和物理因素刺激的作用。越接近表层的细胞结合越疏松并失去弹性而脱落，同时会有新形成的角化细胞来补充。经常受摩擦部位的皮肤的角质层比较厚，如手掌、足底等处；眼睑部的角质层最薄，皮肤比较娇嫩。

角质层的厚薄对人的肤色和皮肤的吸收能力有一定影响。角质层过厚，皮肤会发黄、发黑，缺乏光泽，且吸收能力差。因此，在做皮肤护理时，每月去一次死皮，将过厚的角质细胞去除，不仅使皮肤细嫩而富有光泽，同时也可提高皮肤对营养物质的吸收能力。眼睑部角质层很薄，不能去死皮。如果脱屑次数频繁，脱屑的速度大于皮肤生长的速度，易造成皮肤的损伤。操作时力度要轻，避免损伤皮肤。角质层具有防御和屏障功能。

2. 真皮　真皮位于表皮的深面，为不规则的致密结缔组织，全身厚薄不一。此层由纤维、细胞和基质组成，并含有丰富的血管、淋巴管、神经和立毛肌等。当皮肤划伤深至真皮，会产生疼痛感觉，皮肤会出血。创伤修复过程中纤维组织大量增生，伤愈后会留瘢痕。真皮中有三种纤维，即胶原纤维、弹力纤维和网状纤维。胶原纤维呈条束状交织成网，弹性纤维多盘绕在胶原纤维束上及皮肤附属器和神经末梢周围，网状纤维是胶原纤维的前身。如果真皮中上述三种纤维减少，皮肤的弹性、韧性下降，缺乏支撑，就容易产生皱纹。

基质是无定形的均质胶状物，由成纤维细胞产生，是含有水分、电解质、黏多糖和蛋白质的复合物，也是各种水溶性物质和电解质等的代谢场所，基质有保持皮内水分、提供营养的作用。

真皮由浅部的乳头层和深部的网状层构成(图 1-17)。

(1) 乳头层：乳头层紧贴表皮深面，由薄层的结缔组织构成，大量胶原纤维和少量弹性纤维交织成网，网眼中散布着较多的细胞(以成纤维细胞为主)。此层组织突向基底层的乳头状隆起称为真皮乳头，增加了表皮与真皮之间的接触面，有利于两者的紧密结合和表皮的营养供给与代谢。此层内还可见到一种含有黑素颗粒的载黑素细胞，能吞噬黑素颗粒。乳头层的主要功能有提供营养、感受刺激、形成皮纹、调节体温等。

(2) 网状层：网状层位于乳头层深面，两者分界不明显。此层较厚，是真皮的主要组成部分，大量的粗大呈带状的胶原纤维纵横交织成网状，也有较多的弹性纤维。纤维束的排列方向与体表的张力线一致(相平行)，相邻纤维之间形成一定的角度以适应各方来的拉力。但有少数纤维垂直进入皮下组织，以便进一步固定皮肤，故称此种纤维为皮肤支持带。由于真皮内纤维多且排列特殊，故真皮具有很强的弹性和韧性。网状层具有抗牵拉、进行物质交换与

代谢、防御、再生与修复功能。

3. 皮下组织 皮下组织位于真皮和深筋膜之间，但与真皮之间界限不明显，主要由大量的脂肪细胞和疏松结缔组织构成，故而又称为“皮下脂肪层”。此层含有丰富的血管、淋巴管、神经、汗腺和深部毛囊等。具有缓冲外力冲击、保温、能量储备及参与体内脂肪代谢等功能。若皮下脂肪过度沉积，可造成肥胖，会影响人体的整体美。

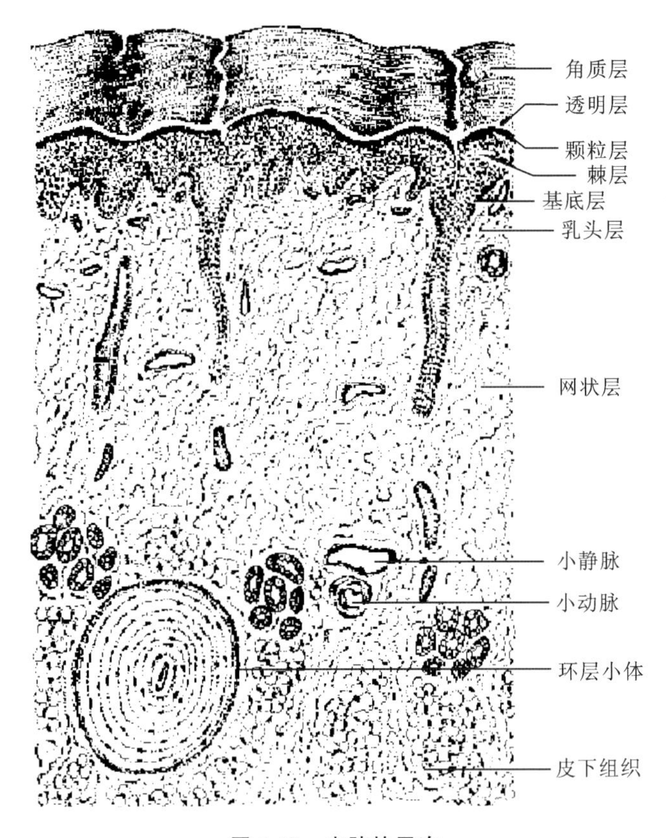

图 1-17 皮肤的层次

4. 皮肤的附属器 皮肤的附属器有汗腺、皮脂腺、毛发和指(趾)甲。

(1) 毛发：毛发由角化表皮细胞构成，可分为终毛和细毛两类。终毛包括头发、胡须、腋毛和阴毛等长毛以及眉毛、睫毛和鼻毛等短毛；细毛包括面部、躯干和四肢的体毛，又称为毳毛。

毛发露出皮肤表面的部分称为毛干，埋于皮肤内的部分称为毛根。毛根末端膨大部分称为毛球，毛球下方凹陷部分称为毛乳头。毛乳头包含结缔组织，为毛发生长提供营养(图 1-18)。毛球下层与毛乳头相接处为毛母基，是毛发的始发点和生长区。每根毛发由同心圆状的细胞排列而成，从中心到外周都可分为三部分：毛髓质，位于毛的中央，是毛发的中轴，由 2～3 层立方细胞组成，不抵达毛的顶端，细毛也无髓质；毛皮质，是毛发的主体，构成毛发的基础包裹髓质，由数层梭形细胞构成，细胞内含黑素颗粒，其含量的多少决定着毛发颜色的深浅程度，若皮质不含黑色素，则为白毛；毛小皮，包于毛皮质外周，由单层呈叠瓦状排列的透明的角化细胞构成。表皮深陷入真皮的上皮小管称为毛囊，由内层的上皮根鞘和外层的纤维根鞘构成，毛囊基部因包裹着毛球而稍显膨大，是毛发生长的场所。毛球、毛囊和毛乳头是毛发生长的三个基本结构因素。毛发具有保温、美化、协助排汗、抗摩擦和缓冲外力的功能。

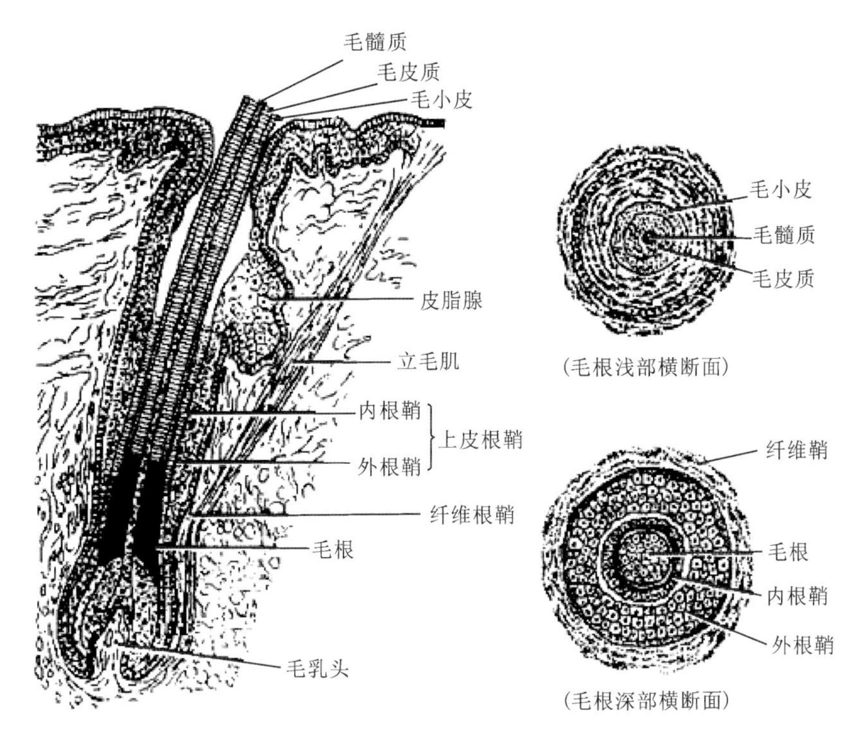

图 1-18　毛发的结构

(2) 汗腺：汗腺为人类特有的皮肤附属器。汗腺的数目随着年龄的增长明显逐渐减少，故在同样条件下老年人出汗明显减少。汗腺因部位不同而存在分布的差异，手掌、足底和腋窝的汗腺最多，随后按头皮、躯干和四肢的次序递减。红唇、阴茎头、阴蒂和小阴唇等处无汗腺(图 1-19)。

根据形态的大小、分泌方式、所在部位和结构的不同，可将汗腺分为大、小两种。

①小汗腺：一般的分泌腺，属单曲管状腺，仅分泌汗液，分布遍及全身。小汗腺包括分泌部和排泄部。分泌部构成腺管，腺管盘曲成丝球状，存在于真皮的网状层和皮下组织的浅部；其腺细胞分泌方式为局浆分泌，即分泌颗粒以出胞方式排入腺腔，故又名局泌汗腺。排泄部即汗腺管，穿过真皮，在表皮内以漏斗状的汗孔开口于体表。

②大汗腺：主要存在于腋窝、乳晕、大阴唇和阴囊等处以及耵聍腺，其形态同小汗腺，也分为分泌部和排泄部。分泌部围成腺管，但管腔较大。

汗腺的主要功能是分泌汗液，其主要成分为水、无机盐和少量尿酸、尿素等代谢产物。分泌物本无臭味，若分泌物受细菌(主要为葡萄球菌)的分解，则产生臭味物质，以腋臭为多见。

(3) 皮脂腺：皮脂腺位于真皮内的毛囊和立毛肌之间，其导管常开口于毛囊。除手掌、足底、口唇和阴茎头外，皮脂腺遍布全身，皮脂腺分泌皮脂，皮脂有润泽皮肤和毛发的作用。若腺体开口阻塞，则皮脂滞留形成皮脂腺囊肿，影响人体美，若感染还可导致疖肿发生。若皮脂腺分泌过少，不足以滋养皮肤，则皮肤会出现皱纹、干裂等现象。性激素有促进皮脂腺发育和分泌的作用，故青年人易生粉刺。雄性激素的促进作用更强，故男性皮脂分泌较女性为多。

(4) 指(趾)甲：指(趾)甲是由角化上皮增厚而成，相当于皮肤的角化层，盖于指和趾末节背侧的远侧 1/2，呈微向背侧隆起的四边形。外露的远侧大部为甲体，埋于皮内的近侧部为甲根，甲体基底部的半月形白色区称为弧影。附着于甲深面的皮肤称为甲床，由生发层和真皮构成，其中甲根深面的甲床称为甲母基，此处角质形成的细胞增生分裂繁殖旺盛，是甲的生长

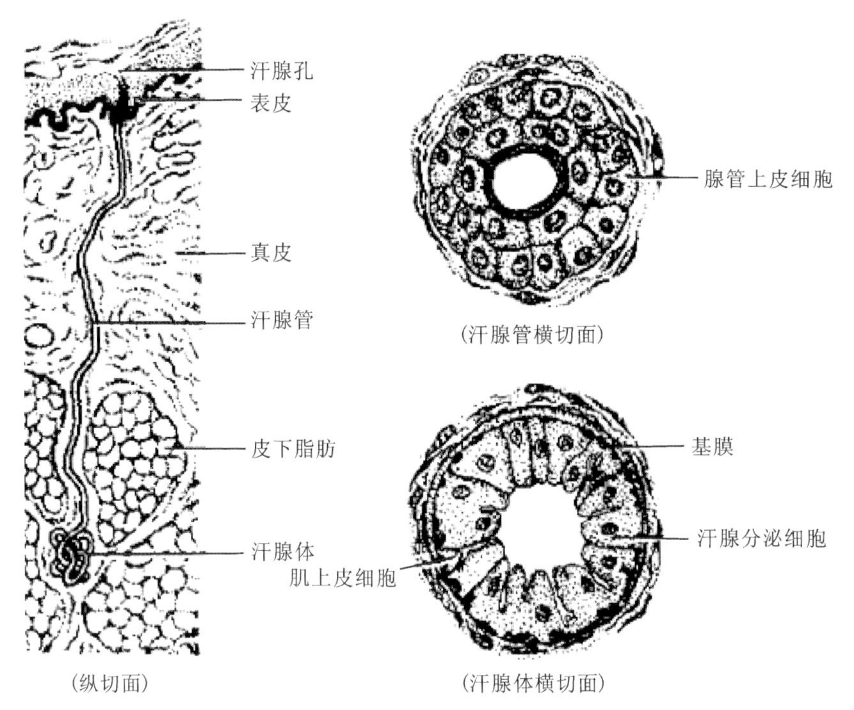

图 1-19　汗腺的结构

点，若将原甲拔去，只要保留甲母基，仍可再生新甲。甲床两侧的皮肤皱襞称为甲侧襞，甲侧襞与甲床之间的浅沟称为甲沟。指(趾)甲具有增加指(趾)间力度、保护指尖端软组织和增加局部美感的功能(图 1-20)。

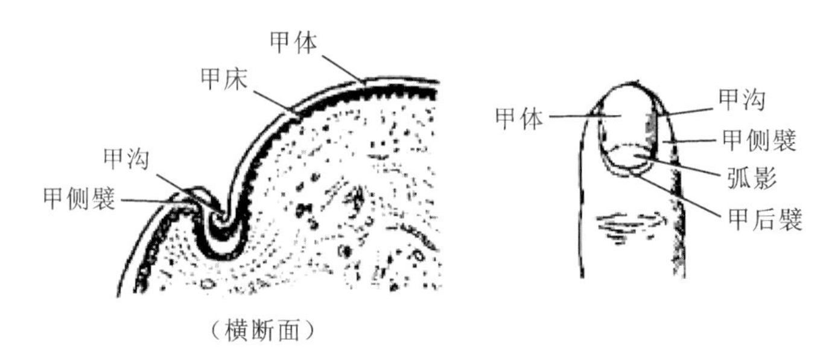

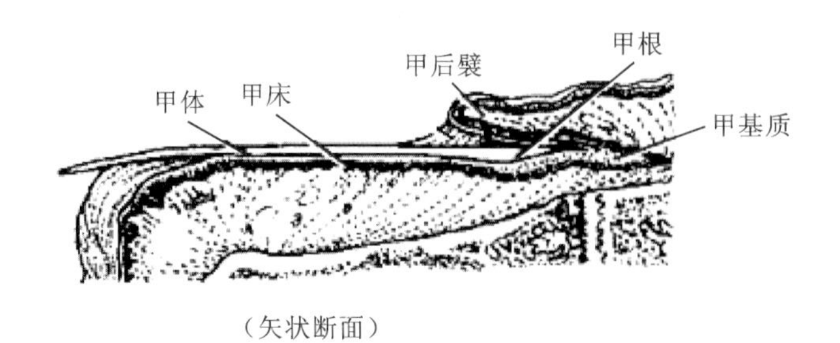

图 1-20　指甲的构造

（三）皮肤的血管、淋巴管、神经与肌肉

1. 皮肤的血管　表皮无血管。动脉进入皮下组织后分支营养皮下组织，上行至皮下组织与真皮交界处形成深部血管网，分支营养毛乳头、汗腺、神经和肌肉。

2. 皮肤的淋巴管 起于真皮乳头层内的毛细淋巴管盲端，沿血管走行，至浅部和深部血管网处，形成浅部和深部淋巴管网，逐渐汇合成较粗的淋巴管，流入所属的淋巴结。淋巴管辅助循环系统，可阻止微生物和异物的入侵。

3. 皮肤的神经 皮肤的神经分为感觉神经和运动神经两种。感觉神经来自脊神经和脑神经，为有髓神经，其末端失去髓鞘，成为游离神经末梢，接受各种刺激。近年来研究表明，神经传导的速度与神经的直径成正比，不同刺激引起的神经冲动，其传导也与此有关，如直径大于 10 μm，神经传导为 30～60 m/s，对触觉、压觉等机械性刺激传导较好；中等直径的神经纤维，传导速度为 10～20 m/s，对温度刺激传导较好；直径小于 5 μm 的神经纤维，传导速度为 1～2 m/s，与痛觉和痒觉的传导有关。运动神经来自植物神经系统，为无髓神经，分布于皮肤的血管、平滑肌和汗腺中，并调节其功能。

4. 皮肤的肌肉 皮肤的肌肉除少数横纹肌外，主要为平滑肌。平滑肌中，主要是立毛肌，收缩时毛发竖立，对皮脂排出等起着重要作用。阴囊和乳晕的肌肉，均为平滑肌，血管壁、汗腺周围也为平滑肌，面部表情肌为横纹肌。

（四）皮肤的生理功能

1. 再生功能 皮肤的再生能力很强，皮肤细胞一般每 10 h 分裂繁殖一次，晚 8:00 至次晨 4:00 的繁殖功能更为活跃，但这必须是在人体消除疲劳、机体处于正常生理功能状态下。副交感神经兴奋，皮肤血供充足，新陈代谢旺盛时皮肤再生才活跃。因此，每个人都应该充分利用每晚的这 8 个小时，使自己处于熟睡状态，即可以为上皮细胞分裂繁殖创造一个最佳的体内环境，坚持下去，有利于延缓衰老，保持皮肤的健康。

皮肤的再生可分为生理性再生和补偿性再生。

（1）生理性再生：皮肤的角质细胞不断死亡脱落，基底层细胞又不断增生分裂繁殖，使皮肤在细胞总量和生理功能上始终保持着相对稳定的动态平衡。这种为平衡生理的现象称为细胞的生理性再生。基底层细胞逐渐向浅层细胞推移演变到表层的角质细胞的周期一般约为 1 个月，即皮肤平均每个月更新一次，但体表各部之间表皮细胞更新周期并不一致，例如额部和头皮的表皮更新速度要比前臂和背部几乎快 1 倍。

（2）补偿性（修复性）再生：当皮肤受损时，表皮细胞分裂繁殖使创口愈合或将创面覆盖，使皮肤恢复其完整性，这种增生繁殖新细胞的现象称为细胞的补偿性（修复性）再生。这种再生是在血痂下进行的，故此时的血痂为完成修复的先决条件。

2. 保护功能 皮肤一方面可以防止体内水分、电解质和营养物质的丧失；另一方面可以阻止外界有害物质侵入，可使机体免受机械性、物理性、化学性和生物性等因素的侵袭，保持机体内环境的稳定。

3. 吸收功能 皮肤的吸收作用是通过表皮至真皮的渗透和腺体导管的吸收两个途径实现的，其吸收的程度与被吸收物质的性质、浓度和剂型有关，也与皮肤角质层厚度、单位面积内所含附属器数量和皮肤的含水量密切相关。水溶性物质不易被吸收，脂溶性物质如维生素 A、维生素 D、维生素 E、维生素 K 和酚类化合物、激素等易被吸收。角质层的厚薄、毛孔的状态、皮肤的温度、皮肤的含水量等都是影响皮肤吸收功能的重要因素。

4. 体温调节功能 皮肤是最重要的体温调节器官，利用很大的体表，通过蒸发、辐射、对流和传导等 4 种方式散热调温；皮肤内血管的舒缩、毛细血管的开闭和汗腺分泌的多少，均对体温调节起着重要的作用；皮肤的导热性能差，有利于保温。

5. 分泌和排泄功能 汗腺的分泌作用受环境温度和情绪变化的影响。出汗除带走大量热外，还随之排出部分新陈代谢产物，维持体内水、盐代谢平衡和减少毒素，故皮肤有“第二肾脏”的美称。

皮脂的排泄受内分泌的控制。雄激素和类固醇激素可促进皮脂排泄。皮脂呈酸性反应，汗液也使皮肤带酸性，均有抑制细菌生长的作用。皮脂还可润滑皮肤和毛发，防止皮肤干燥和皲裂。

6. 感觉功能 皮肤的真皮内有各种功能的感觉神经末梢，可感受痛、温、触、压和痒觉。若感觉神经末梢受损，则皮肤感觉便会减退或消失。

7. 呼吸功能 皮肤通过毛孔、汗孔进行呼吸，从空气中吸收氧气，排出二氧化碳。面部的角质层比较薄且毛细血管丰富，又直接暴露于空气中，因此皮肤的呼吸作用较身体其他部位更为显著。经常化妆、晚上使用较厚重的膏霜都会影响皮肤的呼吸功能，对皮肤造成损害。

本项目重点提示

（1）面部护理的概念和作用。

（2）常见的头面部按摩穴位包括：攒竹、睛明、承泣、四白、巨髎、地仓、迎香、神庭、水沟、印堂、鱼腰、球后、鼻通、承浆、太阳、丝竹空、角孙、耳门、翳风、颧髎、听宫、上关、瞳子髎、听会、阳白、百会、风府、风池、安眠、头维、下关、颊车、大迎和大椎等。

（3）皮肤的生理功能有：再生功能、保护功能、吸收功能、体温调节功能、分泌和排泄功能、感觉功能和呼吸功能等。

能力检测

一、选择题

1. 正常表皮的更替时间为（　　）。

A. 约2个月　　B. 约28天　　C. 约2周　　D. 5～6天

2. 下列不属于皮肤附属器的是（　　）。

A. 竖毛肌　　B. 皮脂腺　　C. 毛发　　D. 毛囊

3. 皮肤的生理功能不包括哪一项？（　　）

A. 抗衰老作用　　B. 呼吸作用　　C. 吸收作用　　D. 防护作用

二、填空题

1. 皮肤的生理功能有：再生功能、________、________、________、________、________、________。

2. 攒竹位于________；承泣位于________。

3. 地仓位于________，用于________；风池位于________，用于________。

三、问答题

1. 影响皮肤吸收的主要因素有哪些？

2. 简述皮肤的结构。

（熊　蕊　阮夏君）

项目二　面部护理服务流程

学习目标

（1）掌握顾客接待与咨询的流程以及皮肤分析方法，具备制订面部皮肤护理方案的能力。

（2）熟悉顾客接待与咨询的内容及顾客档案的基本内容。

（3）了解顾客档案填写的基本要求和顾客档案的制作与使用。

项目描述

本项目主要介绍顾客接待与咨询的流程、顾客档案的基本内容与要求、美容院顾客档案的制作与使用，皮肤分类与皮肤分析方法，面部皮肤护理方案的制订。学生通过本项目的学习，具备顾客接待与咨询、顾客档案制作、皮肤分析、制订面部皮肤护理方案的能力，为以后的工作做准备。

案例引导

美容师小刘，入职第一天，面对顾客不知道该如何接待和服务，十分紧张和困惑。希望得到帮助，尽快融入工作岗位。

问题：

1. 请指导小刘如何接待和服务顾客。
2. 请帮助小刘制订面部皮肤护理方案。

一、顾客接待与咨询

面对日益激烈的行业竞争，美容院怎样才能出奇制胜，赢得更多的客源呢？这关键取决于如下三点：店面形象、接待服务和技术水平，这常常是顾客选择、评定美容院的三大因素。良好的店面形象、优质的接待服务和过硬的专业技术是美容院的最佳广告，能让顾客信赖，也有利于传播美容院的声誉。

美容院工作的第一环节即为接待与咨询，前台是美容院的门面和中枢。前台接待人员热情、礼貌、耐心的咨询态度，准确科学的诊断技术，体现了美容院的技术水平和服务质量。因此，前台接待与咨询工作在美容院起着至关重要的作用。

（一）顾客接待与咨询的内容及要求

1. 顾客接待与咨询的内容 美容院一般均设有接待前台，是顾客进店后接受服务的第一场所，同时还设有接待美容师（图 2-1）。顾客接待与咨询主要包括以下几个方面。

图 2-1 美容院前台

（1）迎接顾客，参观美容院环境。

（2）介绍美容院的服务项目、服务特色及服务流程，使顾客对美容院有一个整体印象；听取顾客美容愿望，回答顾客美容方面的简单咨询。

（3）检测分析顾客皮肤状况，填写顾客资料登记表并慎重保管。

（4）为顾客制订护理方案及计划，对每次护理时间、所用产品、护理美容师等做记录。

（5）安排美容师为顾客做护理。

（6）结算顾客的美容费用。

（7）负责招呼等候的顾客。

图 2-2 美容师形象

（8）送顾客离店。

（9）接听电话，回答咨询，接受预约。

（10）负责随时与顾客保持联络，如顾客生日时致电问候、送祝福或以其他方式表示关心，如送贺卡、鲜花等小礼物。

（11）监督服务人员的表现，注意维持店面形象。

2. 顾客接待与咨询的要求

（1）形象得体：负责接待的美容师必须仪容整洁，化淡妆，身着制服或工作服（图 2-2），也可佩戴署有姓名和编号的工作牌。

（2）熟悉业务：负责接待的美容师对美容院所提供的服务项目及特点、效果、价格等应熟记在心，以便详熟地为顾客介绍。

（3）其他：精神饱满，态度诚恳，举止典雅端庄，谈吐文雅幽默，面带微笑。

知识拓展

接听电话的基本要求

(1) 及时接听电话：电话铃响两声就应当拿起，如果有事拖延，接听时应首先向对方表达歉意。

(2) 主动报名：拿起电话时应先说“您好”，接着报清美容院名称及自己的姓名。

(3) 声音亲切：声音要亲切柔和，语调和缓，语速适中，吐字清楚且面带微笑。

(4) 专心致志：听对方讲话时要专心致志，切忌心不在焉或马虎应对。

(5) 认真记录：在手边准备好纸和笔，对顾客的问题要随时记录，明白其意图。

(6) 表达清晰：讲话要清晰、有条理，口齿清楚、吐字干脆，不要含含糊糊。语言表达尽量简洁，切忌啰嗦。

(7) 有耐心：当顾客讲话比较啰嗦时，仍要耐心对待，切忌打断对方的讲话。通话结束时，应等对方先挂断电话。

(二) 顾客接待与咨询的流程

美容院接待程序一般包括：接待、咨询、分析诊断、建立档案、设计护理计划、沟通护理计划、实施护理计划、效果评价、记录与结账、送客、整理、定期跟踪回访。

1. 接待 热情迎接顾客，引领顾客入座，奉茶，了解顾客需求。

2. 咨询 了解顾客的基本情况、既往美容护理情况、生活习惯与兴趣爱好、面部皮肤基本情况、健康状况等(图 2-3)。

3. 分析诊断 分析顾客面部情况，并诊断皮肤问题(图 2-4)。

4. 建立档案 填写顾客档案，了解需要解决的问题。

图 2-3 顾客咨询

图 2-4 皮肤分析与诊断

5. 设计护理计划 包含护理目的，护理项目，护理产品和护理仪器的选择，护理疗程的确定，护理计划和家居护理方案的设计。

6. 沟通护理计划 与顾客沟通护理计划的目的、实施过程、护理效果及费用等。

7. 实施护理计划 操作前准备好用物，为顾客提供相应的护理服务。

8. 效果评价 美容师引导顾客对比护理前后的变化，询问顾客感受，确认护理效果。

9. 记录与结账 记录顾客服务项目、护理沟通内容、顾客满意度等情况，前台结算付账。

10. 送客 送顾客出门要做到“迎三送七”，即顾客来的时候上前三步去迎接，顾客走的时候送出去七步(图 2-5)。

图 2-5 送客

11. 整理 整理工作区域环境，归还物品。

12. 定期跟踪回访 新顾客在第一次护理后 1～2 天进行电话回访，让顾客感受到关爱之情。7 天左右再次电话回访，用真诚打动顾客，预约下次到店时间。还可沟通家居护理建议等。

二、顾客档案

（一）顾客档案的主要内容

填写美容院顾客档案是美容接待服务工作中一个非常重要的环节，是开展专业护理的第一步，为日后护理服务提供重要依据。美容院通过登记表所建立的详实、可靠的顾客资料库是美容院宝贵的无形资产。因此，精心设计、制作一份内容全面且合理的顾客档案尤为重要。美容院的顾客档案应全面反映顾客个人皮肤情况，包括美容史、皮肤状况、皮肤诊断结果、护肤及饮食习惯、健康状况、护理方案、效果分析、顾客意见等。顾客档案记录的内容为美容师正确地分析皮肤、制订科学的护理方案，提供了准确、详尽的信息。

1. 顾客基本情况

(1) 顾客的联系方式：如姓名、移动电话、固定电话、邮箱、通讯地址、联系时间等。

(2) 顾客的常态情况：如生日、民族、婚姻状况、生育状况、身高、体重等。

(3) 顾客的职业、爱好、性格特征等。

2. 生活习惯 顾客的生活环境、工作环境、饮食情况、运动习惯、喜好音乐、喜好颜色等。

3. 皮肤状况 皮肤类型、皮肤基本状况、皮肤问题、皮肤疾病等。

4. 身体状况 是否有过重大疾病史、手术史，是否孕期、月经期，月经情况，精神、睡眠等身体健康状况。

5. 既往美容护理情况 常用护肤品、化妆品，皮肤养护习惯、常做的护理项目及每次所购买产品名称等。

6. 护理方案与计划 本次护理方案与下次护理计划的设计。

7. 护理记录与效果评价 护理程序及方法、护肤品选择与建议、家居护理情况、护理效果对比与顾客满意度等。

8. 备注或顾客意见 记录顾客的要求、评价等。

（二）填写顾客档案的基本要求

(1) 让顾客明白填写的目的，以便积极配合。

(2) 填写字迹要清楚，不可随意涂改。顾客资料由顾客本人填写，皮肤分析由美容顾问填写或美容师通过问话形式协助填写。

(3) 填写内容要及时、真实、准确、详实，详细记录每次顾客到店的情况。

(4) 美容师需注明顾客护理过程中的注意事项，当次向顾客介绍的产品和项目，下次美容师应主推的产品和项目，顾客在护理中所关心的话题和顾客对护理的喜好，如皮肤受力程度等。

(5) 美容师需将顾客的最新资料送到前台汇总，及时更新记录，同时记录顾客每次护理后的感受，护理效果前后对比确认。

(6) 顾客档案应按照一定的顺序编辑，如按姓氏笔画、汉语拼音、制卡时间顺序或皮肤情况等，装订成册或制成电脑数据库。

(7) 档案由专人管理，电脑数据库设定密码，以防遗失泄密。

(8) 填写顾客档案要尊重其意愿，切忌强制记录。

(9) 要为顾客保守秘密，如顾客的年龄、住址或美容项目、消费金额等，不可随意让人翻看。

(三) 顾客档案的制作与使用

1. 顾客档案的制作 顾客档案封面一般包含美容院名称、顾客姓名、档案编号、填写日期等，封底一般包含美容院地址、乘车方式、联系电话等，档案具体内容见表 2-1。

表 2-1 美容院顾客档案

(1) 顾客基本情况

姓名		性别		年龄		出生年月	
籍贯		身高		体重		婚姻状态	
血型		星座		兴趣爱好			
会员卡号		职业		工作单位			
联系方式			家庭地址				

(2) 生活习惯

每周工作日	□周一 □周二 □周三 □周四 □周五 □周六 □周日
每日工作时间	上午：时 分 — 时 分 下午：时 分 — 时 分 晚上：时 分 — 时 分
工作环境	□喧闹 □适宜 □安静
工作压力	□很大 □一般 □偶尔 □没有
饮食习惯	□咖啡/茶 □煎炸食物 □高糖分食物 □烟/酒
食欲	□偏酸 □偏辣 □偏甜 □偏苦 □清淡 □油腻
睡眠习惯	□早睡(22:00 前) □晚睡(22:00 后) □贪睡
睡眠时间	□充足睡眠(超过 8 h) □睡眠不足(4～8 h)
睡眠质量	□好 □一般 □差
最合适的美容时间	□周一 □周二 □周三 □周四 □周五 □周六 □周日
	上午：时 分 — 时 分 下午：时 分 — 时 分

续表

请勿打扰时间	□周一 □周二 □周三 □周四 □周五 □周六 □周日
	上午： 时 分 — 时 分 下午： 时 分 — 时 分

（3）面部皮肤基本情况

皮肤类型	□干性 □中性 □油性 □混合性
皮肤肤色	□白皙 □黝黑 □暗沉 □偏黄 □无光泽 □色素型黑眼圈 其他：
皮肤弹性	□好 □一般 □不佳
皮肤含水量	□正常 □缺乏 □偏低
皮肤保水量	□高 □中 □差
皮肤油脂量	□旺盛 □局部过量 □正常 □缺乏
痤疮类型	□脓肿型 □结节型 □脓包型 □丘疹型 □黑头粉刺 □白头粉刺
皮肤色素分布	□额头 □面颊 □鼻梁 □下颌 □口周 其他：
皮肤色素状态	□雀斑 □晒斑 □黄褐斑 □老年斑 □色素痣 □胎记 其他：
皮肤老化状况	□松弛 □无弹性 □干瘪 □皱纹 □细纹 其他：
皮肤皱纹状态	□全面皱纹 □额横纹 □眼角皱纹 □眉间皱纹 □眼下皱纹 □鼻唇纹 □局部细纹
皮肤红血丝状况	□微血管扩张 □静脉怒张 □静脉曲张 □血管性痘印
皮肤敏感度	□特别敏感 □偶尔敏感 □从不敏感 敏感源： 敏感现象： 好发部位： 好发时间：
曾使用产品	□国内 □国外 □日化线 □专业线 □医疗线
产品类型	□补水保湿型 □美白淡斑型 □紧致抗皱型 □控油清爽型 □其他：
家居护理意识	□强烈 □一般 □没有
家居护理工作	□每天 □经常 □偶尔 □从不
最希望解决的皮肤问题	皮肤问题： 原因：
曾经是否接受过专业护理项目	□是 护理项目： 效果满意度： □否 原因：
关注面部护理的方面	□安全 □疗效 □人员专业度 □品牌 □肤色 □肤质 其他：

（4）身体健康状况

身体抵抗力	□好 □一般 □差 □很差
体型状态	□偏瘦 □正常 □偏胖 □肥胖
肥胖原因	□饮食习惯 □遗传 □荷尔蒙失调 □缺乏运动
脂肪类型	□硬脂 □软脂 □蜂窝组织 □水肿
最不满意的部位	□颈部 □胳膊 □腰部 □腿部 □臀部 □其他： 不满意原因：

续表

曾经是否做过瘦身项目	□是　瘦身部位： □否　原因：
身体亚健康症状	□神经衰弱　□失眠多梦　□皮肤敏感　□过敏体质　□头晕目眩 □体虚感冒　□颈椎疼痛　□烦躁不安　□月经失调　□胸闷气短 □食欲下降　□乳房胀痛　□小便泛黄　□疲劳无力　□胃痛便秘
排便情况	□每天定期　□隔三差五　□经常便秘
月经生理状况	□顺调　□不顺调　□轻微疼痛　□剧痛
月经生理周期	□规律　□偶尔失调　□不规律
妇科疾病状况	□严重　□较严重　□一般　□没有
曾经是否做过大手术	□是　□否
喜欢的按摩力度	□偏重　□适中　□偏轻
最希望解决的身体问题	皮肤问题： 原因：
曾经是否接受过专业护理项目	□是　护理项目：　　效果满意度： □否　原因：
关注身体护理的方面	□安全　□疗效　□人员专业度　□品牌　□肤色　□肤质　其他：

（5）美容院护理疗程设计

护理目的	
护理项目	
护理产品	
护理疗程设计	
家居护理	
家居产品	

顾问签名：　　　　美容师签名：　　　　顾客签名：

（6）消费记录

项目	次数	日期	储值金额	充值	余额	顾客签名	备注
储值记录	1						
	2						
	3						
项目	次数	日期	项目卡名称	金额	护理项目	顾客签名	备注
办卡记录	1						
	2						
	3						

续表

项目	次数	日期	产品名称	金额	数量	顾客签名	备注
购买产品记录	1						
	2						
	3						
项目	次数	日期	活动名称	金额	活动项目	顾客签名	备注
沙龙活动记录	1						
	2						
	3						

(7) 护理服务记录

日期	护理项目	护理记录	美容师签字

(8) 意见与建议

项目	非常满意	满意	一般	待改善
服务态度				
床位布置				
环境卫生				
专业形象				
操作熟练度				
护理效果				
手法力度				
专业知识				
家居护理指导				
护理产品				
意见与建议	顾客签字： 日期： 年 月 日			

2. 顾客档案的使用

(1) 美容师需了解的内容：

①顾客的基本情况：姓名、年龄、地址、职业、学历、收入、所使用产品或护理项目、生活习惯、平均月收入、兴趣爱好、感兴趣的沙龙活动等。

②通过一问一答形式和顾客的双向沟通，更准确地了解顾客需求。

③顾客皮肤护理规划以及每次护理记录。

④顾客感兴趣的话题以及在护理中的喜好和注意事项。

⑤顾客到店情况，长时间没有到店的原因，积分情况，参与活动记录情况。

⑥每次向顾客介绍的产品和项目，当次应主推的产品和项目。

(2) 前台需要与顾客联系的内容：

①提前一周给顾客送生日礼物，生日当天致电祝贺。

②定期给顾客短信或微信发送新产品资料、促销方案等。

③提醒顾客定期护理。

④建议顾客换季产品的购买与正确使用。

(3) 店长需要做到的内容：

①定期检查顾客消费和护理情况，检查美容师的服务执行情况。

②定期通过顾客档案，了解顾客的到店情况。注明长时间没有到店的顾客，了解原因并安排人定期跟进。

③及时了解顾客对店内各方面的要求和建议，以便及时改进工作。

④了解顾客感兴趣的主题沙龙活动，做出相应的活动策划。

⑤根据顾客的积分记录，准确掌握顾客的消费情况。

⑥将存在的问题在例会中和美容师沟通，以便及时调整工作方案。

三、皮肤分析

(一) 皮肤分类

人的皮肤按其皮脂腺的分泌状况，一般可分为四种类型，即中性皮肤、干性皮肤、油性皮肤和混合性皮肤。但在实际操作过程中敏感性皮肤、痤疮性皮肤也是常见的皮肤类型。各类皮肤具有各自不同的特点。

1. 中性皮肤 中性皮肤是健康理想的皮肤，多见于青春发育期前的少女或幼儿。皮脂分泌量适中，皮肤既不干也不油，肌肤红润细腻，富有弹性。皮肤纹理不粗不细，毛孔较小，厚薄适中，对外界刺激不敏感。皮肤的 pH 值为 5～5.6(图 2-6)。

2. 干性皮肤 皮肤白皙，毛孔细小而不明显。皮脂分泌量少，皮肤比较干燥，容易产生细小皱纹。毛细血管表浅，易破裂，对外界刺激比较敏感，易生红斑。干性皮肤可分为缺水和缺油两种。干性缺水皮肤多见于 35 岁以上人群及老年人，干性缺油皮肤多见于年轻人。干性皮肤的 pH 值为 4.5～5(图 2-7)。

3. 油性皮肤 肤色较深，毛孔粗大，皮纹较粗，皮脂分泌量多，皮肤油腻光亮，不容易起皱纹，对外界刺激不敏感。由于皮脂分泌过多，易患痤疮，常见于青春期年轻人。油性皮肤的 pH 值为 5.6～6.6(图 2-8)。

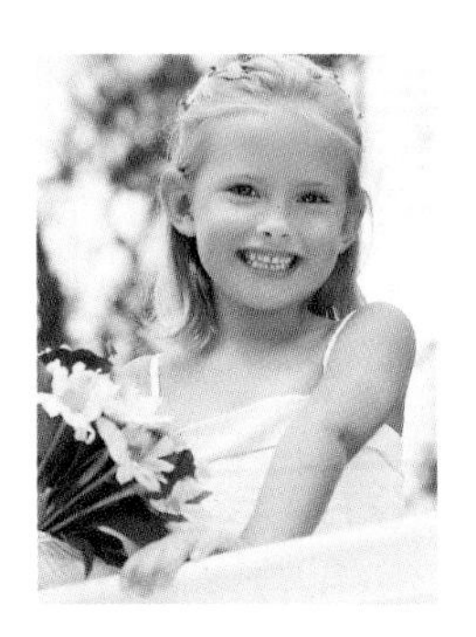

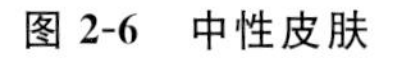

图 2-6 中性皮肤

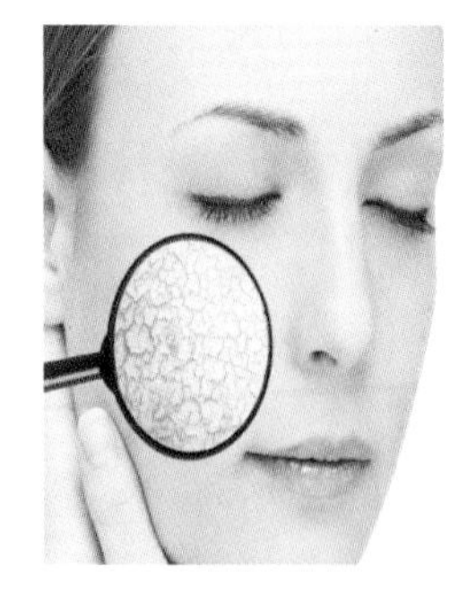

图 2-7 干性皮肤

图 2-8 油性皮肤

4. 混合性皮肤 兼有油性皮肤和干性皮肤的特征。在面部 T 区(前额、鼻、口周、下巴)呈油性状态,眼部及两颊呈干性状态。混合性皮肤多见于 25～35 岁的年轻人(图 2-9)。

5. 敏感性皮肤 可见于上述各种皮肤,其皮肤较薄,对外界刺激很敏感。当受到外界刺激时,会出现局部微红、红肿、包块及刺痒等症状(图 2-10)。

6. 痤疮性皮肤 多见于青春期,大部分人 30～35 岁以后可以自愈。皮脂腺分泌过多,不能及时排出,积于毛囊内,皮肤油腻,毛孔粗大,易出现黑头粉刺、白头粉刺甚至痤疮。黑头粉刺是由于皮脂积于毛囊内,毛囊口处的皮脂与灰尘及角化死细胞混合,凝成小脂栓,堵塞毛孔而形成的;白头粉刺是由于皮脂积于毛囊内,毛孔被角化细胞覆盖而形成的。痤疮是由于皮脂堵塞毛孔,导致皮肤内缺氧,皮脂中含有大量的营养,使痤疮杆菌大量繁殖,堵塞毛囊,使毛囊发炎而形成的(图 2-11)。

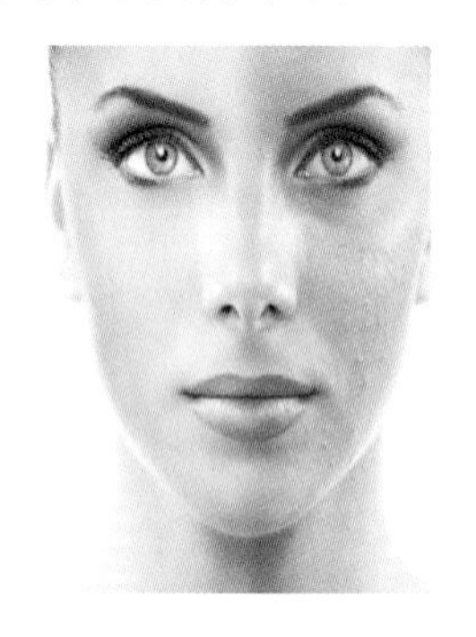

图 2-9 混合性皮肤

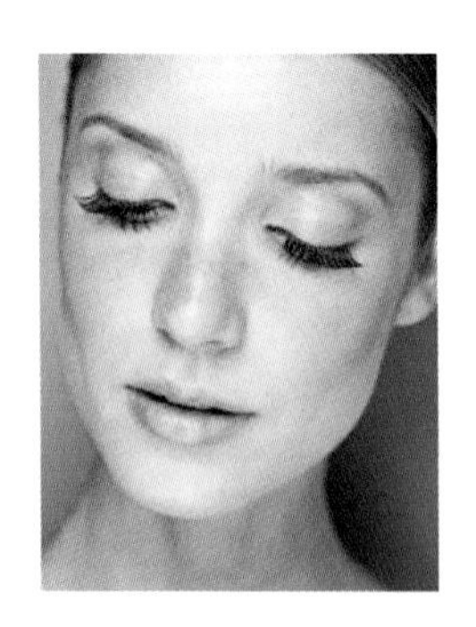

图 2-10 敏感性皮肤

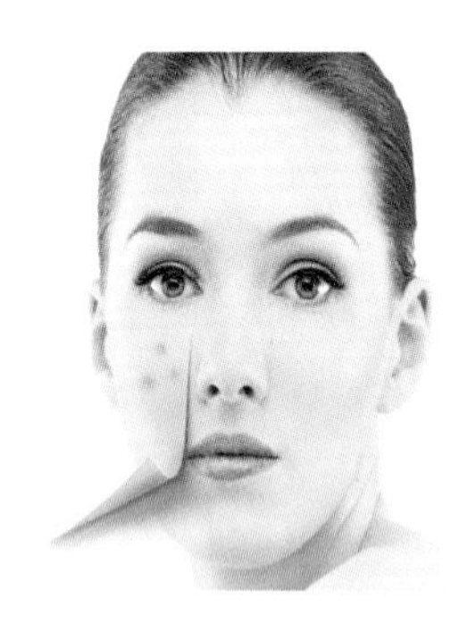

图 2-11 痤疮性皮肤

(二) 皮肤分析方法

1. 目测法 面部清洁后,用毛巾将水擦干,皮肤会逐渐出现紧绷感,不用任何护肤品,静静观察皮肤的状况,计算紧绷感消失的时间。同时观察其肤色、皮脂分泌的情况,皮肤的湿润度、毛孔状态、纹理、肤质、瑕疵、血液循环状况、敏感情况和特殊病变。

2. 纸巾擦拭法 彻底清洁皮肤后,不用任何护肤品,2 h 后用干净的吸油纸分别轻按颈部、面颊、鼻翼和下颌等处,观察纸巾上油污的多少。此法只适合家庭自我检测,参考使用。

3. 美容放大镜法 洗净面部,待皮肤紧张感消失以后,用放大镜仔细观察皮肤的纹理及毛孔状态,操作时注意用棉片盖住眼睛。

4. 美容透视灯观察法 美容透视灯又称为滤过紫外线灯,滤出的紫外线照射在不同类型的皮肤上呈现不同颜色。可帮助美容师了解皮肤表面和深层的组织情况。清洁面部,用湿棉片遮住顾客双眼，以防紫外线刺伤眼睛，待皮肤紧张感消失以后再进行检测。

5. 美容光线显微镜检测仪观察法 该方法是应用微电脑皮肤显示器来观察。该仪器利用光纤显微镜技术，采用新式的冷光设计，清晰、高效的彩色或黑白电脑显示屏，使顾客亲眼目睹自身皮肤或毛发状况。由于该仪器具有足够的放大倍数，一般为 50～200 倍，可直接观察皮肤基底层。

6. 触摸法 用手触摸顾客的皮肤，测试皮肤的柔润度、角质层的厚薄、皮肤的弹性、皮表温度等。

7. 虹膜观察法 它是一种新型的肌肤与身体亚健康的检测方法。专业检测分析系统利用专用虹膜检测仪将显微测试图输入电脑，然后进行分析，可以观察到先天体质的强弱，推测身体在生化上的需求、目前的健康程度以及药物、色素、毒素累积的情况。美容师可根据分析出的健康状况为顾客提供相应的护理方法。

（三）皮肤分析的基本程序

1. 询问 按美容院顾客皮肤诊断、分析表所要求填写的内容，以交谈询问的方式让顾客自我介绍并做最基本的资料记录，为准确分析皮肤提供参考资料。

2. 肉眼观察 可以直接用肉眼观察判断皮肤的大致情况，然后进行触诊，比如用拇指和食指在局部做推、捏、抹动作，仔细观察皮肤毛孔、弹性及组织情况；或用手指掠过皮肤，感觉其粗糙、光滑、柔软程度等。如果顾客有化妆，一定要先卸妆，彻底清洁面部皮肤，并且等皮肤的 pH 值完全恢复正常后，再进行皮肤分析。

3. 仪器观察 借助专业仪器检测皮肤，更加准确地判断皮肤的状况。

4. 分析结果与制订护理方案 根据皮肤检测结果分析皮肤，确定护理目的，制订护理方案。

四、面部皮肤护理方案的制订

面部皮肤护理方案是美容师实施护理操作的重要依据，能够对不同类型的皮肤及皮肤问题进行有针对性的护理，做到有的放矢。上述顾客档案登记表中记录的详细内容，可以帮助美容师为顾客制订系统的护理计划。

（一）护理方案的基本内容与要求

1. 制订护理方案的依据

（1）分析皮肤类型：分析皮肤，判断皮肤类型和状况特征。

①通常在美容院采用目测法、美容放大镜法、电脑皮肤分析仪分析法和皮肤测试仪检查法进行皮肤分析。

②通过分析皮肤的油脂和水分辨别皮肤的类型。

③观察皮肤的其他状况，如敏感程度、肤色、弹性、瑕疵等，寻找需要进行护理和改善的方法。

（2）确定护理目的：面部护理的任何方法和手段都是围绕护理目的展开的，确定护理目的需要根据以下几点。

①顾客的需求。

②皮肤的类型。

③需要改善的皮肤状况。

案例引导

顾客1:肉眼难以看见毛孔,皮肤呈现哑光状态,有小细纹,在秋冬季节常出现脱皮情况,顾客希望使皮肤变得滋润些。

分析:

顾客的需求:使皮肤变得滋润

皮肤类型:干性肌肤

护理目的:改善皮肤干燥,补充皮肤水分和营养

案例引导

顾客2:皮肤油脂分泌旺盛、有大量红色丘疹型痤疮,有褐色色素沉着,有细纹和脱皮现象,时常面部发现有红、肿、发痒等过敏现象。顾客需要改善痤疮及祛除色素沉着。

分析:

顾客的需求:改善痤疮及祛除色素沉着

皮肤类型:油性皮肤

需要改善的皮肤状况:敏感、痤疮、色素沉着

护理目的:最近数次护理为防敏修复护理

2. 护理方案的内容 护理方案包括沙龙护理计划和家居护理方案。

(1)沙龙护理计划:

①护理产品设计。

②护理仪器选择。

③护理手法选择。

④护理程序选择。

⑤护理疗程设计。

(2)家居护理方案:

①护肤品的选择:

a.根据顾客皮肤类型和沙龙护理计划,为顾客选择合适的家居护理产品。

b.与顾客沟通家居护理疗程,并告知家居产品使用方法。

②注意事项:

a.明确自己皮肤类型,选择合适的护肤品。

b.按照正确的护肤程序使用护肤品。

c.掌握护肤品的正确用法和用量。

d.坚持早、中、晚做好常规护肤。

e.特殊护肤产品按照疗程使用。

③起居生活配合:

a.保证充足的睡眠,生活有规律。

b.保持良好心态,疏导压力。

c.膳食营养均衡,适当运动。

3. 护理方案的要求

(1) 分析皮肤类型,明确顾客需求,根据顾客需求制订护理方案。

(2) 在有多个护理目的时应该选择顾客最需要解决的问题为当下护理目的。

(3) 与顾客有效沟通沙龙护理计划,配合家居护理。

(二) 护肤品的选择与使用

1. 根据皮肤性质选择合适的化妆品

(1) 中性皮肤:毛孔细致,有光泽,油脂分泌适中。①皮肤困扰:夏季T区略微油腻,冬季偏干,随着年龄的增长逐渐转变为干性皮肤。②保养重点:适度清洁与保护,选用性质温和、滋润型的护肤品。

(2)干性皮肤:皮肤无光泽,缺乏弹性,出现皮屑。①皮肤困扰:干燥紧绷。②保养重点:适度清洁、保护,选用高保湿的洁面产品,乳液或乳霜。

(3) 油性皮肤:毛孔粗大,油脂分泌旺盛。①皮肤困扰:满面油光,易生黑头粉刺甚至痤疮。②保养重点:重清洁,抑制油脂。选用清洁能力强、含有控油成分的泡沫洁面乳,结构清淡含控油成分的乳液。

(4) 混合性皮肤:T区油、两颊偏干或中性。①皮肤困扰:T区油性强,易生痤疮。②保养重点:分区护理,T区选择油性皮肤适用的护肤品,两颊选择中性皮肤适用的护肤品,偏干部位选用干性皮肤适用的护肤品。

2. 日常护肤品的选择与使用 日常护肤需要使用洁面产品、化妆水、眼霜、精华液、乳液、面霜、防晒与隔离霜等。周期性护肤需要使用面膜、按摩霜、调理霜、去角质产品、眼膜等。

(1) 清洁类护肤品:即洗面奶,包括洁面乳、洗颜霜、清洁霜、磨面膏、清洁啫喱等。主要用于清洁、营养、保护皮肤。早、中、晚各使用1次。

(2) 化妆水:也称紧肤水,分为爽肤水和柔肤水,它的作用在于再次清洁以恢复肌肤表面的酸碱性,调理角质层,使肌肤具有更好的吸收功能,为使用保养品做准备。由于洗面奶洗后皮肤呈弱酸性,爽肤水或者柔肤水的弱碱性可以调节皮肤的酸碱度。清爽型的化妆水适合偏油性肤质,保湿型的化妆水则适合偏干性肤质。面部清洁后"5点法"爽肤2~3遍。"5点法"即将所需护肤品分为5份,分别置于额头、两颊、鼻子、口周,然后用美容指(即中指和无名指)点拍均匀,指腹弹拍至吸收。每次使用1~2 mL。

(3) 眼霜:用来保护眼睛周围比较薄的一层皮肤,对眼袋、黑眼圈、鱼尾纹等均有一定的效用,但是,不同的眼霜有不同的作用。可以根据自身年龄、季节、肤质突出问题选择使用。爽肤后,用无名指将"绿豆粒"大小的眼霜由外眼角向内眼角,以打圈的方式涂抹眼周至吸收。

(4) 精华液:成分精致,功效强大,效果显著。一般含有微量元素、胶原蛋白、血清,具有抗衰、抗皱、保湿、美白、祛斑等作用。精华素比较浓稠,精华液比较稀薄,在选择时遵循三点:首先根据功效而定,选择美白、保湿或抗皱型精华素;其次是按照皮肤性质而定,如干性皮肤选择保湿滋润型精华素,而油性皮肤选择能收缩毛孔的精华素;最后视季节而定,一般在夏天可选用水质精华素,而冬天则偏向油质精华液。爽肤后"5点法"使用,每次用量为5~8滴。

(5) 乳液:含水量较大,能为皮肤补充水分。其含有少量油分,可以滋润皮肤,保湿的同时能够锁住水分。乳液主要有去污、补充水分(锁水)、补充营养的作用。在精华液后"5点法"使用,每次用量为"黄豆粒"大小。

(6) 霜:类似乳液功能,乳液相对较稀薄,而面霜质地浓稠,保湿作用更好,一般适用于秋冬季节。面霜、日霜具有抵抗外界刺激和抗氧化的作用,晚霜则注重夜间修复。在乳液后"5

点法"使用,每次用量为"黄豆粒"大小。

(7) 面膜:敷在面部的美容护肤品,利用覆盖在面部的短暂时间,暂时隔离外界的空气与污染物,提高肌肤温度,使皮肤毛孔扩张,促进汗腺分泌与新陈代谢,皮肤适度地收紧,增加张力,使皮肤皱纹舒展开来。在清洁和爽肤后使用,一般一周使用1~2次。

(8) 隔离防晒霜:对皮肤起到防护作用,能够防晒并隔离被污染的空气和防辐射作用。在面霜之后使用,每次用量为"黄豆粒"大小,用美容指轻拍均匀。

知识拓展

护肤品选择的原则

选择护肤品一般应坚持皮肤适应性原则,选择购买和使用时应"一看、二闻、三涂抹"。

1."一看" 看护肤品外包装、生产企业、经营企业、产品成分、有无卫生许可证号和批准文号标识,看护肤品颜色和形态。

2."二闻" 闻化妆品的气味。

3."三涂抹" 手臂内侧涂抹,感觉滑润舒适,质地细腻。

总之,选择护肤品要避开三个误区,即广告、价格、追求快速高效,特别注意四个方面,即安全、品质、性能和服务。如果皮肤出现异常状况,如过敏、红肿等,就要立即停止使用任何护肤品,并向专业人士咨询。

3. 根据年龄、地区和季节选择适合的护肤品

(1) 年龄:随着年龄的增长,皮肤所需要的养分有所不同。①青春期:油脂分泌旺盛。保养应重清洁和控油,选用清爽含有控油成分的乳液。②20~25岁:皮肤不缺乏养分,保养重点是补水。③25~30岁:25岁是皮肤衰老的分水岭,女性过了25岁皮肤就逐渐衰老,应选用高水分的乳液或乳霜,使皮肤保持水润健康的状态。④30~40岁:30岁以后皮肤衰老速度加快,不仅要补水,还要适当补充营养。应选用含有胶原蛋白的营养霜或乳液,也可每周敷营养面膜。⑤40岁以后:属于衰老性皮肤,保养重点是补充营养,选用有高营养、抗衰老的面霜。

(2) 地区和季节:南方四季潮湿,选用具有收敛、控油、清爽功能的产品。北方四季分明:春天空气干燥,植物花粉易引起敏感,选用含有防敏、补水成分的产品;夏天天气炎热,紫外线强,皮肤油脂分泌旺盛,选用清洁力度强、控油和具有防晒功能的产品;秋天空气干燥,选用补水产品为主;冬天天气寒冷,皮肤油脂分泌减少,血液循环慢,选用补水、滋养的产品。

本项目重点提示

(1) 顾客接待与咨询的流程一般包括:接待、咨询、分析诊断、建立档案、设计护理计划、沟通护理计划、实施护理计划、效果评价、记录与结账、送客、整理、定期跟踪回访。

(2) 顾客档案登记表的主要内容包括:顾客基本情况、生活习惯、面部皮肤基本情况、身体健康状况、美容院护理疗程设计、消费记录、护理服务记录、意见与建议。

(3) 皮肤一般可分为四种类型,即中性皮肤、干性皮肤、油性皮肤和混合性皮肤。但在实际操作过程中敏感性皮肤、痤疮性皮肤也是常见的皮肤。各类皮肤具有各自不同的特点。

(4) 皮肤分析的基本程序包括询问、肉眼观察、仪器观察、分析结果与制订护理方案。

(5) 面部皮肤护理方案的基本内容：定沙龙护理计划、家居护理方案。

能力检测

一、选择题

1. 下列哪项不属于美容院前台接待美容师的职责要求？（　　）

A. 形象得体　　B. 熟悉业务　　C. 精神状态饱满　　D. 制订护理方案

2. 下列哪项是面部皮肤护理方案的基本内容？（　　）

A. 分析护理目的　　B. 销售计划　　C. 确定护理目的　　D. 沙龙护理计划

3. 干性皮肤的特征不包括下列哪项？（　　）

A. 不易敏感　　B. 皮肤比较干燥　　C. 缺乏弹性　　D. 皮脂分泌量少

4. 油性皮肤的 pH 值是多少？（　　）

A. 5.6～6.6　　B. 5～5.6　　C. 4.5～5　　D. 4.5～6

5. 下列哪项是乳液具有的作用？（　　）

A. 去污　　B. 防晒　　C. 补充水分(锁水)　　D. 增白

二、填空题

1. 沙龙护理计划包括________、________、________、________和________。

2. 皮肤一般可分为四种类型，包括________、________、________和________。

3. 皮肤分析程序包括________、________、________、________与________。

三、问答题

1. 美容院前台接待美容师有哪些岗位职责？

2. 如何制作顾客档案、指导顾客填写并有效使用？

3. 常见的皮肤类型有哪几种？各有何特征？

4. 如何为顾客制订面部皮肤护理方案？

5. 如何指导顾客选择合适的护肤品？

（张　颖　王　艳）

项目三　面部皮肤护理

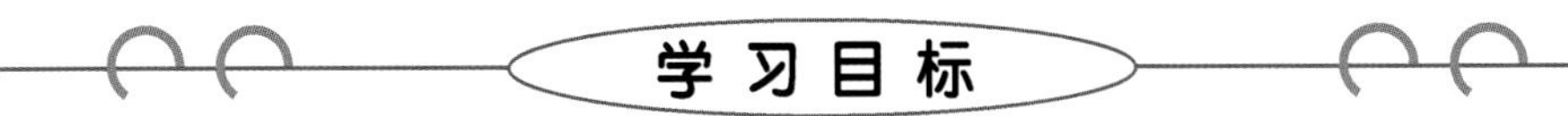

学习目标

(1) 掌握面部基础护理的操作程序,具备为顾客设计个性化面部护理方案的能力。

(2) 熟悉面部刮痧与面部芳香疗法的操作程序及注意事项。

(3) 了解面部刮痧与面部芳香疗法的基础知识。

项目描述

面部皮肤保养是现代女性追求美丽与健康的首要工作。不管是污染的空气,还是面部化妆,都会增加皮肤负担,如不做好面部皮肤护理,面部皮肤问题便会接踵而来。本项目主要介绍面部基础护理、面部刮痧和面部芳香疗法。学生通过本项目的学习,能了解如何养护面部,掌握面部护理的程序,具备为顾客设计个性化面部护理方案的能力。

案例引导

王小姐,32岁,白领,165 cm,58 kg。喜欢化妆,额头可见少许痤疮,两侧颧骨皮肤少许细小斑点;生活和工作压力大,身体疲惫,面部气色欠佳。想做面部护理,放松身心,缓解压力。

问题:

1. 作为美容师,你认为应该给这位顾客做什么护理项目?有哪些注意事项?
2. 请你给顾客介绍一下家居保养面部的方法。

面部皮肤长年累月暴露在空气中,紫外线照射及空气中尘埃、细菌等有害物质刺激,加上自身分泌的油脂、汗液,代谢的死细胞等,这些因素都会影响皮肤正常功能的发挥,甚至引起皮肤感染及痤疮,导致皮肤提前衰老。正确的面部皮肤护理可以清除皮肤表面的污垢、皮肤分泌物,保持汗腺、皮脂腺分泌物排出畅通,防止细菌感染。洁肤可使皮肤得到放松、休息,以便充分发挥皮肤的生理功能,呈现青春活力;亦可调节皮肤的 pH 值,使其恢复正常的酸碱度,保护皮肤。

一、面部基础护理的操作程序

面部皮肤护理的每个程序都有其不同的目的、作用及效果，操作程序应该根据护理目的不同而设定，各程序之间相辅相成，但又不是一成不变的，应进行优化整合。完整的面部基础护理程序包括准备工作、清洁皮肤、面部按摩、敷面膜和后续工作。有序的工作是完美服务的基本保证，美容师应严格按照护理程序实施操作。

（一）准备工作

1. 美容师准备 化淡妆、着工作服、穿工作鞋、戴口罩、去首饰、修剪指甲、洗手、消毒双手。

2. 用物准备

（1）床位准备：①铺床单：将两条床单重合平铺在美容床上并覆盖枕头；再铺上被子，被子近床头端与枕头近床尾端平齐，将上层床单多余部分向上反折，再将上层床单和被子一角向上反折(45°)。②铺毛巾：将毛巾平铺于枕头上，第一条毛巾上边缘齐枕头上边缘；第二条毛巾下边缘齐枕头下边缘，用于包头；第三条毛巾横折两次，放于枕头右边，作肩巾待用；第四条毛巾挂在护理车上，护理时擦拭美容师手上的水渍(图 3-1)。

（2）护理产品准备：美容师将顾问开具的护理方案交给配料员准备护理用品，主要包括卸妆液、棉签、棉片、洁面乳、爽肤水、按摩膏、面盆、面巾纸、乳液、精华液、眼霜、面霜和防晒霜等。护理用物放于护理车上，摆放整齐(图 3-2)。

（3）护理设备准备：检查美容仪器、设备的电路是否通畅，运转是否正常。

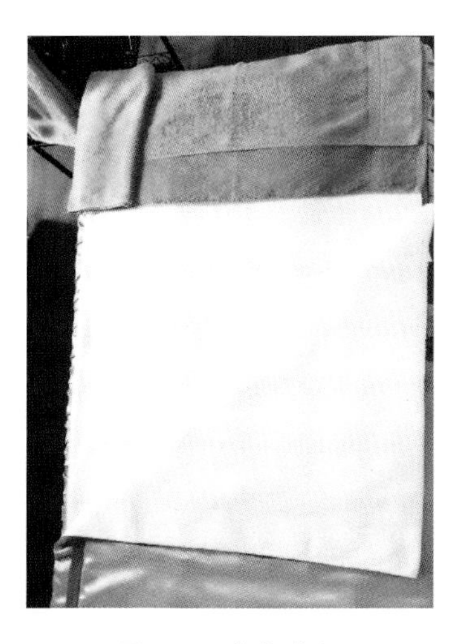

图 3-1 床位准备

图 3-2 护理产品准备

3. 环境准备 保持护理间空气清新，点香熏灯，灯光明暗度适宜；播放舒缓音乐；室温一般控制在 24～26 ℃。

4. 顾客准备 协助顾客换拖鞋，告知顾客将饰物，如戒指、项链、手镯等取下放进衣柜锁好。必要时协助顾客沐浴，更换美容服，让顾客取仰卧位躺在美容床上，包头，铺肩巾。

（1）包头：

①顾客平躺，将毛巾的长边向下折叠 2 cm 左右，折边在下与后发际平齐(图 3-3(a))。

②左手拿住折边左角，沿发际从耳后往上右方拉紧至额部压住头发，右手配合将头发收拢至毛巾下包住。用同样方法拉起毛巾右角往上左方压住发际头发，然后将毛巾右角塞进折

边内固定好(图 3-3(b))。

③将顾客耳朵抚平,双手四指扣住毛巾边缘,轻轻将包好的毛巾向后拉至发际。

④检查毛巾松紧是否适度(图 3-3(c))。

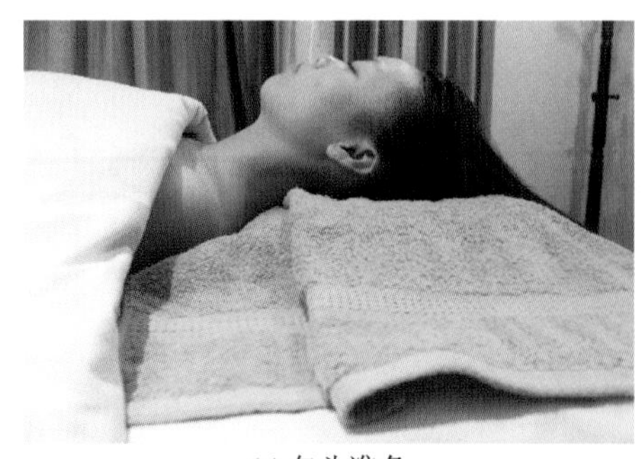

(a) 包头准备

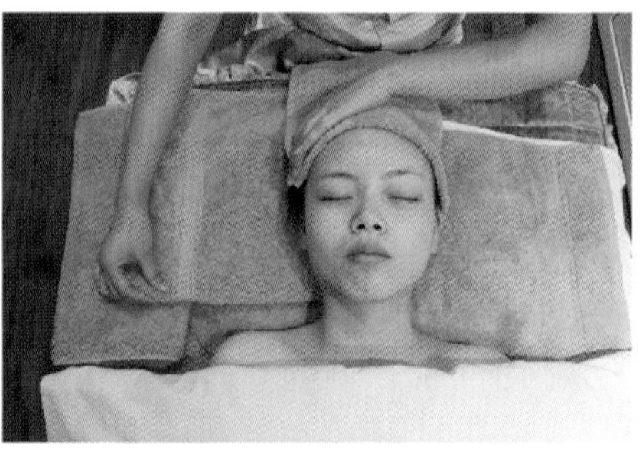

(b) 包住头发

(c) 发尾整理

图 3-3　包头

(2) 铺肩巾:①"V"字法:双手分别捏住毛巾长边的两个角,沿颈部左侧平铺,将胸前远侧角向上反折,铺于颈部右侧。注意盖住顾客衣领,保护顾客衣领不被污染(图 3-4(a))。②"一"字法:毛巾平铺在顾客胸前,近侧边缘向内折约 2 cm(图 3-4(b))。

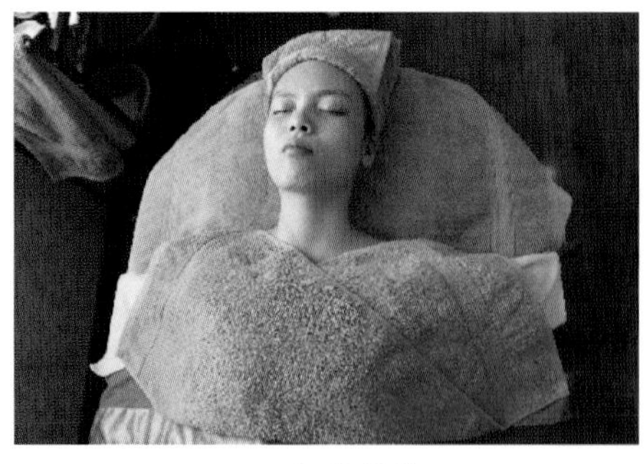

(a) "V"字法

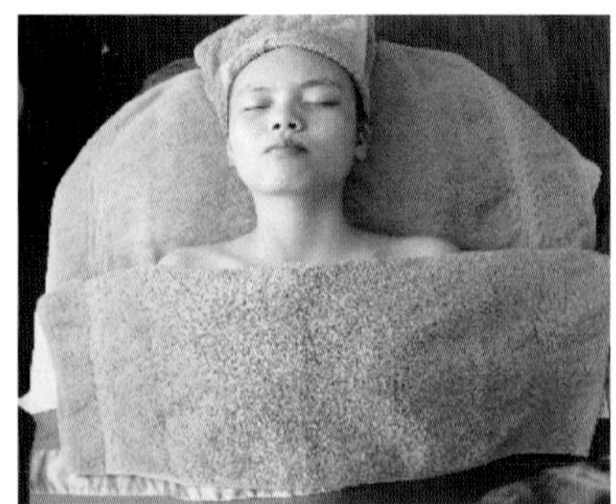

(b) "一"字法

图 3-4　铺肩巾

(二) 清洁皮肤

清洁皮肤主要是为了清除皮肤表面的污垢,如灰尘、细菌、残留化妆品、汗液、油脂、老化角质等,保持汗腺和皮脂腺分泌通畅,促进新陈代谢和营养物质的吸收,是顾客在接受皮肤分析前必须完成的步骤。皮肤清洁是护肤品吸收、取得良好护理效果的前提，具体包括卸妆、表层清洁和深层清洁。

1. 卸妆 彩妆中的粉底、色素多为油性，附着于皮肤表面，不易脱落，难以清洗，使用专业卸妆产品才能清除。

卸妆产品有卸妆油、卸妆乳、卸妆水等(图 3-5(a)、图 3-5(b)、图 3-5(c))。卸妆用物包括面巾纸、小棉片、棉签等。

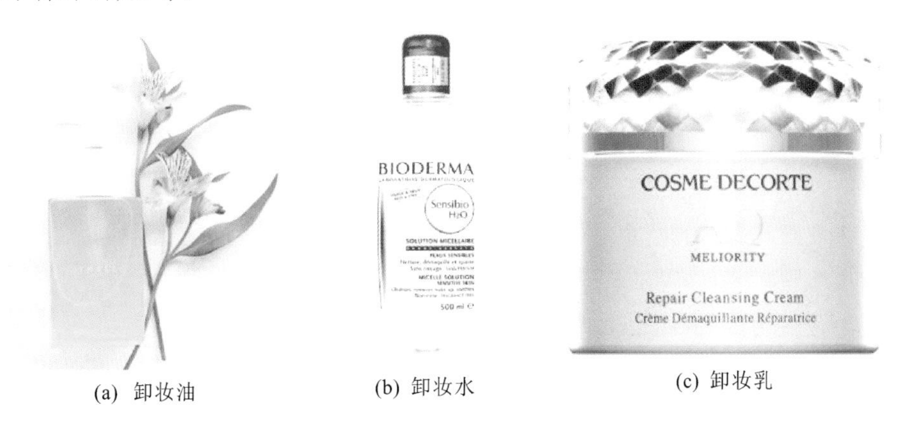

(a) 卸妆油 (b) 卸妆水 (c) 卸妆乳

图 3-5 卸妆产品

(1) 卸妆操作程序：依次卸除睫毛膏、眼线、眼影、眉色、唇膏(口红)、腮红、粉底。

①卸睫毛膏：取一块小棉片，滴上卸妆产品，嘱顾客闭眼，将小棉片置于顾客睫毛下。用棉签蘸取卸妆产品，从睫毛中间向两边卸除睫毛膏，一手轻提上眼睑，另一手用棉签从睫毛根部以向内打圈的方式卸除睫毛膏，直至卸除干净(图 3-6(a))。

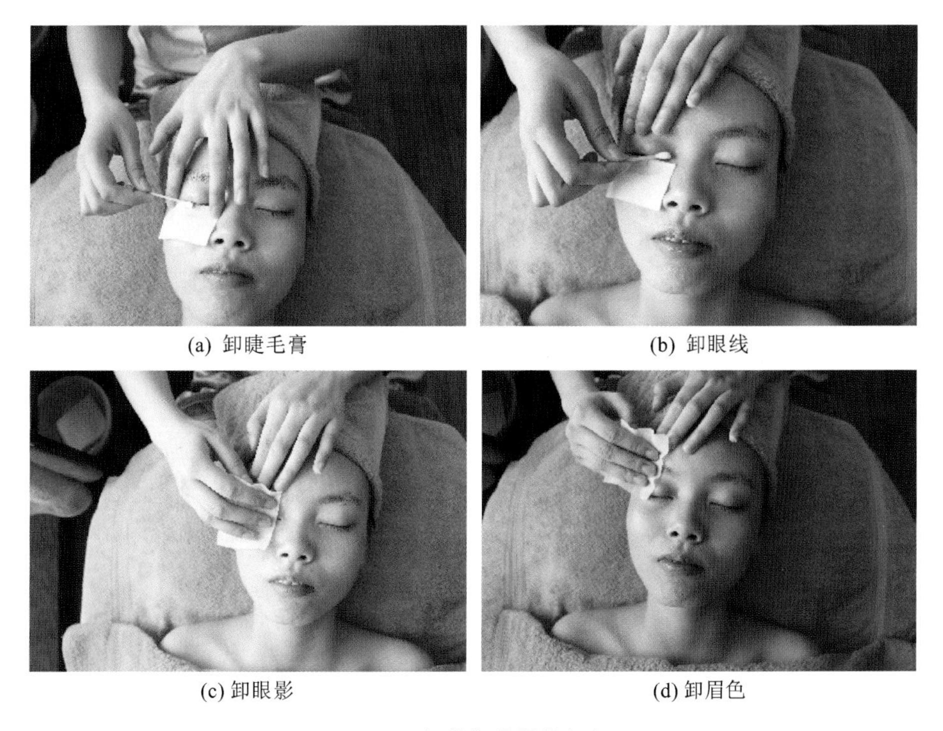

(a) 卸睫毛膏 (b) 卸眼线

(c) 卸眼影 (d) 卸眉色

图 3-6 卸妆部分操作程序

②卸眼线：更换棉签蘸取卸妆产品，从内眼角向外眼角拉抹，卸除眼线（图 3-6(b)）。

③卸眼影：将眼睑下的小棉片上翻覆盖上眼睑，由内眼角擦至外眼角，卸除眼影（图 3-6(c)）。

④卸眉色：取棉片滴上卸妆产品，由眉头往眉尾拉抹，卸除眉色（图 3-6(d)）。

⑤卸唇色：左手轻轻固定左侧嘴角，右手用清洁棉片（滴上卸妆产品），从固定侧擦拭至嘴角另一侧，分别卸除上、下唇部妆容（图 3-7）。更换右手固定右侧嘴角，重复上述动作。

⑥卸腮红和粉底：取卸妆产品依次置于面部"5 点"：下巴、鼻尖、额头、两侧面颊（图 3-8）。用美容指将产品抹开，顺肌肤纹理打圈，涂抹全脸，待卸妆产品充分溶解腮红、粉底后，用面巾纸擦拭。

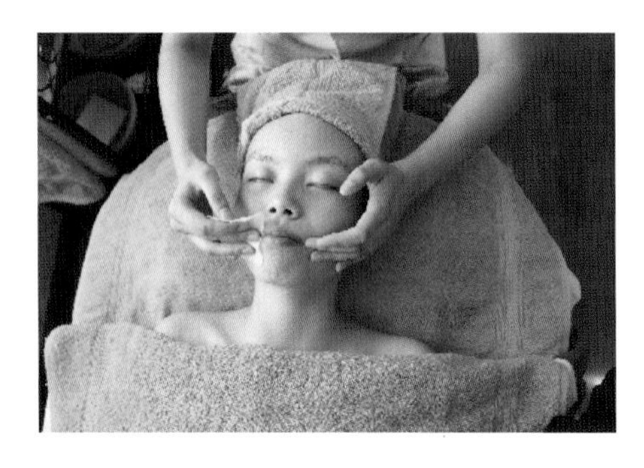

图 3-7　卸唇色

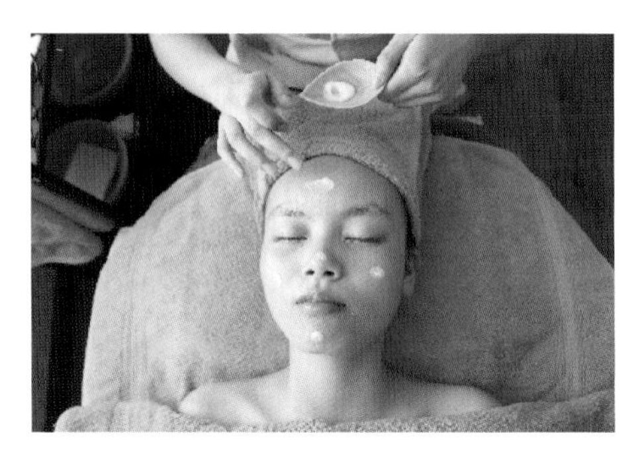

图 3-8　卸腮红和粉底

面巾纸主要用于清洁面部，用时需绕在手指上。缠绕时要求整齐、牢固、迅速。面巾纸大小约 15 cm×8 cm。①折叠方法：横折一次，将重合短边向内折约 1 cm（图 3-9(a)）；掌心向下，用食指和中指夹住纸巾一侧，折面朝向自己（图 3-9(b)）；将面巾纸另一端向下绕过食指、

(a) 短边内折

(b) 将面巾纸缠绕于手指

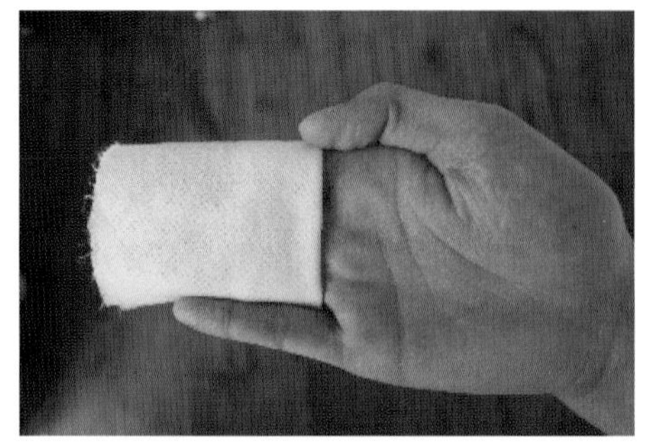

(c) 缠绕好的面巾纸

图 3-9　面巾纸折叠方法

中指、无名指，夹在小指和无名指之间(图 3-9(c))。②使用方法：按照“面部 12 条线”进行擦拭：上眼睑至太阳；下眼睑至太阳(图 3-10(a))；鼻子三条线(图 3-10(b))；额头三条线(图 3-10(c))；鼻翼至耳上(图 3-10(d))；人中、嘴角至耳中；下巴和下颌分别至耳下(图 3-10(e))。

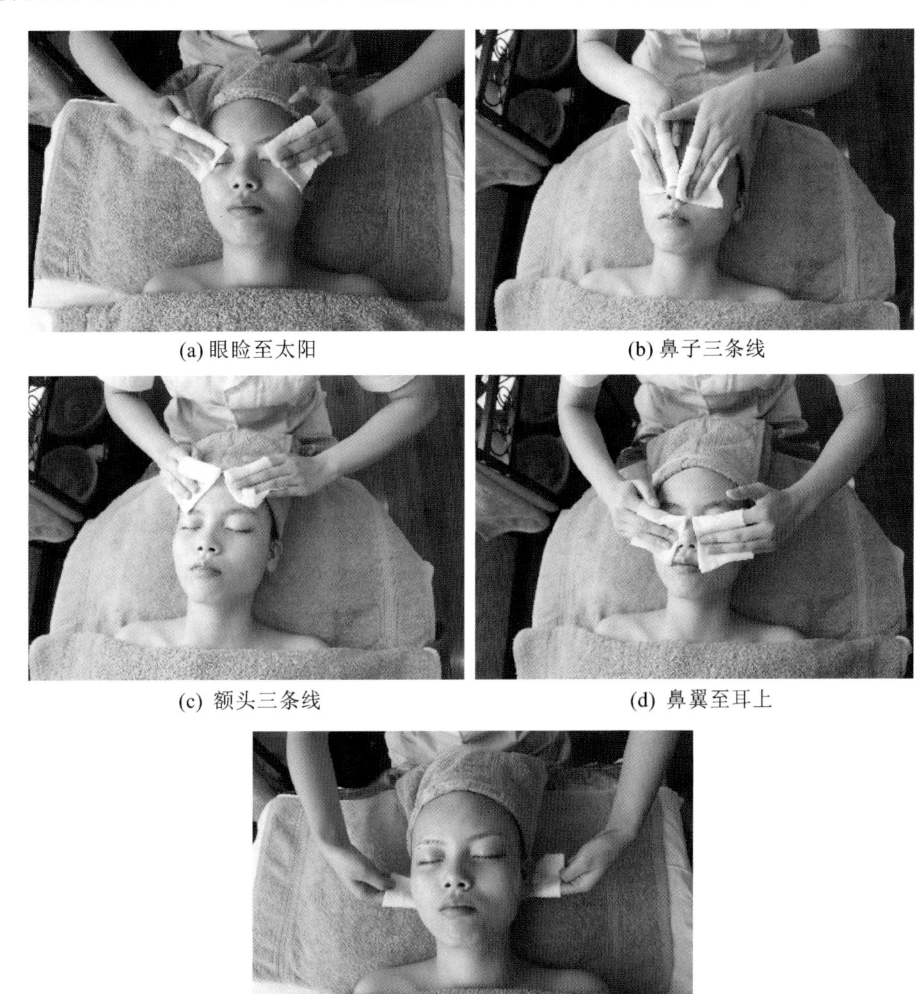

(a) 眼睑至太阳 (b) 鼻子三条线

(c) 额头三条线 (d) 鼻翼至耳上

(e) 人中、嘴角、下巴和下颌至耳部

图 3-10 面巾纸使用方法

(2) 卸妆要求与注意事项：

①卸妆要彻底。

②眼部皮肤比较敏感，操作要轻柔。

③卸妆时注意不要让产品流入顾客的口、眼、鼻中。

2. 表层清洁 卸妆主要清除面部的彩妆及污垢。卸妆后，面部有少许彩妆和卸妆产品残留，还有细菌、灰尘附着在面部表层的皮脂膜，也需彻底清洗去掉，因此需用洁面产品清洗面部皮肤，即为表层清洁。常用的表层清洁产品有洁面皂(图 3-11(a))、洗面奶(图 3-11(b))、洁面啫喱、泡沫洁面乳等。表层清洁用物有面巾纸、面盆、水。

(1) 表层清洁的操作程序：

①上洁面乳：取适量洁面乳置于掌心，另一手掌蘸取少量清水，两手相互揉搓起泡后均匀

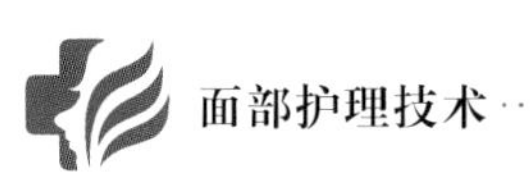

(a) 洁面皂　　(b) 洗面奶

图 3-11　表层清洁产品

涂抹至面部。

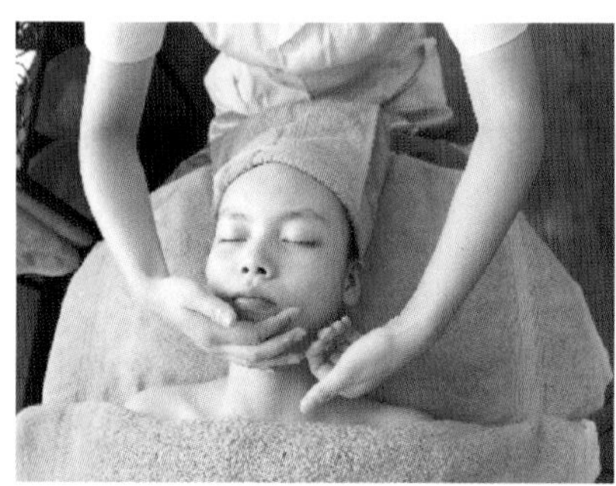

(a) 清洁下巴

(b) 清洁口周

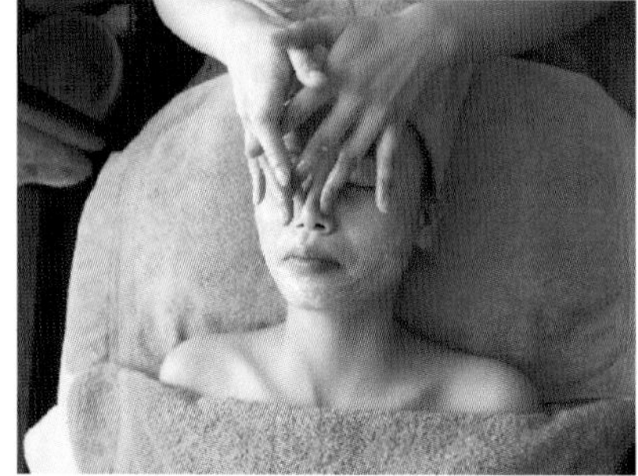

(c) 清洁鼻子

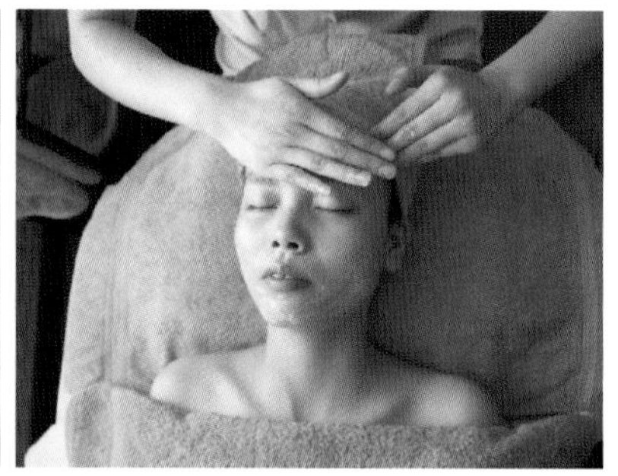

(d) 清洁额头

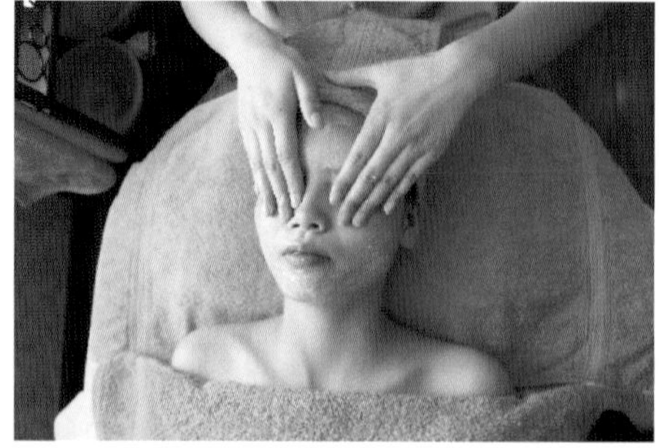

(e) 清洁眼周

图 3-12　面部清洁操作程序

②面部清洁：用美容指以打圈方式依次清洁下巴、口周、鼻子、额头、眼周(图 3-12)、脸颊及耳部。清洁时间为 1～2 min。

③用面巾纸擦拭干净。

(2) 洁肤水及产品的选择：

①水质的选择：自然界的水有软水和硬水两大类。清洗皮肤时选择软水最佳。软水是不含或仅含少量可溶性的钙盐、镁盐的水，性质温和，对皮肤无刺激，如自来水、蒸馏水等。硬水是指含有钙盐、镁盐较多的水，长期使用硬水清洁面部，会使皮肤脱脂、干燥。

②水温的选择：皮肤护理最适宜的水温为 34～38 ℃，对皮肤有镇静作用，有利于皮肤休息和解乏，也便于洗净油污，对皮肤无伤害。水温过冷(低于 20 ℃)对皮肤有收敛作用，可锻炼肌肤，使人精神振奋，但用过冷的水洁肤，不易清除皮肤上的油污，油性、痤疮性皮肤不适用。水温过高(高于 38 ℃)对皮肤有镇痛和扩张毛细血管作用，但经常使用会使皮肤脱脂，血管壁活力减弱，导致皮肤淤血、毛孔扩张，皮肤容易变得松弛无力，出现皱纹。对于正常肌肤，可冷热水交替使用，水温的冷热变化可以使皮肤浅表血管扩张和收缩，增强皮肤的呼吸能力，促进血液循环。

③清洁产品的选择：根据顾客皮肤状况，选择合适的清洁产品。干性、中性皮肤应选择乳液状的洗面奶，其性质温和，清洁效果良好，清洁皮肤的同时，在皮肤上留下滋润保护膜，对皮肤刺激性小。油性、混合性皮肤应选择泡沫型洁面乳、洁面啫喱，其表面活性剂能够产生丰富的泡沫清洁皮肤，清洁力度较好，含有润肤剂，使用后皮肤清爽而不紧绷。

(3) 注意事项：

①洁肤时力度柔和，注意清洁面部死角部位，如鼻翼旁。

②擦拭时面巾纸不宜过干或过湿。

③护理时沿肌肤纹理走向，不可上下来回反复。

④清洁皮肤时间不宜过长，一般不超过 3 min，以免刺激皮肤。

3. 蒸面 蒸面不仅可以起到冷、热效应，也可达到辅助清洁皮肤的效果。一般情况下，蒸面分为冷喷和热喷。

(1) 热喷：常选择奥桑喷雾仪。

(2) 冷喷：常用仪器有冷喷仪(图 3-13)。适用于任何皮肤，尤其是黑斑、敏感性皮肤。其作用有：

①收敛毛孔。

②抑制黑色素细胞，淡化色斑。

③使皮肤血管收缩，降低皮肤表面温度，能消除炎症、红肿，使组织充血减轻。

④降低皮肤的敏感性，起抗过敏作用。

冷喷的操作方法如下。

①水箱注满蒸馏水，接通电源。

②打开冷喷仪电源开关，即有水雾产生，调节雾量大小。

③将喷雾对准顾客面部，操作时间一般为 15 min 左右。

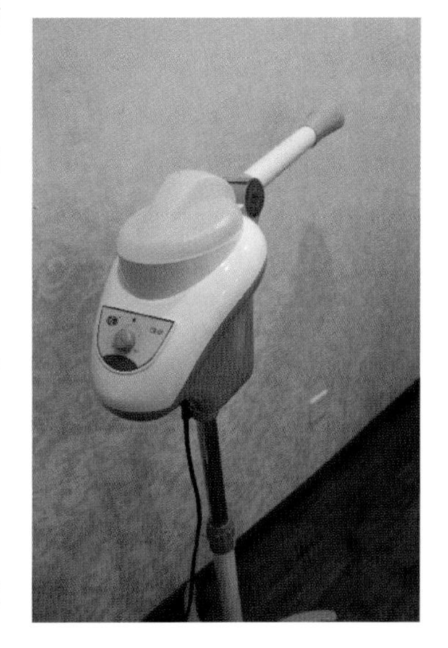

图 3-13 冷喷仪

使用仪器注意事项如下。

①定期清洗水箱并消毒。

②控制好注水量，保证不漏水、不溢出。

③避免碰撞机体，远离高热源。

④水箱无水时，仪器保护系统会使仪器自动关闭或无法启动。

4. 深层清洁 深层清洁也称脱屑、去角质或去死皮，即去除皮肤角质层内衰老死亡的细胞，是常见的皮肤护理方法之一。随着皮肤的不断自我更新，最外层的死细胞会不断脱落，由新生细胞来补充。在某些因素影响下，皮肤的新陈代谢功能减退，死细胞的脱落过程放缓，在皮肤表层堆积过厚，皮肤会显得粗糙、发黄、无光泽，甚至出现痤疮，影响皮肤正常生理功能发挥。脱屑就是借助人工去死皮的方法，去除堆积在皮肤表面的死细胞，以使皮肤更好地吸收各种营养。

(1) 脱屑的分类：

①自然脱屑：由皮肤自身正常的新陈代谢过程来完成。表皮细胞经过一定时间(28 天左右)由基底层逐渐生长到达皮肤表面，变为角化死细胞而自行脱落。

②物理性脱屑：不通过任何化学手段，只使用物理的方法使表皮的角质层发生位移、脱落的方法。常用产品有磨砂膏(图 3-14(a))、撕拉型深层清洁面膜。这种脱屑方法对皮肤刺激较大，适用于健康皮肤。

(a) 磨砂膏

(b) 去角质膏

图 3-14 脱屑常用产品

③化学性脱屑：将含有化学成分的去死皮膏、去死皮水、去死皮啫喱、去角质膏(图 3-14(b))或者果酸涂于皮肤表面，使附着于皮肤表层的角质细胞变软(可擦拭去掉)的方法。此脱屑方法适用于干性、衰老性皮肤和敏感皮肤。

(2) 脱屑的操作方法：

①物理性脱屑的操作方法：

a. 表层清洁。蒸面结束后，取适量磨砂膏分别点涂面部 5 点，然后均匀抹开。

b. 用双手美容指蘸水，以指腹打圈，类似洁面打圈方法。干性、衰老性皮肤脱屑时间短；油性皮肤、T 区脱屑时间稍长；眼周皮肤不做脱屑。整个过程以不超过 3 min 为宜。

c. 将磨砂膏清洗干净。

②化学性脱屑的原理与操作方法：去死皮膏、去死皮液、脱屑水的主要成分是聚合乙烯、有机酸，还含有润肤剂和胶合剂。其中，有机酸可溶解和剥离角质。去死皮膏性质温和，对皮肤刺激小。有的去死皮膏用酵素作为角质溶解剂，性质更加温和，适合敏感性皮肤使用。操

作方法如下：

a. 取适量去死皮膏，5 点法均匀涂于面部，然后抹开。

b. 停留片刻（时间长短参照产品说明，约半分钟）。

c. 将纸巾垫于面部皮肤四周。

d. 一手美容指微微撑开局部皮肤，另一手美容指以打圈方式轻轻地揉搓，方向由下往上，从中间向两边。最后用湿面巾纸将去角质膏和角质细胞擦拭干净。

（3）脱屑的注意事项与禁忌：

①脱屑前，一般先蒸面，可使毛孔张开，有利于清除毛孔深层污垢。

②脱屑一般以 T 区为主，两颊视情况而定，眼周禁止脱屑。

③根据顾客的皮肤性质选用脱屑的方法与产品。

④皮肤发炎、外伤、严重痤疮、特殊脉管状态等问题皮肤均不适合脱屑。

⑤脱屑的间隔时间，根据季节、气候、皮肤状态而定，不可过勤，以免损伤皮肤。一般 1～2 次/月。

⑥手法不宜过重，脱屑后的皮肤需要彻底清洁干净。

（三）面部按摩

按摩是皮肤保养中最重要的一个环节，它不是简单的揉搓，而是在掌握一定技巧的基础上，顺应肌肤纹理走向操作。在中医理论中，面部皮肤与各脏腑相呼应，同时也与全身经脉相连，可以体现脏腑的功能状态。面部按摩也能使面部气血充盈，肤色红润，同时结合面部穴位的点压（点按），给人以舒适感，改善微循环的同时，可辅助治疗头面部病痛。

1. 面部按摩的定义及作用 面部按摩，是在整个面部涂上按摩介质，用轻柔的手法在面部进行揉、捏、弹、拍、压等，不仅能使人面部的肌肉、神经得以放松，而且能消除疲劳，还能使面部轮廓更加清晰，皮肤更加光润。面部按摩的作用主要如下：

（1）增加血液循环，促进新陈代谢：面部按摩手法加速了面部血液的流动，促进血液循环，从而增加面部皮肤的养分供应，加快了代谢废物的排出，进而促进皮肤的新陈代谢，使皮肤焕发光彩，减缓衰老。

（2）提高皮肤温度，增加皮肤的保湿能力：通过面部按摩，可以提高皮肤温度，皮肤温度升高，皮脂膜和汗腺的分泌会增加，毛孔亦会张开，蓄积于毛囊的污垢更易排出，分泌的皮脂可以滋润皮肤，有利于皮肤水分的保持，使皮肤更加柔润。

（3）放松肌肉和神经，消除疲劳：面部按摩促进血液循环的同时带给面部肌肉神经营养，排除其代谢废物，可以有效地减轻肌肉紧张，安抚神经，消除疲劳，使人放松，恢复肌肤活力。

（4）去除死皮，清洁皮肤：面部按摩是美容师用双手在面部进行揉、捏、弹、拍、压等操作，皮肤最外层的角质细胞在外力作用下松动剥离，提高了皮肤的清洁度。

2. 面部按摩介质 面部按摩需要使用按摩介质，其主要作用是润滑皮肤，减少按摩过程中的摩擦阻力。按摩介质中可添加不同物质，如面部所需的各种营养成分，具有美白、保湿、活肤、抗衰老、抗敏感等有效作用的因子。按摩介质根据性状可分为按摩膏（图 3-15）、按摩油、按摩啫喱，每种性状的按摩介质，根

图 3-15 按摩膏

据添加的营养因子适用于不同种类的皮肤。

3. 面部按摩的基本原则 根据面部皮肤的特点，在面部按摩过程中尽量减少面部皮肤位移，要做到力达深层，而表皮基本不动。

(1) 按摩走向从下向上：随着年龄增长，生理功能减退，肌肤会出现松弛现象。由于重力作用，松弛的肌肉会下垂而显现出衰老状态。因此，按摩方向应由下向上。

(2) 按摩走向从里向外，从中间向两边：在进行面部抗衰老性按摩时，应尽量将面部的皱纹展开，并推向面部两侧。

(3) 按摩方向与肌肉走向一致，与皮肤皱纹方向垂直：因为肌肉的走向一般与皱纹的方向是垂直的，因此，按摩时走向与皱纹方向垂直，就能保证与肌肉走向基本一致。

(4) 按摩时尽量减少肌肉位移：当肌肉发生较大位移时，肌肉运动方向的另一侧肌纤维紧绷，过度牵拉，持续的张力会使肌肤松弛，加速其衰老。因此，在进行按摩时，要尽量减少肌肉位移。

4. 面部按摩的基本手法

(1) 按压手法：用手指或手掌按压面部皮肤肌肉。手指按压多在按摩中用以刺激腧穴，行气活血，消除疲劳。操作时注意按压力度逐渐加深，到达一定刺激深度时，停顿 3 s 左右，再慢慢减压，等力度完全放松之后移动到下一位置。

(2) 深压摩擦手法：利用手指或手掌在皮肤组织上施加压力并摩擦的动作。如用拇指划拉额头，可以促进血液循环和腺体分泌。操作时注意手指指腹或手掌紧贴面部，施压划拉，用力均匀渗透，动作有韵律。

(3) 揉捏手法：用手指揉动或提捏某一部位的皮肤、肌肉，包括揉、捏、挤等动作。如夹划眉筋，用食指与中指夹住眉筋，并慢慢划拉，可放松肌肉，消除疲劳。操作时注意力度轻、稳，指腹紧贴皮肤，用力均匀，动作连贯。

(4) 安抚手法：用手指或手掌做轻柔缓慢而有节奏的连续按摩动作。面颊、额头等宽大的地方用手掌操作，眼周、口周等面积窄小的地方用手指操作，可以放松肌肉和神经，镇静皮肤。操作时多采用拉抹动作，指腹或手掌服帖，用手腕带动手指或手掌运动。

5. 面部按摩的手法要求 熟练的面部按摩手法，需要动作连贯而有节奏，能满足不同需求的顾客。因此还需做到以下几点要求。

(1) 持久：每步操作可重复 3～5 遍，双手力度在按摩操作中能持续保持。点压腧穴时手指需按而留之，力度应遵循由轻到重、由重到轻的原则。

(2) 有力：按摩手法必须具有一定力度，才能刺激到深层肌肉。力度大小根据顾客感受及皮肤状况，及时调整。

(3) 均匀：操作手法应有韵律感，可配合背景音乐，调整节奏，不能时快时慢；用力平稳，不能忽轻忽重。

(4) 柔和：美容师在进行面部按摩时，手掌或手指应该柔软而服帖，手法转换应流畅连贯。

(5) 得气：在点穴时有酸、麻、胀、重等感觉。

6. 面部按摩的方法 面部按摩手法应根据顾客皮肤特点和实际情况灵活运用，面部按摩时间为 10～15 min。在按摩过程中，注意手法连贯、力度沉稳，手感柔软服帖，全部动作以舒缓的节奏进行。顾客仰卧于美容床，提前完成表层清洁、深层清洁、爽肤。

(1) “四四三二一”(图 3-16)：依次按压地仓(四指，除去拇指)、迎香(四指，除去拇指)、颧髎(三指,除去拇指和小指)、瞳子髎(二指,中指和无名指)、太阳(一指,即中指)。

(2)"五点二指"(图 3-17):以美容指按压地仓、迎香、颧髎、听宫和太阳。

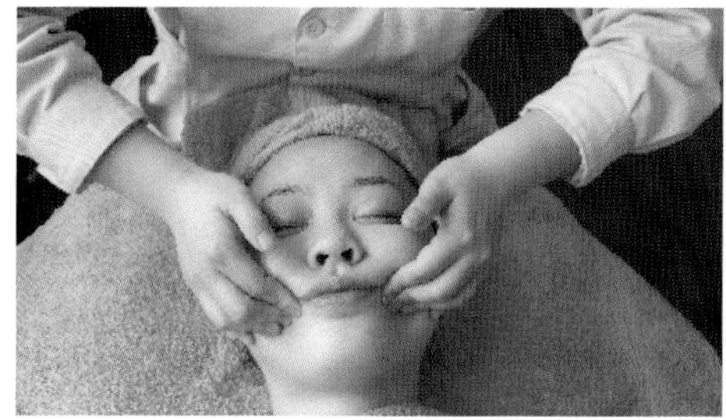

(a) 按压地仓

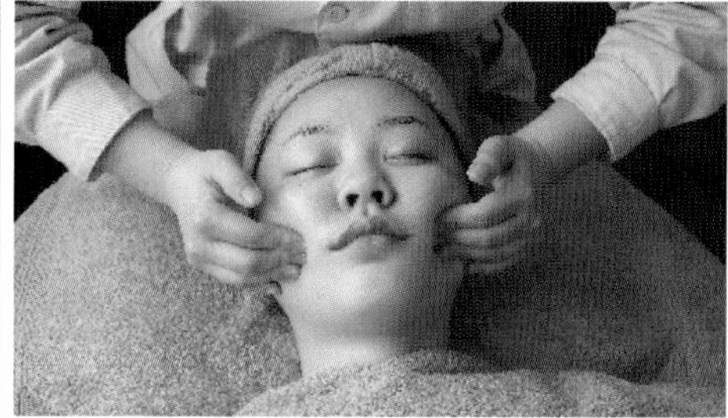

(b) 按压颧髎

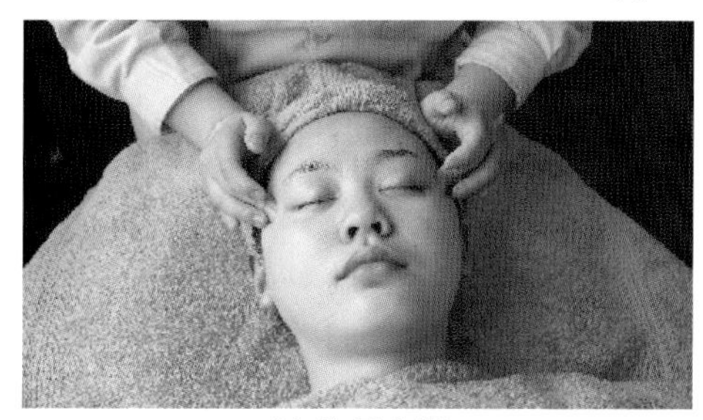

(c) 按压太阳

图 3-16 "四四三二一"操作示意图

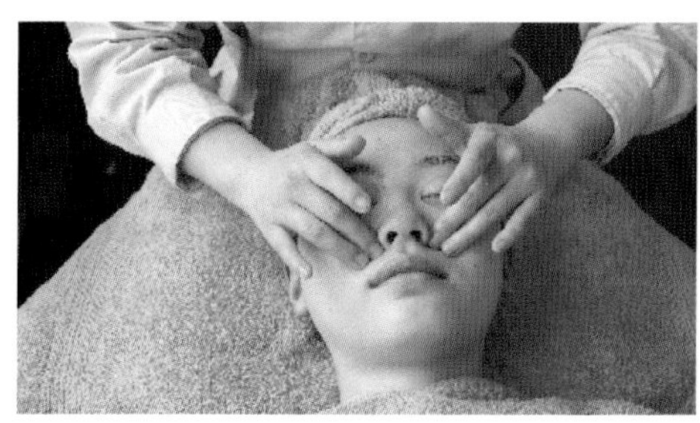

(a) 按压迎香

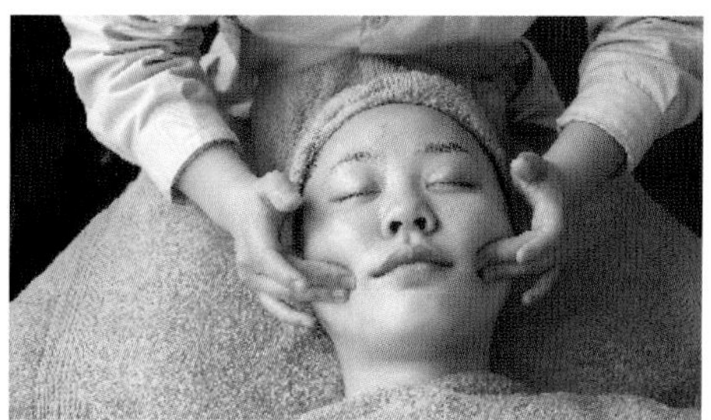

(b) 按压颧髎

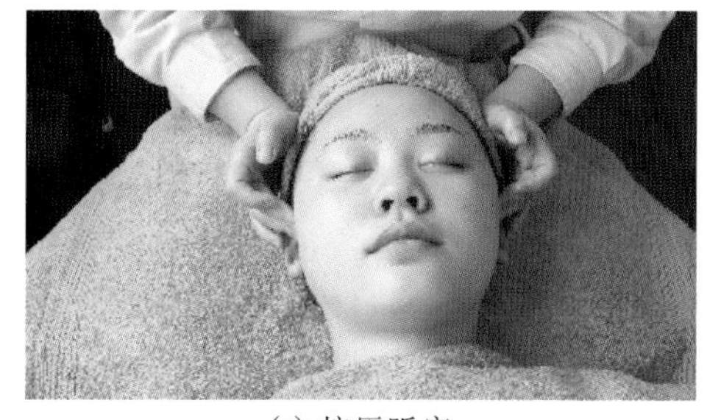

(c) 按压听宫

图 3-17 "五点二指"操作示意图

(3)"下二上三"(图 3-18):双手下划至下颌骨下重叠,抬下颌两次;放松后重叠的手掌上划至下巴,在承浆穴向下按压三次。

(4)"美容指划口周"(图 3-19):用美容指划口轮匝肌,力度上重下轻。

(5)"四指五点"(图 3-20):以四指(除拇指)按压,由迎香穴开始,沿颧骨下缘至耳后翳风，每隔两横指按压一次，共按压五个点。

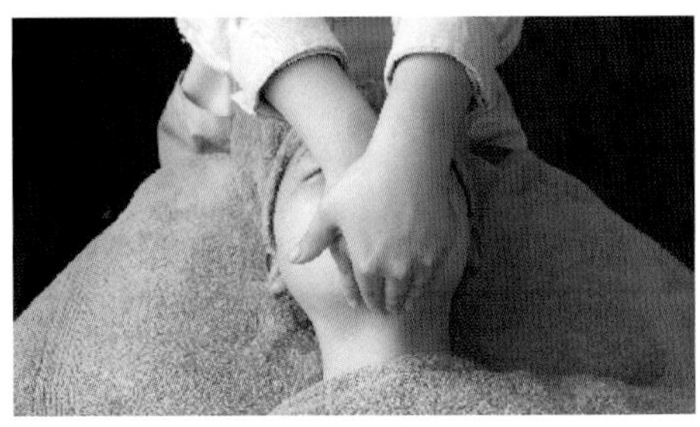

(a) 抬下颌

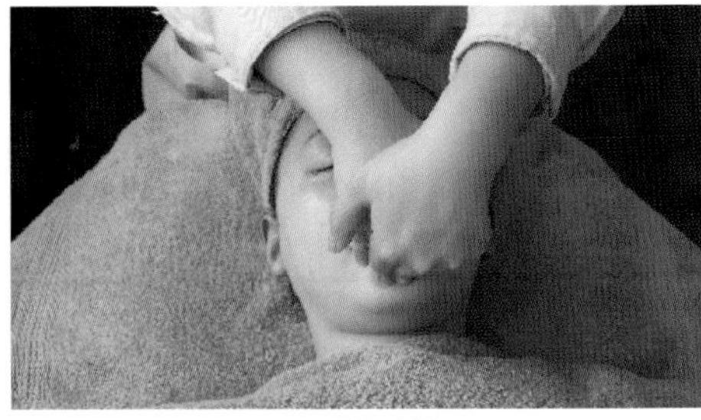

(b) 压下颌

图 3-18 “下二上三”操作示意图

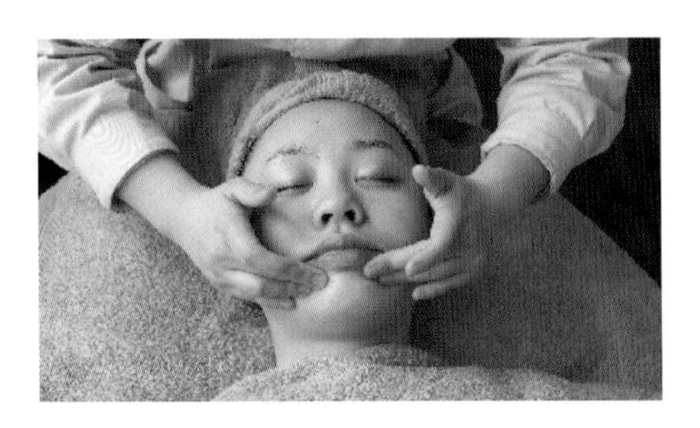

图 3-19 “美容指划口周”操作示意图

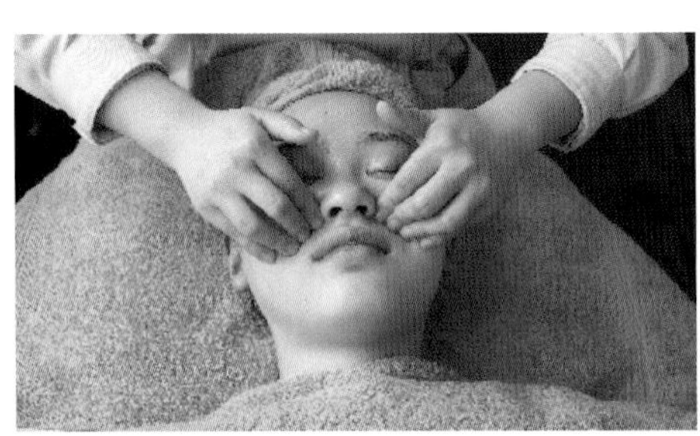

(a) 按压迎香

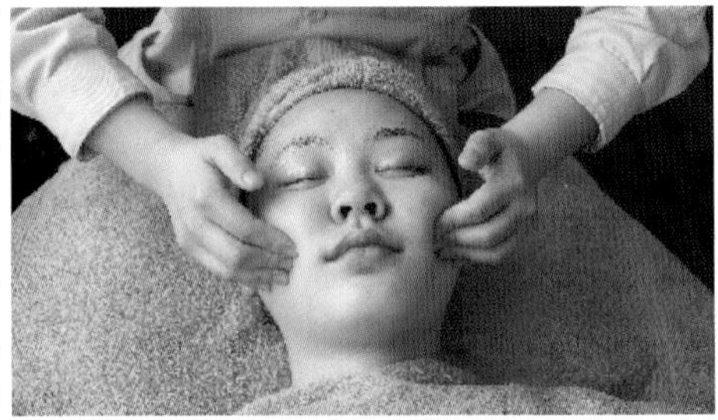

(b) 按压颧髎

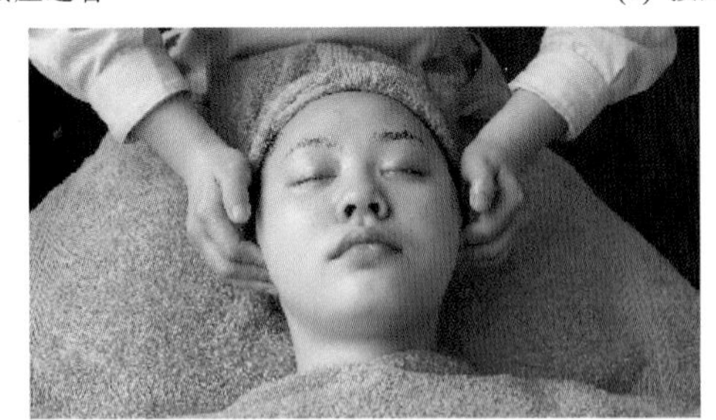

(c) 按压翳风

图 3-20 “四指五点”操作示意图

(6) “蝴蝶飞”(图 3-21)：大拇指由耳下沿脸颊外缘划至下巴，反转手掌，以食指指背带力拉至耳后。

(7) “边提边排”(图 3-22)：四指(除拇指)沿下颌内缘由下巴至耳下上提下颌然后向后下划拉，淋巴排毒。

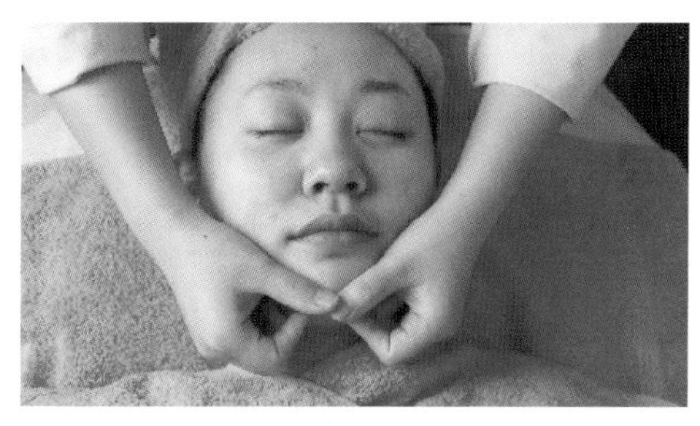

(a) 蝴蝶飞1

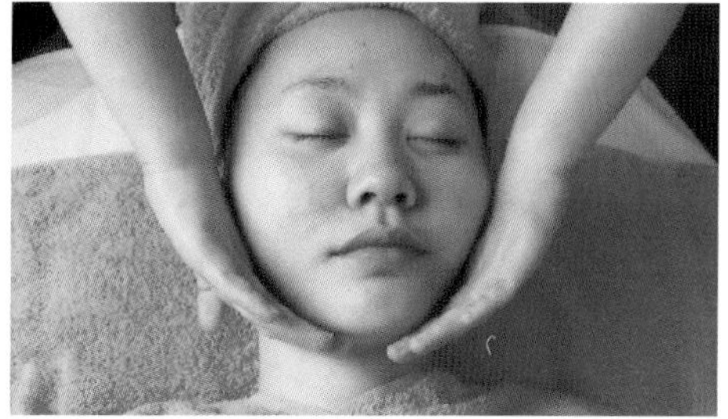

(b) 蝴蝶飞2

图 3-21 “蝴蝶飞”操作示意图

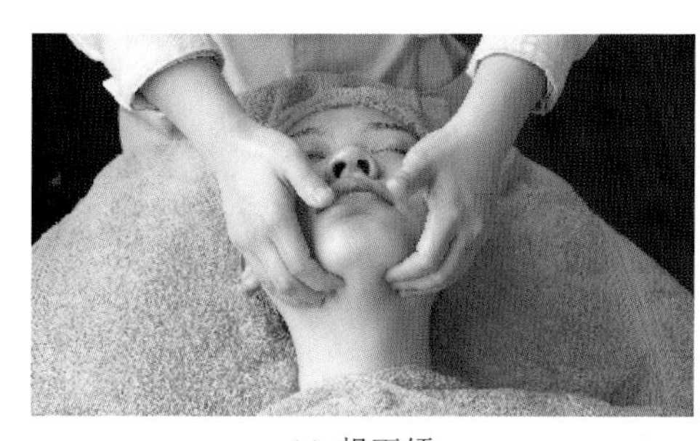

(a) 提下颌

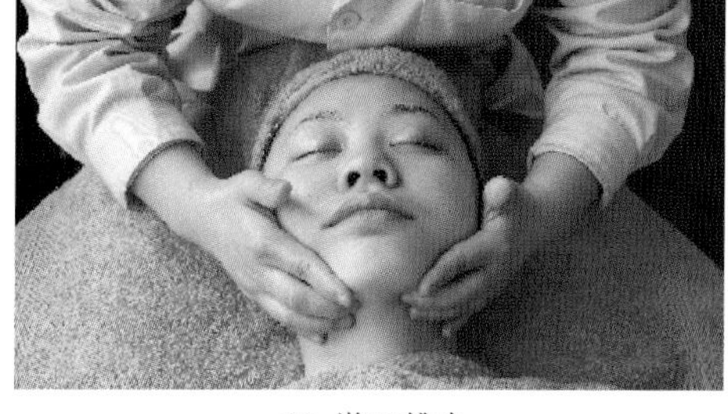

(b) 淋巴排毒

图 3-22 “边提边排”操作示意图

(8)“拇指划口周”(图 3-23):大拇指以上重下轻的方式划口轮匝肌。

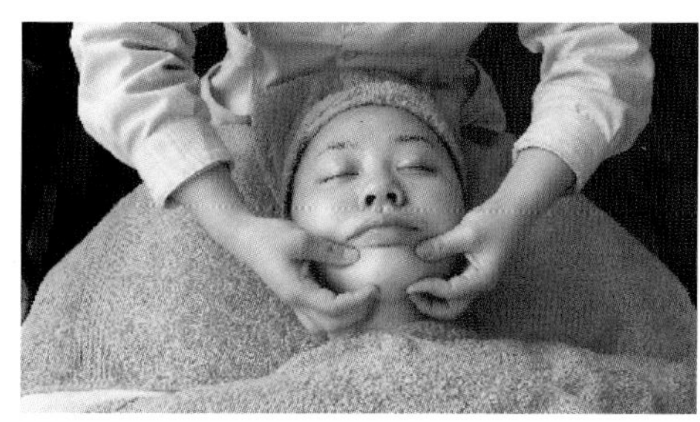

(a) 拇指划下口周

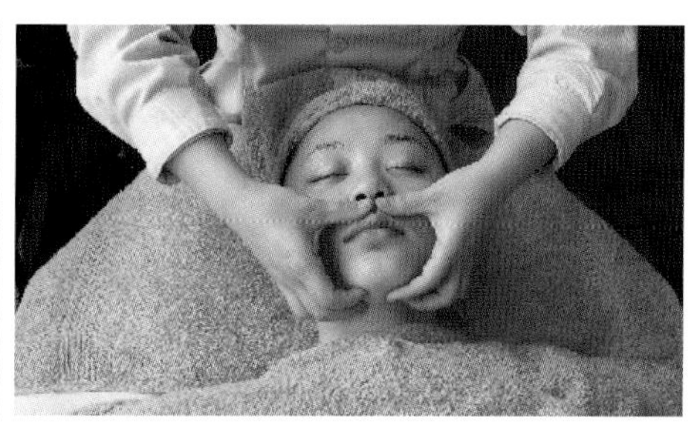

(b) 拇指划上口周

图 3-23 “拇指划口周”操作示意图

(9)“塑鼻型”(图 3-24):大拇指划口轮匝肌至鼻梁,再由鼻头划出至承浆,塑造鼻型。

(10)“按印堂”(图 3-25):大拇指划口轮匝肌至鼻梁、鼻头,再上划至印堂按压,沿眉骨往两侧划出,安抚鼻根肌。

(11)“额头三条线”(图 3-26):重复(8)、(9)、(10)步骤,在额头中间和发际线处分别往两侧划出,于发际线中间按压神庭穴。

(12)“过山车”(图 3-27):一手中指按压印堂,划至对侧睛明、迎香、地仓,然后手掌包下巴,沿同侧脸颊带力拉至太阳,双手交替进行。

(13)“横向按摩额头”:用双手掌依次横抹额头。

(14)“斜向安抚额头”(图 3-28):双手掌交替安抚额头,即掌根从对侧眉骨斜拉至另一侧发际线处。

(15)双手掌交替依次向上安抚额头。

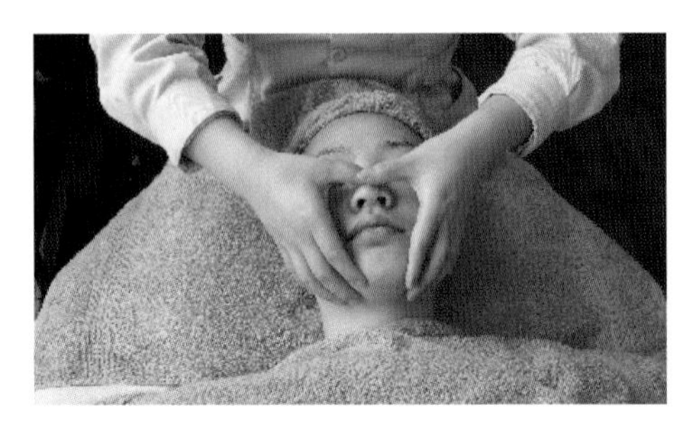

图 3-24 “塑鼻型”操作示意图

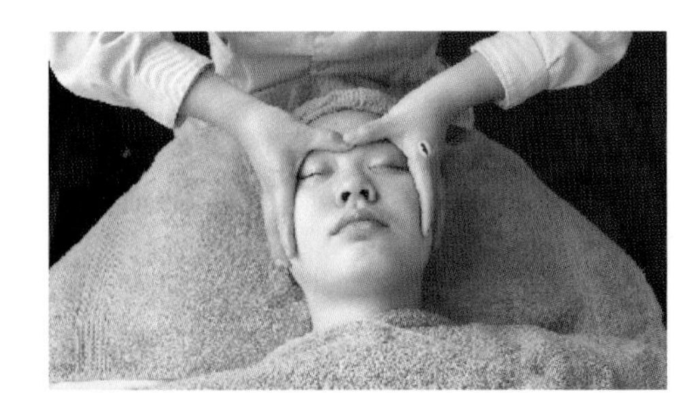

图 3-25 “按印堂”操作示意图

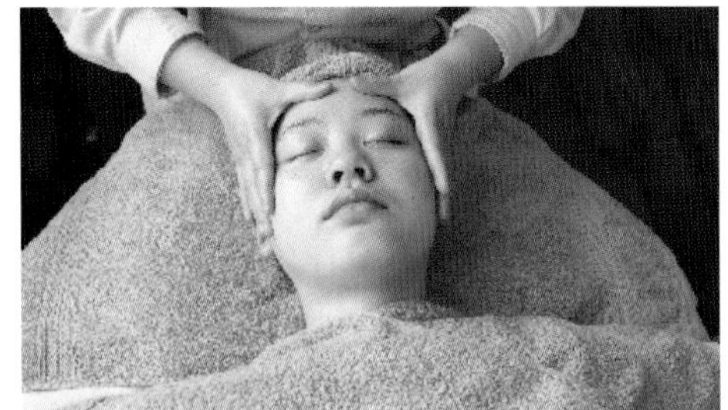

(a) 按压额头中间线

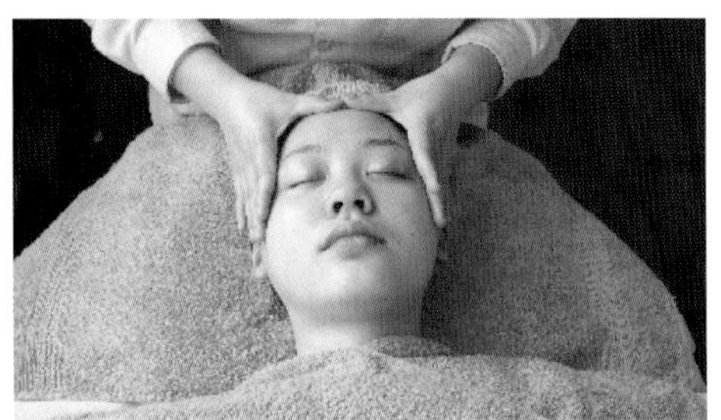

(b) 按压神庭

图 3-26 “额头三条线”操作示意图

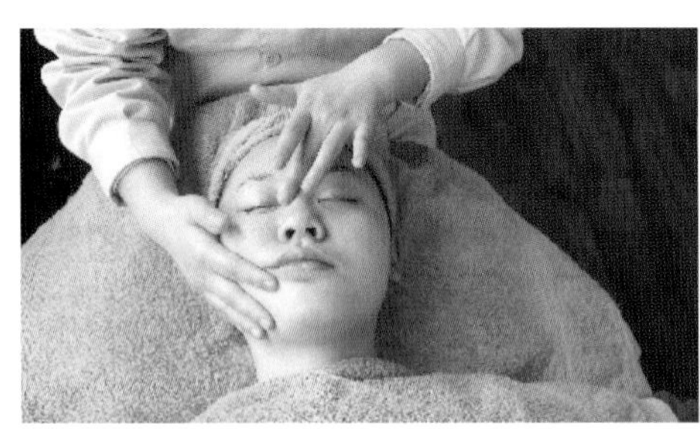

(a) 划至睛明

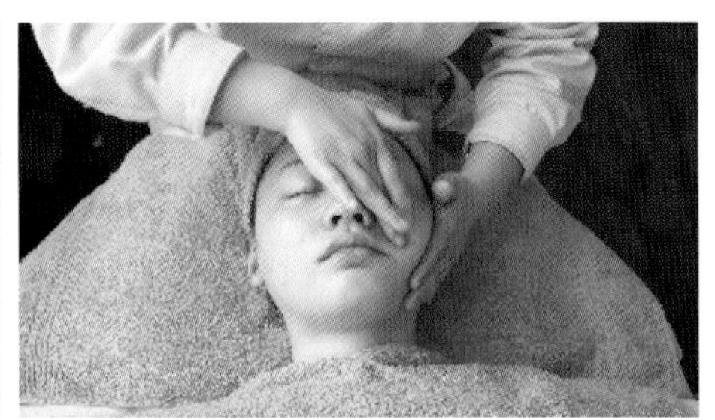

(b) 划至迎香

图 3-27 “过山车”操作示意图

(16)“推眉”(图 3-29):一手掌平放额头上,另一手美容指按压同侧攒竹,划至对侧眉骨,然后手掌包额头向上划拉;另一手美容指重复操作。

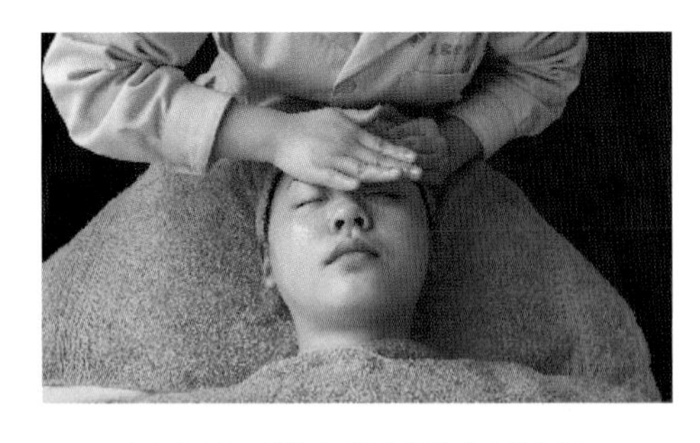

图 3-28 “安抚额头”操作示意图

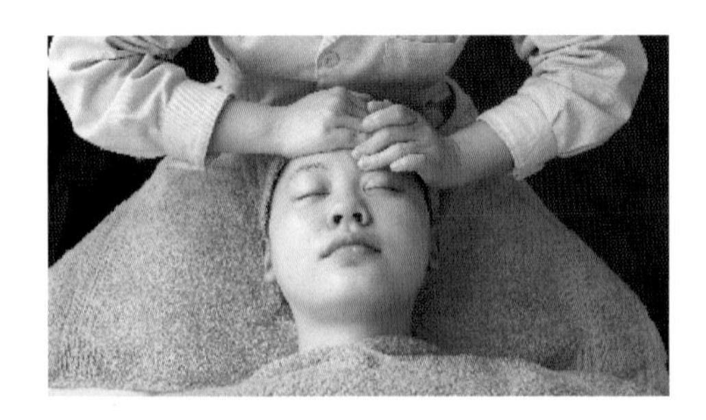

图 3-29 “推眉”操作示意图

(17)“爽眉”(图 3-30):以食指、中指夹住眉筋向眉尾划出至太阳,于太阳划三圈,再由下眼圈划回眉头。

(18)“点压三穴”(图 3-31):用美容指依次按压攒竹、鱼腰、丝竹空。

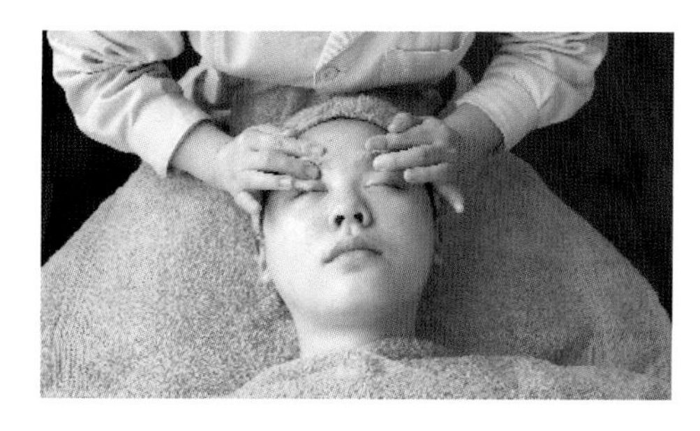

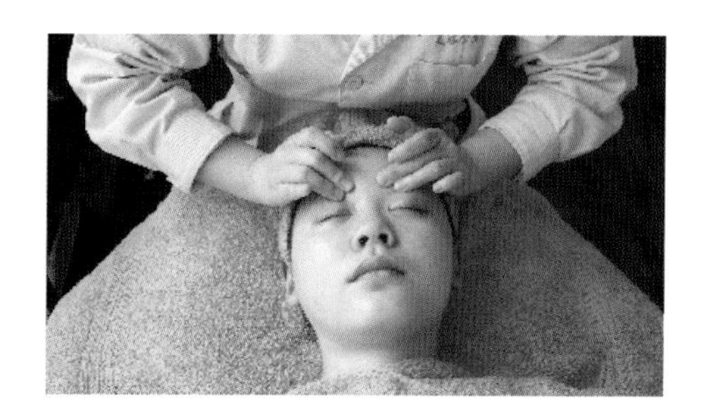

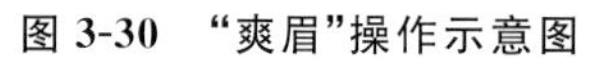

图 3-30 “爽眉”操作示意图

图 3-31 “点压三穴”操作示意图

(19)“拨眉”(图 3-32):在眉尾处双手交替向上拨划眉筋。

(20)“划 8 字”(图 3-33):大拇指在眉尾、眼尾处划“8”字,力度上重下轻。

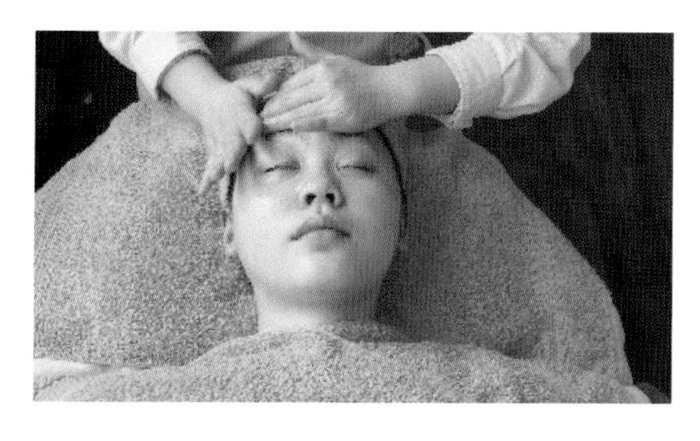

图 3-32 “拨眉”操作示意图

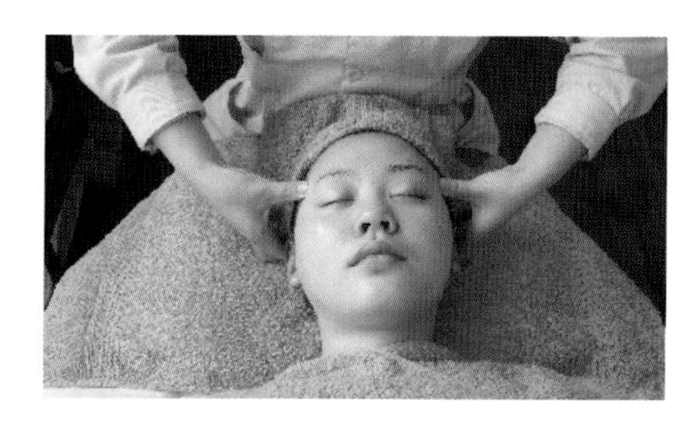

图 3-33 “划 8 字”操作示意图

(21)“全脸安抚”(图 3-34):双手掌安抚全脸,双手掌合十,小指从印堂出发,依次经过睛明、迎香,于地仓打开双手,安抚口周下巴,沿脸颊外缘划回额头,重复上述操作,再分别于迎香、睛明打开双手,安抚脸颊、眼部。

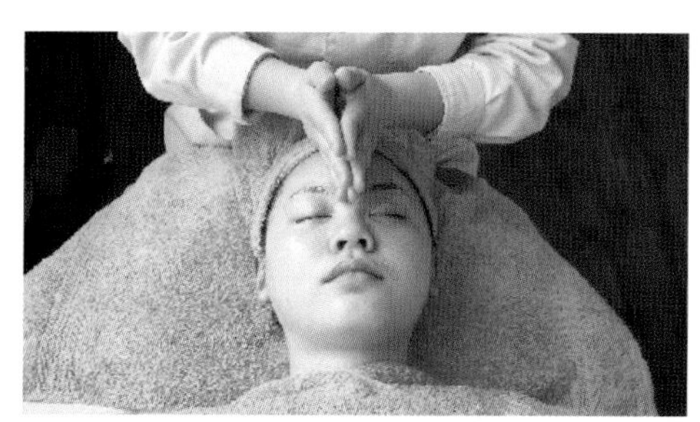

(a) 印堂出发

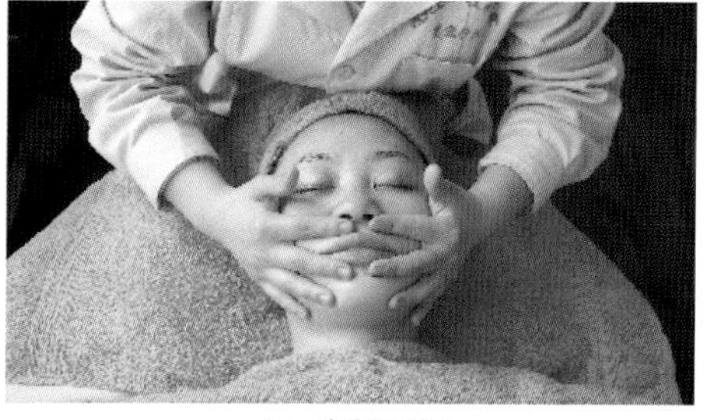

(b) 安抚下颌

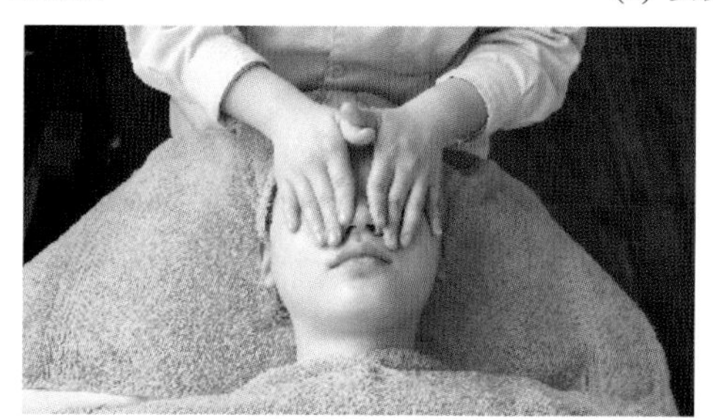

(c) 安抚面部

图 3-34 “全脸安抚”操作示意图

(22)“按压额头”(图3-35):用双手掌依次横抹额头,然后双手掌心向下重叠,按压额头。

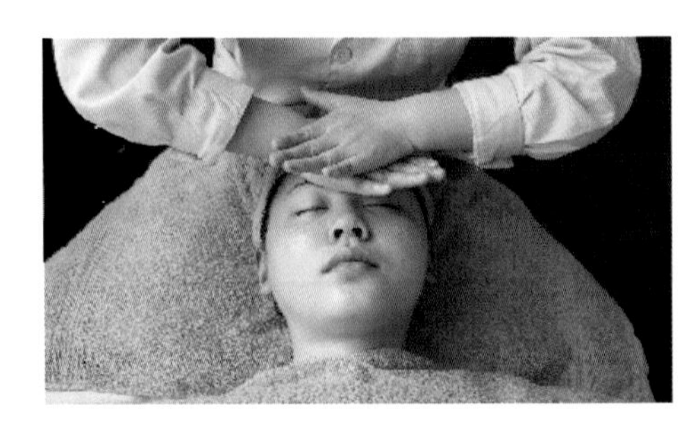

图3-35 “按压额头”操作示意图

(23)“划拉耳根”(图3-36):双手沿面部轮廓,下划至耳下,用食指、中指划拉耳前耳后。

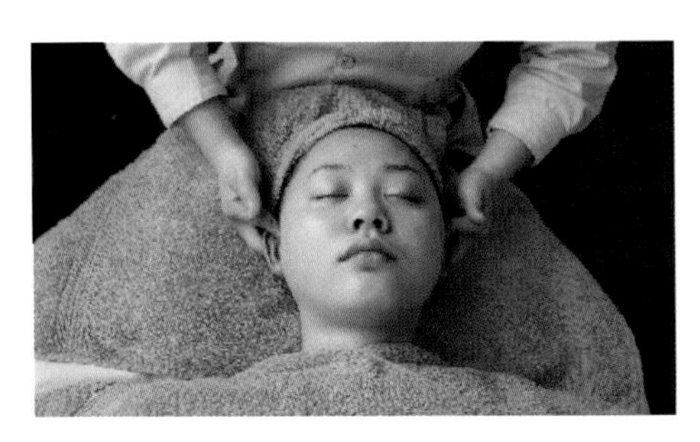

图3-36 “划拉耳根”操作示意图

7. 面部按摩的注意事项

(1) 根据顾客皮肤特点灵活选择按摩介质、按摩方法,按摩动作要熟练、准确,能够配合不同部位的结构特点变换手型。

(2) 按摩时间不可太长,以10～15 min为宜。长时间的按摩易导致皮肤疲劳甚至擦伤,影响按摩效果;长期、长时间按摩容易导致皮肤老化,皮肤抵抗力下降。

(3) 按摩过程中双手不可同时离开顾客面部,如需暂时离开,动作要轻柔,重新开始也应如此。

(4) 严禁让按摩产品进入顾客的眼、鼻和嘴中。

(5) 敏感皮肤按摩时间不宜超过5 min,避开敏感部位,点穴要轻,少弹脸、摸脸。用加有扑尔敏的水冷喷,配合使用抗敏系列产品作为按摩介质操作。

(6) 不能进行常规按摩的情况:过敏皮肤、面部红血丝、痤疮性皮肤、急性皮炎、哮喘、气管炎、鼻炎患者等。

(四) 敷面膜

1. 面膜的作用原理 面膜是集洁肤、护肤和美容为一体的多用途保养品。通过在面部敷、抹面膜并停留一定时间,使其与肌肤充分接触,形成一层薄膜,然后将膜取下或用清水洗掉。原理是:利用覆盖在脸部的短暂时间,暂时隔离外界的尘埃和污染的空气,提高肌肤温度,使皮肤毛孔扩张,促进汗腺分泌与新陈代谢,使肌肤的含氧量上升,以利于肌肤排出皮肤细胞新陈代谢的产物和表皮累积的油脂类物质。面膜中的水分渗入表皮的角质层,使肌肤变得柔软、光亮,有弹性。

2. 面膜的作用

(1) 清洁:由于面膜对皮肤表面物质的吸附作用,在剥离或洗去面膜时,可以将皮肤上的分泌物、皮屑、污垢等随面膜一起除去，达到彻底清洁肌肤的效果。

（2）营养：面膜覆盖在皮肤表面，将皮肤与外界空气隔离，使皮肤温度上升，减少皮肤水分丢失，软化角质层，扩张毛孔，促进血液循环，使皮肤有效地吸收面膜中的活性营养成分，达到护肤效果。

（3）紧致肌肤：在面膜形成和干燥的过程中，由于表面张力的作用，可以收紧松弛的肌肤，有助于消除和减少面部细小皱纹。

（4）特殊作用：面膜中添加的不同成分，可用于不同问题皮肤，解决不同肌肤问题。

3. 面膜的分类及特点

（1）按功能不同可分为：

①清洁面膜：最常见的一种面膜，可以清除毛孔内的污物和多余油脂，去除老化角质，使肌肤清爽、干净。

②补水面膜：含保湿剂，将水分锁在膜内，软化角质层，帮助皮肤吸收营养，适合各类肌肤。

③美白面膜：彻底清除死皮细胞，兼具清洁、美白双重功效，使肌肤重现柔嫩光滑，白皙透明。

④抗皱面膜：紧致肌肤，减少皱纹，特别适用于没有时间去美容院做护理的女性。

⑤修复面膜：内含植物精华，软化表皮组织，促进肌肤新陈代谢，适用于干性或缺水性肌肤。

⑥滋养面膜：含有多种维生素与胶原蛋白，补充肌肤所需营养，令肌肤新生，焕发活力。

⑦舒缓面膜：迅速舒缓肌肤，消除疲劳感，恢复肌肤的光泽和弹性，适用于敏感性肌肤。

（2）按形态不同可分为：

①泥膏型：常见的有矿物泥面膜。

②撕拉型：黑头粉刺专用鼻贴。

③冻胶型：以睡眠面膜最常见。

④湿纸巾型：很常见，一般是单片包装、浸润着精华液的面膜纸。

常见的面膜如下：

①粉状面膜：一般分软膜和硬膜，硬膜较少使用，软膜在美容院和家居中使用较多。

软膜：美容院经常使用的面膜。使用时与液体（一般为纯净水、纯露、爽肤水、牛奶等）混合调至糊状，涂敷在面部（图 3-37）。15～20 min 后形成质地细软的薄膜。温和补水，吸附皮肤的分泌物，可整张膜取下。软膜粉中添加不同成分而具有不同作用，因此适合不同类型的皮肤。常见的有维生素 E 软膜，具有抗衰老作用；当归软膜，可改善肤色、去皱抗老化；芦荟、洋甘菊、薰衣草软膜，温和补水，适用于敏感性肌肤。

硬膜：主要成分是熟石膏，能形成很坚硬的膜，由于熟石膏遇水会放热，可以使膜体温度持续渗透，能够燃烧皮下脂肪，故而也常用于减肥。当硬膜粉中添加冰片、薄荷等具有收敛、消炎作用的成分时，则是一种冷膜，通过对皮肤的冷渗透从而达到抑制皮脂分泌、清热消炎、镇静肌肤的作用，能形成坚硬的膜，不易整张膜取下，容易形成碎末。因此，在涂敷硬膜前，需覆盖一层与脸部大小一致的纱布，露出眼睛和嘴巴，将硬膜迅速涂敷在纱布上。由于硬膜的吸水性和收敛作用较强，有一定压迫性，故一般 1 个月使用 1 次。

使用面膜粉时需注意：

a. 调制面膜时，注意水量控制。加水过多，会使面膜太稀，不易成形，敷面膜时会因重力

(a) 取面膜粉

(b) 加水调和

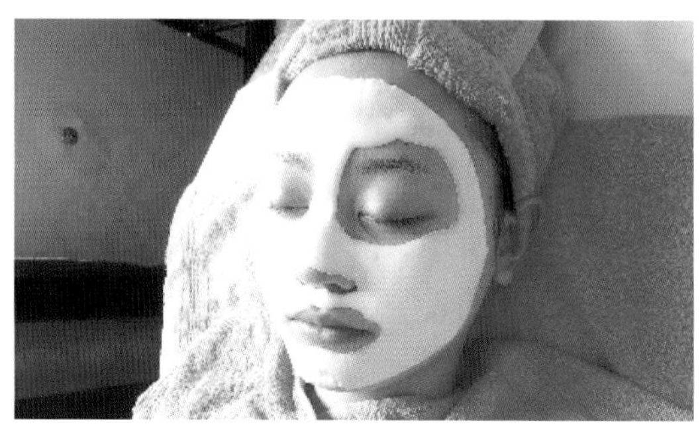

(c) 涂敷面膜

图 3-37 粉状面膜

流至顾客颈部引起不适；加水太少，会使面膜迅速凝固成型而不易涂敷至全脸，影响效果。

b. 涂敷面膜时，按照先 U 区再 T 区的顺序涂敷，控制好厚度，太薄，会影响效果，揭膜时难以完整取下；太厚，成型时间长，且易造成面膜粉的浪费。

c. 注意把握取膜时间。约 15 min 便可取下面膜，取面膜的时间应根据具体情况而定。如空调房内敷膜时间不宜太久，否则面膜会受环境影响而迅速干燥，倒吸面部皮肤水分。

d. 取膜之后，要注意下巴、耳后、发际等位置是否残留膜渣，可用湿面巾纸擦拭干净。

②膏状面膜：膏状面膜含油性成分，更具滋润效果，易调整面膜厚度，可局部用，不受脸型限制，服帖度好。根据添加的有效成分不同，可分为美白面膜、舒缓面膜、控油面膜、营养面膜等。根据使用之后的状态不同，分为可干型面膜和保湿型面膜。

可干型面膜：面膜涂于皮肤之后，逐渐凝固干燥，可整体揭除。膜体与皮肤的亲和力较强，涂膜之后，随着膜体干燥，皮肤紧绷感越来越明显，收敛性强。揭膜时将毛孔深层污垢及老化角质一起带下，具有较好的清洁作用，适用于油性皮肤，敏感性皮肤禁用。

保湿型面膜：面膜敷于皮肤一直保持湿润状态，面膜中的有效成分在湿润环境中发挥作用，常用于眼部或干性皮肤的护理，有较好的滋润作用。

使用膏状面膜需要注意：

a. 涂敷顺序与软膜涂敷顺序相似。由皮肤温度最低的地方开始，依次涂敷脸颊、唇周、鼻翼、额头，温度低处需要花较长时间渗透，依照此顺序涂敷，就能确保渗透率相同。

b. 先用化妆水滋润皮肤。若皮肤没有充分滋润，面膜中的油脂容易阻塞毛孔。

c. 面膜涂敷厚度适中，过薄无法形成一个封闭渗透的“护肤场”。

③海藻面膜：海藻面膜的有效成分是从海藻中提炼的海藻胶，具有凝胶作用，能增加肌肤

的锁水性、紧缩性及弹性,达到除皱目的。还有平衡油脂分泌,消炎杀菌之功效,对痤疮有很好的治疗效果。

海藻面膜是深褐色颗粒状(图 3-38),用温水调制,轻轻搅动,1 min 内析出很多海藻胶(图 3-39),即可敷于面部,待干,取下面膜。

图 3-38 海藻面膜

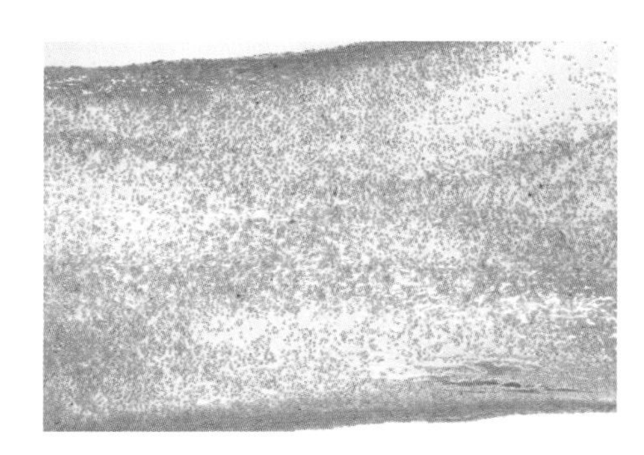

图 3-39 析出海藻胶

④骨胶原面膜:骨胶原面膜属于抗衰老面膜的一种。骨胶原含有多种细胞因子,可以使细胞再生,一次可以修复上万个肌肤细胞,补充大量透明质酸和胶原蛋白,能提供皮肤所需要的水分,形成很好的保护水层,修复皮肤红血丝,减轻皱纹,分解色斑,祛除色素沉着,提升面部轮廓。适合缺水、疲惫、衰老肌肤,以及敏感脆弱肌肤和晦暗肌肤。

使用骨胶原面膜时需注意:

a. 足够量的水:骨胶原面膜运用冷冻干燥法保存,活性成分处于休眠状态,是水溶性产品,只有加入足够量的水,活性成分才能完全释放出来。骨胶原面膜具有超强的吸水功能,吸水量是自身重量的 30 倍,即 1 张面膜 2 g,注入 60 mL 水才可以发挥出更好的效果。

b. 使用时间:由于面膜的活性成分遇水后释放以及皮肤对骨胶原面膜吸收均需要一定的时间,因此,面膜停留的最佳时间为 30～45 min。

c. 水质选择:使用纯度较高的水。普通水中的杂质和某些成分容易破坏骨胶原的成分,影响效果。建议使用纯净水。使用玫瑰纯露、橙花纯露、薰衣草纯露效果更佳。

d. 清洁皮肤:敷面膜前彻底清洁皮肤,将水溶性产品清洁干净,不使用带油分的产品按摩。因含油物质遇骨胶原会影响骨胶原的吸收,且易引起皮肤过敏。

e. 注意事项:取下面膜后不要用清水清洁,用化妆棉蘸取爽肤水擦拭即可。调制好的骨胶原面膜处于弱酸性,对肌肤有保护作用,遇水后影响其效果。

⑤片状面膜:片状面膜是人们最常用的一款家居面膜。使用方便,操作简单。片状面膜中的片状物质只是作为一种载体,真正有效物质是混入其中的高浓度精华液。作为载体的片状物质有蚕丝的、全棉的、涤纶的、混纺的、天丝的、生物纤维的等,其中生物纤维面膜最好,混纺面膜质量最次。

无纺布面膜是贴式面膜的一种,以无纺布为精华液载体。纯棉无纺布面膜的感觉柔润舒服,密封性好,透气性一般,精华液少时会翘起来,不是很服帖。它使用广泛的原因是成本低,价格低廉(图 3-40)。

蚕丝面膜轻薄,服帖性好,透气性好,因材质薄承载精华液有限,精华液多留在面膜袋中,在敷膜过程中精华液容易流失(图 3-41)。

生物纤维面膜服帖性好,透气不滴水,具低敏性,产品中富含类似人体表皮细胞的生物活性体,与人体肌肤细胞亲和性极高,营养物质极易通过毛孔吸收,到达肌肤深层,效果好。美

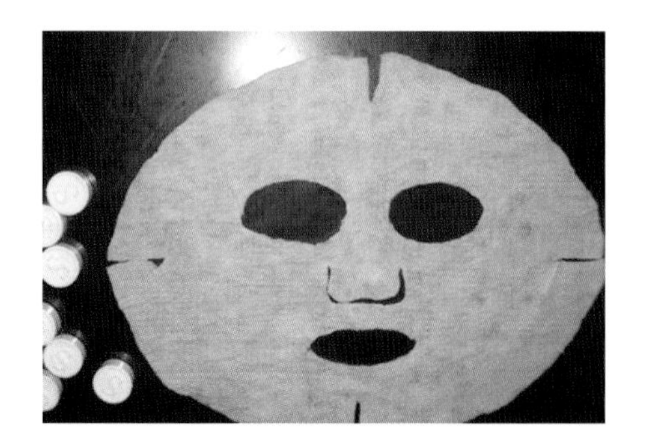

图 3-40 无纺布面膜

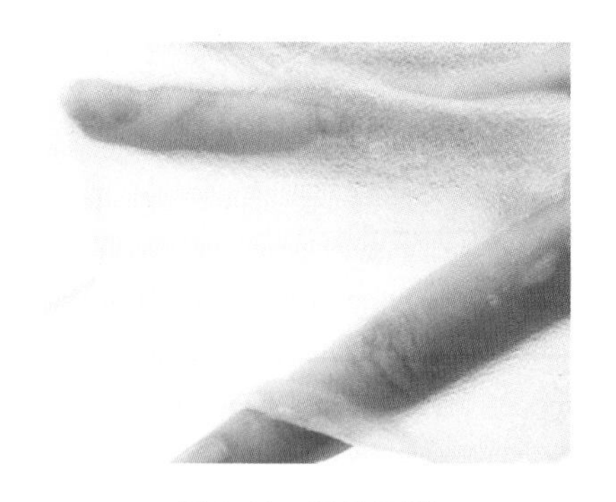

图 3-41 蚕丝面膜

容会所已在逐步推广这种面膜,但由于制作成本较高,目前使用不广泛(图 3-42)。

片状面膜的功能多以保湿补水、美白肌肤和滋养修复为主。对皮肤深层清洁的效果没有膏状面膜和面膜粉明显。根据外观,片状面膜分为压缩面膜和非压缩面膜。

a. 压缩面膜约五角硬币大小,厚度为 6 mm 左右。使用时放在液体中,自然膨胀,打开即是一片适合脸型大小的面膜纸。外出时携带方便。浸泡的液体可根据需要选择相应功能的精华液。

b. 非压缩面膜使用比较方便。有适合脸型面膜和可覆盖颈部面膜,可满足大众的需求。

⑥睡眠面膜:睡眠面膜是做完基础护肤之后,将面膜敷在脸上直接睡觉的一种面膜,一般第二天早晨正常洁面即可。睡眠面膜是啫喱或乳霜质地,像涂了一层护肤品,其特点是可以保留在面部不用立即清洗,能有效舒缓身心疲劳,提升睡眠质量,更好地促进肌肤夜间新陈代谢,备受女性欢迎。睡眠面膜一般在 6~8 h 后清洗,一周做 2~3 次。

⑦超导面膜:一款创新型面膜,它由 PET 薄膜与丝质纤维膜经高科技热融合技术复合而成,因其具有极佳的吸收度、保湿度、亲肤性及超强的导入性等特征,故被称为"PET 超导面膜"(图 3-43)。面膜外层有许多细小气孔,厚度只有 2 μm;面膜内层为超轻的 40 g/m^2 的 Rayon 纤维膜巾。

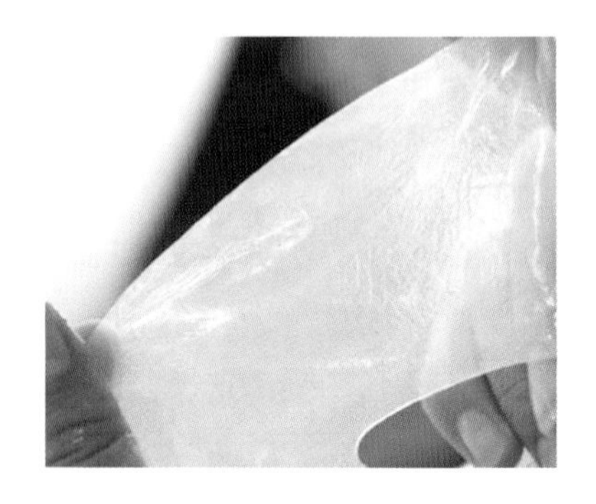

图 3-42 生物纤维面膜

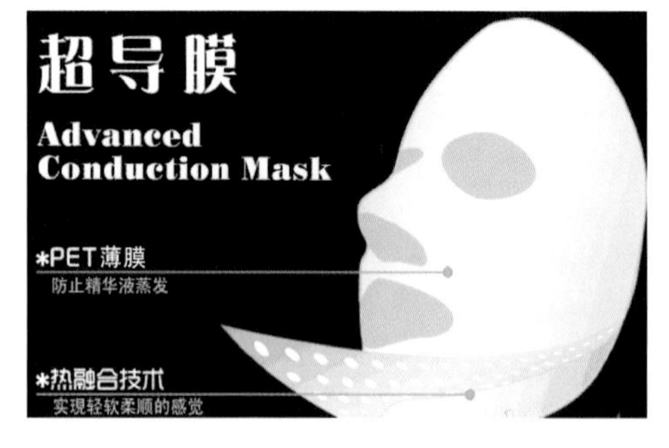

图 3-43 超导面膜

超导面膜的特点是其外层 PET 薄膜的特殊材质及细小气孔,让面膜透气不透水。可以有效阻隔空气反吸,防止精华液蒸发,能利用大气负压原理增加面膜附着力和传导作用,让精华液加倍渗透和吸收。

使用超导面膜时注意,不要撕开 PET 薄膜,将丝质面贴敷在脸上,10~15 min 取下面膜,无需洁肤。

知识拓展

自制面膜

自制面膜是指选用天然材料自己动手制作的面膜。新鲜水果、蔬菜、鸡蛋、蜂蜜、中草药和维生素等富含皮肤所需营养，副作用少，物美价廉，是现代女性的美容佳品。制作方法如下。

1. 苹果面膜　将苹果磨成泥，加入柠檬汁和少许盐后搅拌，敷于脸上。具有清洁、去角质效果。

2. 黄瓜、胡萝卜、蛋清面膜　将黄瓜和胡萝卜搅碎，与蛋清混合敷于面部。具有滋养肌肤、改善肌肤粗糙的功效。黄瓜、胡萝卜打成汁加入蜂蜜饮用，会配合面膜发挥更好的效果。

3. 蜂蜜、蛋黄、橄榄油面膜　三者混匀敷于面部即可。为肌肤补充养分。

4. 陈醋、蛋清面膜　将鸡蛋浸入醋内，72 h 后捞出，取蛋清备用。每晚睡觉前以蛋清敷面，一周两次。可以控制痤疮生长，有杀菌消炎之功效。

4. 使用面膜的注意事项

(1) 使用面膜前，先清洁面部，也可按摩后用热毛巾敷面 2 min 再敷膜。温热水让毛孔打开，排出皮脂和污垢，面膜营养成分更易吸收。

(2) 干性皮肤或气候干燥时，需先拍柔肤水，再敷面膜。

(3) 涂敷面膜时，用手指将面膜均匀涂于脸部和颈部，注意面膜必须距离眼睛和嘴唇 0.5 cm左右，避免眼睛和嘴唇受到刺激引起不良反应。距离发际 0.5～1 cm，以免面膜黏附于头发而不易清洗。

(4) 敷面膜后脸部不宜做表情动作，以免面膜与皮肤接触不紧密而影响吸收与效果。

(5) 面膜涂敷时间根据其性质来定。水分含量适中，约 15 min 即可清洗，以免面膜过干而吸收肌肤水分；水分含量高，30 min 即可清洗。因此，敷面膜时间不是越长越好，时间“超支”，会导致肌肤失水、失氧。

(6) 面膜应由下往上轻轻揭下。

(7) 揭去面膜之后，面部清洗与否，根据面膜的使用要求而定。清洁类面膜需要清洗，保养类面膜不需要清洗。滋润或补水面膜取下后建议用手按摩 2～3 min，使营养充分吸收。

(8) 敷面膜频率依据年龄和皮肤情况而定，年轻人 1 次/周。若皮肤过于粗糙、松弛，可 2 次/周。如果使用太频繁，易引起角质层增厚，改变皮肤的正常代谢，导致红肿、敏感等不良反应。滋养类面膜频繁使用，容易导致痤疮产生。

(9) 补水类面膜在极度缺水的状态下使用，面部会有刺痛感，属正常现象。面膜简易测试方法：可先在前臂内侧皮肤上涂抹少量面膜，20 min 后若无过敏反应，即可敷在脸上。

(10) 面膜是周期性护理品，为皮肤补充水分和营养，因此，敷完面膜后要涂抹乳液或面霜，锁住水分，滋养肌肤。

(五) 后续工作

1. 基本保养　面膜养护结束之后，整个面部护理的主要护理操作就基本结束，此时需要滋养皮肤，做好防护工作。依次使用爽肤水、眼霜、精华液、乳液、面霜，涂擦防晒隔离霜，做好防晒工作。

2. 协助顾客 拆开包头巾，取下肩巾，协助顾客起床，整理其衣物、头发。必要时为顾客化淡妆。

3. 结账与送客 协助顾客结账，送顾客离店。

4. 环境整理 护理车上的用物整理归还至配料间；污物放到指定位置；美容床上用物、顾客浴衣回收清洗消毒，更换干净物品备用；关闭仪器，切断电源，做好仪器养护；打扫护理间的卫生。

5. 记录 认真记录护理及沟通内容，如护理项目、顾客喜好、皮肤受力程度、顾客感兴趣的沙龙活动及满意度等。

6. 跟踪回访 美容师可通过电话、微信等方式咨询顾客护理后的感受，提醒顾客家居护理注意事项，预约下次到店时间。

二、面部刮痧

（一）刮痧疗法基础知识

1. 面部刮痧的概念 面部刮痧是根据刮痧治病的原理派生出来的一种新颖的面部护理方法，是运用特质刮痧板，沿面部经络实施一定手法，刺激面部腧穴，疏通经络、平衡阴阳、调整面部生物信息，从而达到行气活血、舒缓皱纹、美白消斑、排毒养颜、护肤美容的目的。

2. 面部刮痧的材料

（1）刮痧板：刮痧板是刮痧的主要工具，可在人体各部位的经络与腧穴使用。通常选用天然的水牛角、磬石、玉石、木材等制成刮痧板。

刮痧板根据身体部位和面部结构差异而被设计成不同的形状，如矩形、三角形、鱼形等。面部刮痧选择用鱼形刮痧板(图 3-44)；矩形、三角形刮痧板多用于躯干和四肢(图 3-45)。

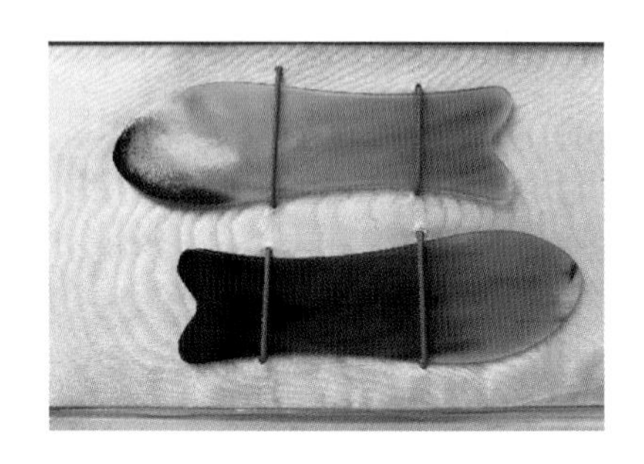

图 3-44 鱼形刮痧板

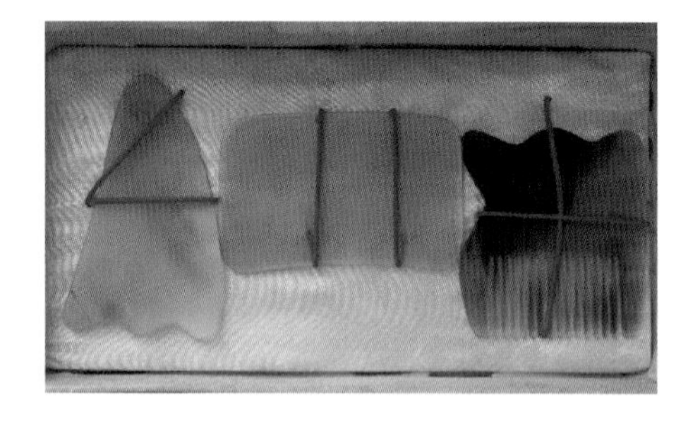

图 3-45 其他形状刮痧板

鱼形刮痧板，结构包括鱼头、鱼身、鱼腰、鱼尾、鱼背、鱼尾曲线状凹口六个部分。面部刮痧操作常用两块鱼形板，左右手各一个配合使用。刮痧操作时，鱼头与鱼尾部专门用于点穴，鱼身、鱼尾处的曲线状凹口多用于面部经络刮拭，鱼背多用于安抚皮肤。

（2）刮痧介质：刮痧时，为保护皮肤或减少皮肤的摩擦损伤，除借助某些药物的辅助作用外，常在刮痧区涂抹脂类、霜类、油类等润滑物质，这些物质统称为刮痧介质。常采用天然植物油添加红花、紫草等天然中草药提炼加工而成，也可直接选用芳香精油作为刮痧介质。

3. 面部刮痧的作用与原理

（1）排出代谢产物：刮痧操作可使局部组织血管扩张，黏膜渗透性增强，血液循环和淋巴循环加快，细胞的吞噬作用增强，改善肌肤深部微循环，使肌肤代谢产物经血液循环和淋巴循环途径迅速排出，改善面色，有效消除、淡化色斑。

（2）促进和恢复细胞的自身功能：面部刮痧有刺激表皮神经末梢、增强传导功能的作用。可以激活受损的细胞，促进和恢复细胞自身分泌和再生等功能。

（3）疏通细胞营养的供给渠道：通过鱼形刮痧板对面部皮肤的刮拭，能有效疏通细胞营养的供给渠道，恢复和增强皮肤细胞自身吸收和利用营养的功能。刮拭产生的热效应，增加了血液、淋巴液和体液的流量，使供应皮肤营养的血液循环持续加快，及时补充营养物质，滋养肌肤，减少皱纹，延缓皮肤衰老。

（4）增强皮肤的免疫功能：面部刮痧可以促进正常免疫细胞的生长、发育，提高其活性，可促进淋巴细胞、白细胞和其他免疫细胞对病毒、细菌的吞噬作用。

（二）面部刮痧的操作程序

1. 准备工作 面部刮痧属于面部护理的一部分，准备工作基本同面部基础护理，另备鱼形刮痧板两块、刮痧介质。

2. 基本步骤

（1）清洁：常规清洁皮肤。

（2）消毒：用75%酒精棉片消毒鱼形刮痧板。

（3）上油：取刮痧介质，由下至上，均匀涂抹全脸。

（4）开穴：

①面部点穴（图3-46）：用鱼头以十字法，依次刺激以下穴位。

a. 承浆、大迎、听宫、听会、太阳。

b. 地仓、颊车、听宫、听会。

c. 人中、迎香、巨髎、太阳。

d. 鼻通、颧髎、太阳。

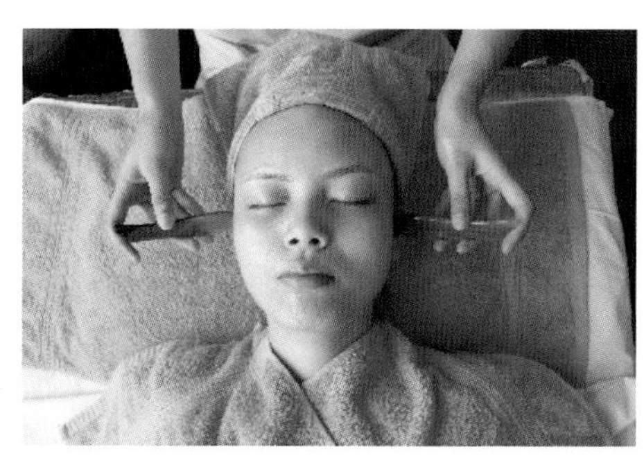

(a) 十字按压听宫

(b) 十字点按听宫

图3-46 面部点穴

②拉抹鼻梁：鱼头朝下，自太阳穴，以鱼身侧曲线下划至下巴，鱼头调转向上，继续划至鼻梁处，该过程称为“走板”，立起刮痧板重叠，鱼尾曲线凹处跨在鼻梁上，双板来回拉抹鼻梁，力度上重下轻（图3-47）。

③将重叠刮痧板呈扇形打开，用下压的鱼尾点压眼周穴位（睛明、承泣、四白、球后、瞳子

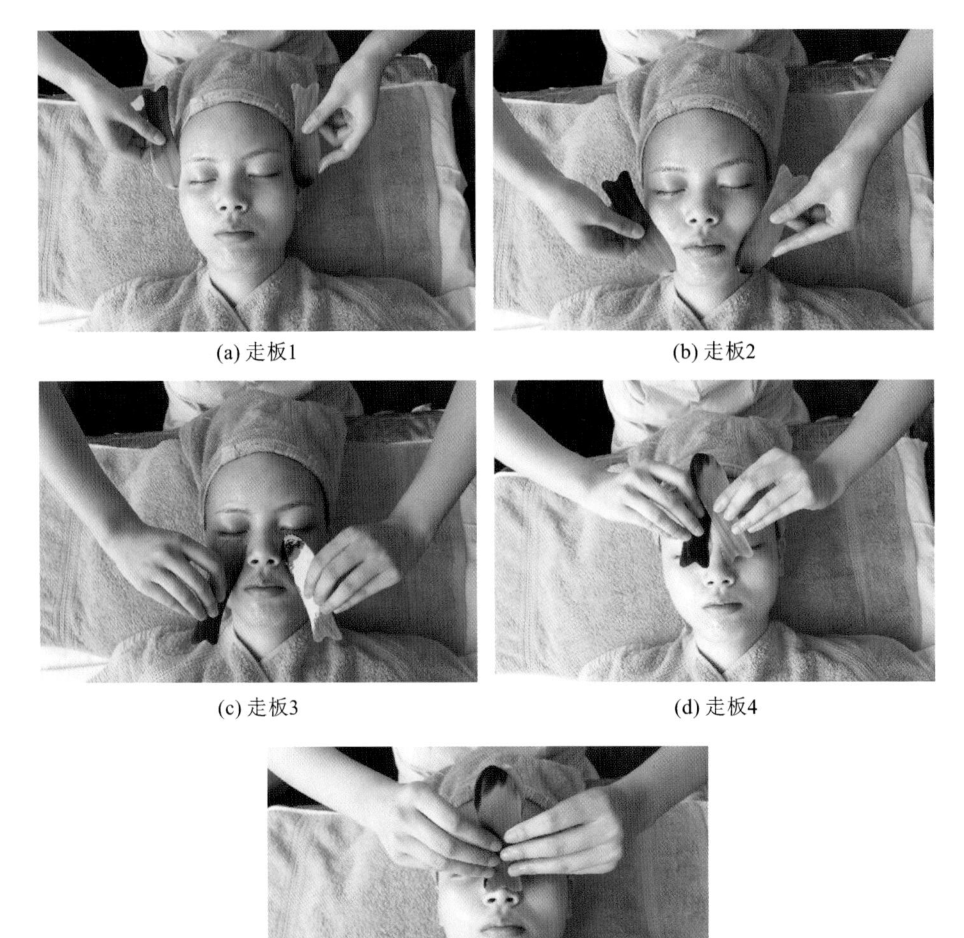

(a) 走板1　(b) 走板2　(c) 走板3　(d) 走板4　(e) 拉抹鼻梁

图 3-47　拉抹鼻梁操作示意图

髎)，鱼尾曲线凹处跨立眉骨，划回至鼻梁，来回拉抹，再呈扇形打开，继续以鱼尾点压攒足、鱼腰、丝竹空等穴位(图 3-48)。

④拉抹额头：自眼尾处，走板，自印堂处重叠两个鱼形板(一板鱼头与另一板鱼尾重叠)，横立刮痧板，用鱼身曲线拉抹额头(图 3-49)。

⑤分开拿好鱼形板，用鱼头点压额部印堂、神庭、头围、太阳等穴位，鱼头朝下，板面平行，用鱼身曲线拉抹耳前，最后用鱼头点压翳风、风池(图 3-50)。

(5) 刮拭全脸：

①上油安抚面部。

②利用鱼身曲线，鱼头向下，分 8 条线刮拭面部(图 3-51)：

a. 下颌骨、耳前。

b. 承浆、耳前。

c. 地仓、耳前。

d. 人中、耳前。

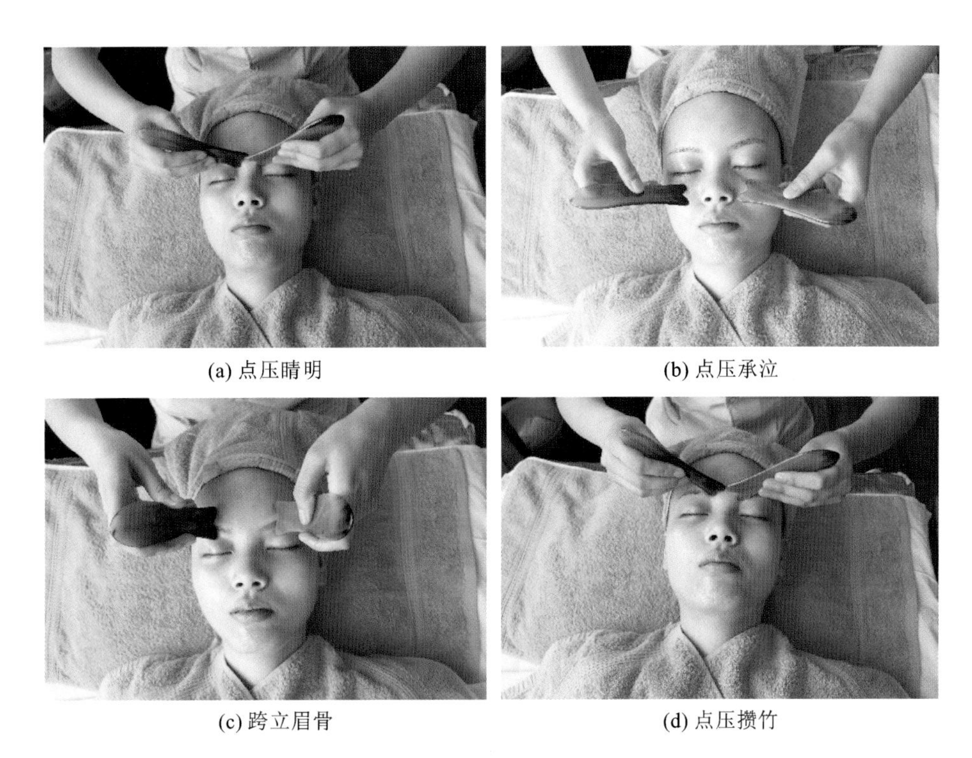

(a) 点压睛明　(b) 点压承泣

(c) 跨立眉骨　(d) 点压攒竹

图 3-48　鱼尾点压眼周穴位

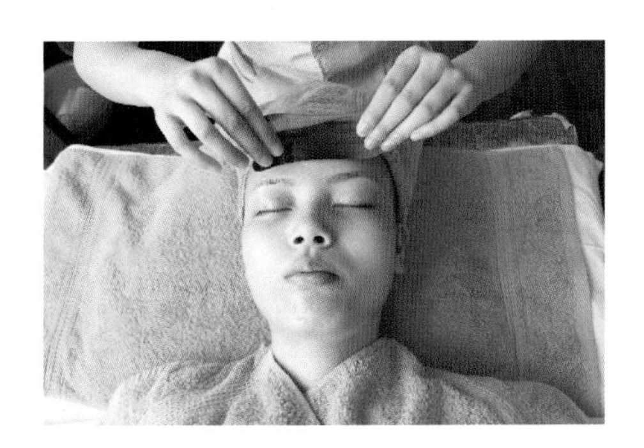

图 3-49　拉抹额头

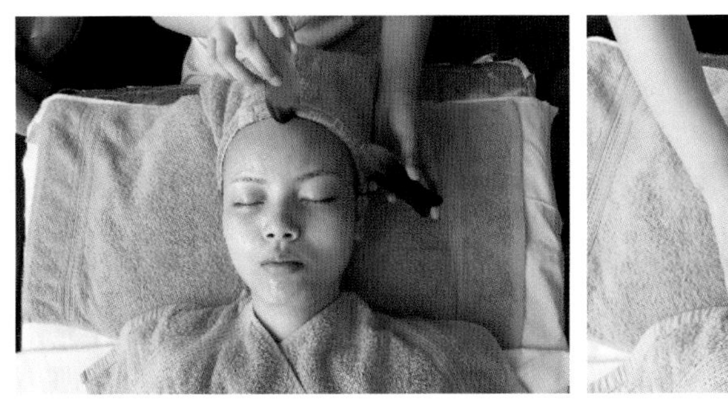

(a) 点压神庭

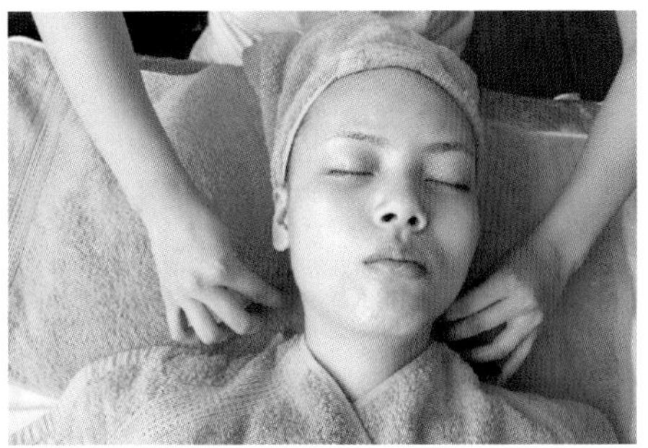

(b) 点压风池

图 3-50　鱼头点压神庭、风池

e. 迎香、耳前。

f. 鼻通、耳前。

g. 眼部、太阳，分两步刮拭：第一步刮拭眼袋，鱼头向上，从内眼角刮至太阳；第二步刮拭上眼睑，鱼头向下，从内眼角刮至太阳。

h. 走板，刮拭额头。

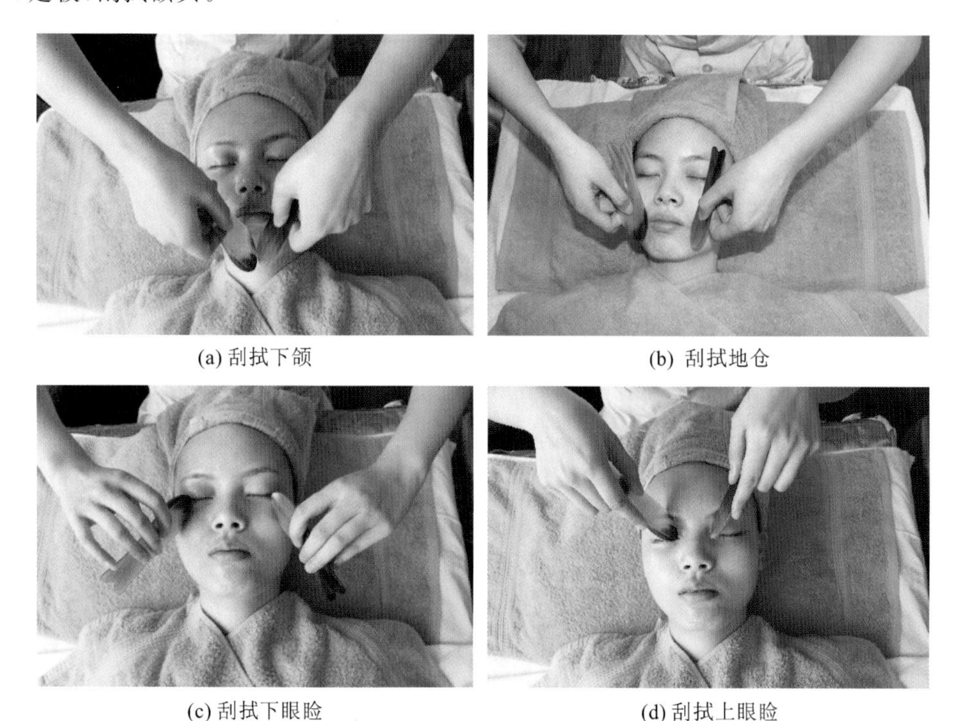

(a) 刮拭下颌　(b) 刮拭地仓

(c) 刮拭下眼睑　(d) 刮拭上眼睑

图 3-51　刮拭脸部

③拉抹法令纹：分开鱼形板，走板至鼻翼两侧，鱼尾下压，在法令纹处上下划拉(图 3-52)。

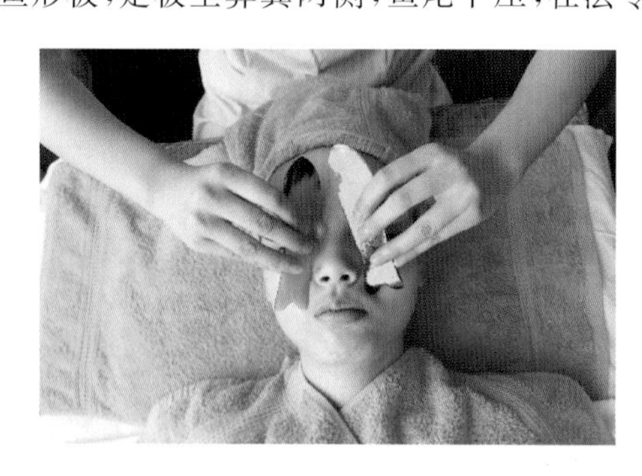

图 3-52　拉抹法令纹

④提升笑肌：鱼身曲线沿着笑纹，利用鱼形板提升肌肉至太阳(图 3-53)。

⑤眼部刮拭(图 3-54)：

a. 眼部点压：一手鱼形板立起，鱼尾自神庭穴划至鼻梁，用鱼尾点压与操作手同侧眼周穴位：睛明、承泣、四白、球后、瞳子髎；另一手拿板搁置一旁。

b. 眼部排毒：用鱼头侧曲线顺操作眼的眼袋刮拭，向下排毒，经耳前至颈下。

c. 切割黑眼圈：双板鱼头侧曲线交叉在下眼睑处划拉。

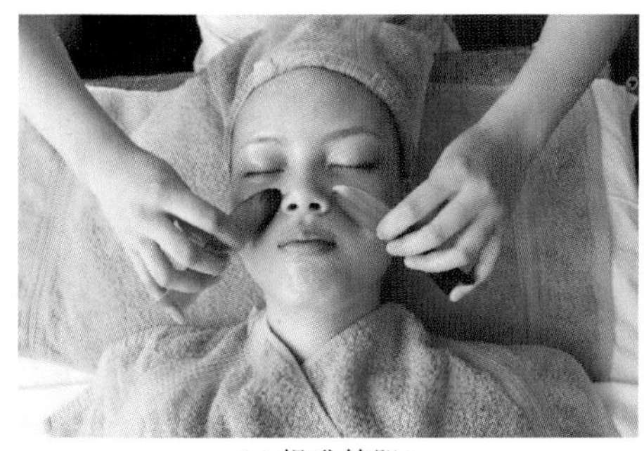
(a) 提升笑肌1

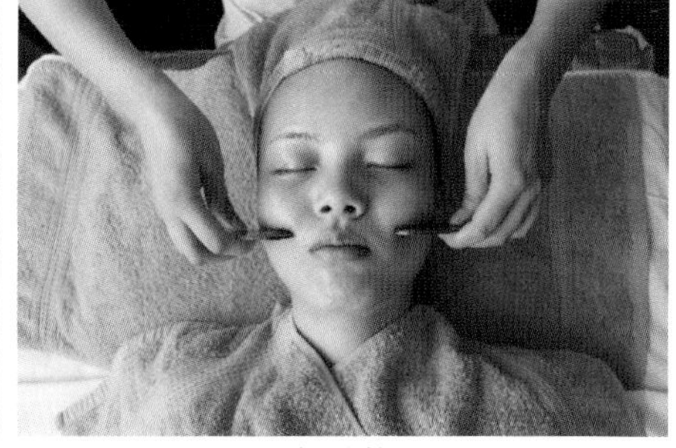
(b) 提升笑肌2

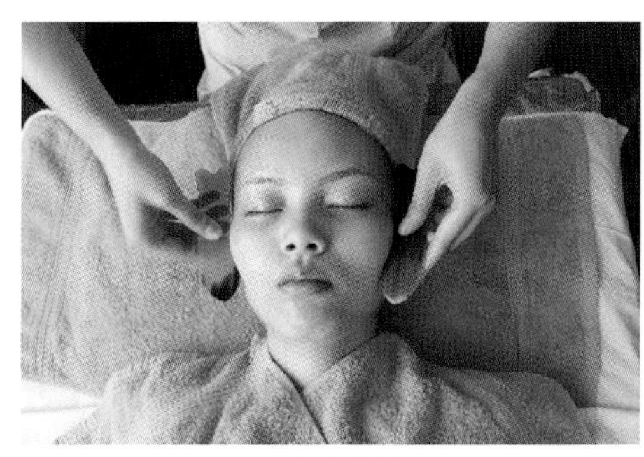
(c) 提升笑肌3

图 3-53 提升笑肌

d. 拉抹鱼尾纹：双板鱼头板面在眼尾交替拉抹外眼角。

e. 同法刮拭另一只眼。

f. 眉骨排毒：双板鱼身曲线同时夹住眉骨由内向外作眉骨排毒。

⑥重复第②步 8 条线刮拭全脸，第一条线用鱼尾刮拭。

⑦放松面部肌肉：用鱼尾轻点刺激面部各个穴位，起到放松作用，线路走向参考刮拭全脸的 8 条线（图 3-55）。

⑧塑造面部轮廓：走 S 路线刺激面部轮廓，一手先做，另一手重复上述动作（图 3-56）。

⑨安抚面部。

3. 后续工作 用湿面巾纸擦拭全脸，爽肤润肤，结束操作。

（三）面部刮痧的注意事项

1. 面部清洁要求 刮痧前先用清水洗面，若用洗面奶清洁皮肤，选择不含软化角质和去角质成分的洗面奶。

2. 刮拭用具选择 面部刮痧工具应为水牛角、木材、玉石等天然材料制作的鱼形刮痧板，不宜使用塑料材质。面部刮痧前要先涂敷刮痧介质，如本草刮痧油或芳香精油等，保持足够的润滑度，不可未涂抹刮痧介质直接刮拭，以免造成皮肤损伤。

3. 刮痧时间和力度 面部刮拭动作要求慢、柔、匀、稳，刮拭速度宜缓慢，时间和力度因人而异。

4. 禁忌证

（1）患皮肤传染性疾病者忌用。

（2）皮肤有创面或有严重痤疮者慎用，以免操作不慎引发感染。

（3）过敏性皮肤急性过敏期忌用。

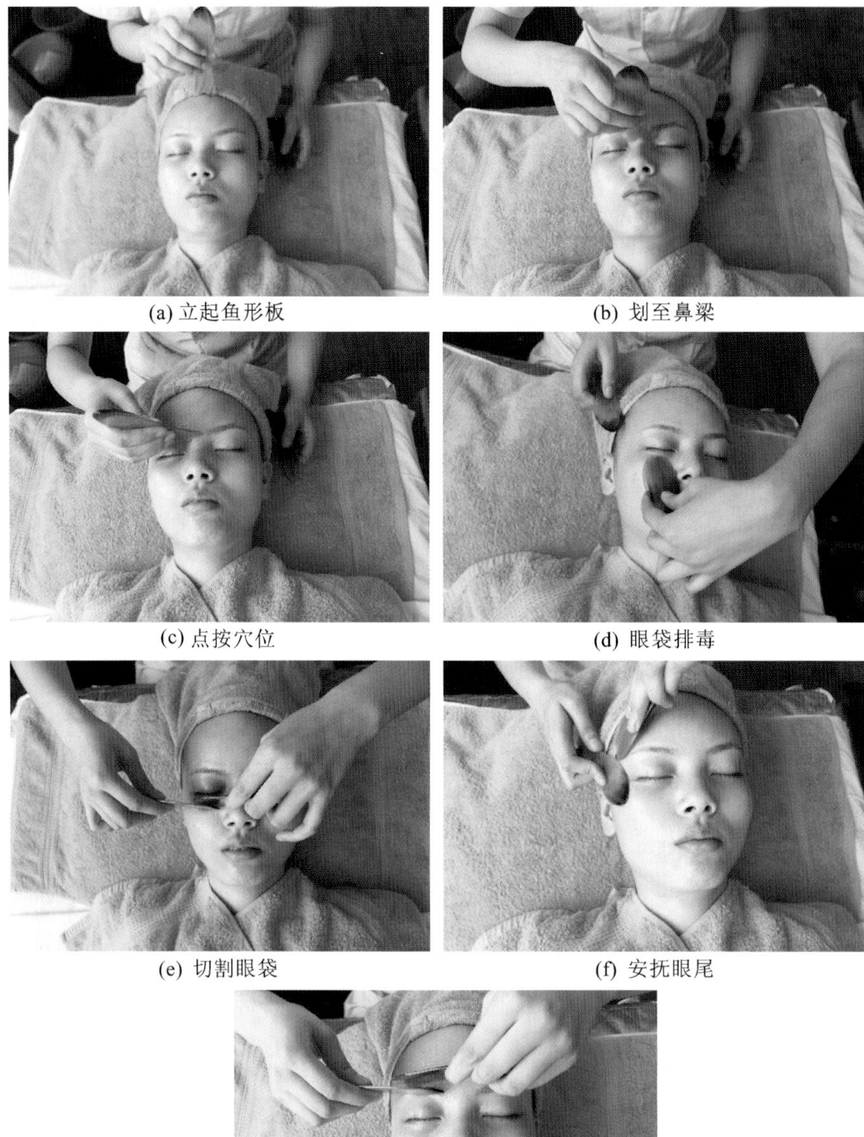

(a) 立起鱼形板　(b) 划至鼻梁

(c) 点按穴位　(d) 眼袋排毒

(e) 切割眼袋　(f) 安抚眼尾

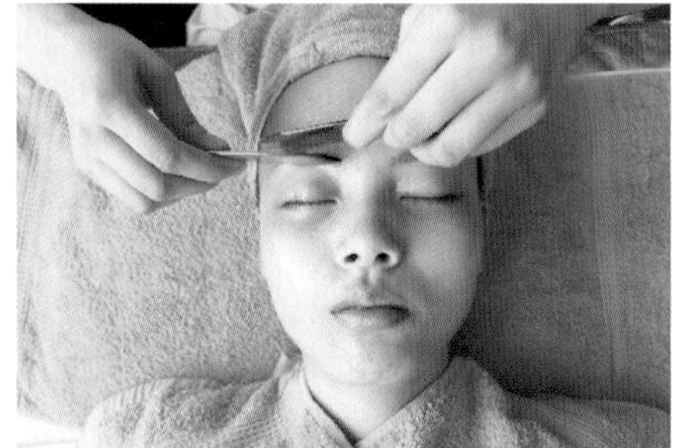

(g) 眉骨排毒

图 3-54　眼部刮拭

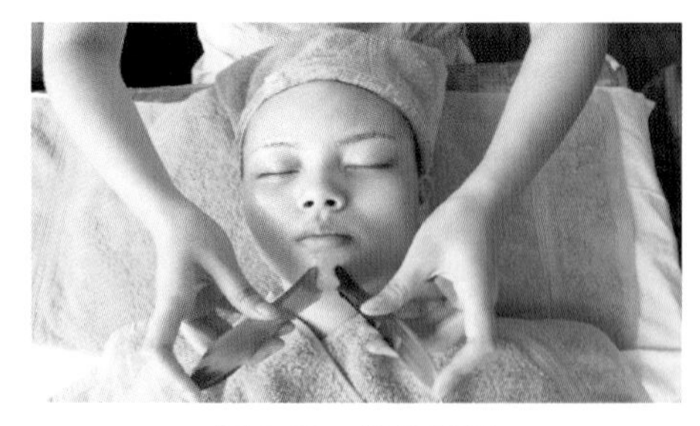

图 3-55　放松面部

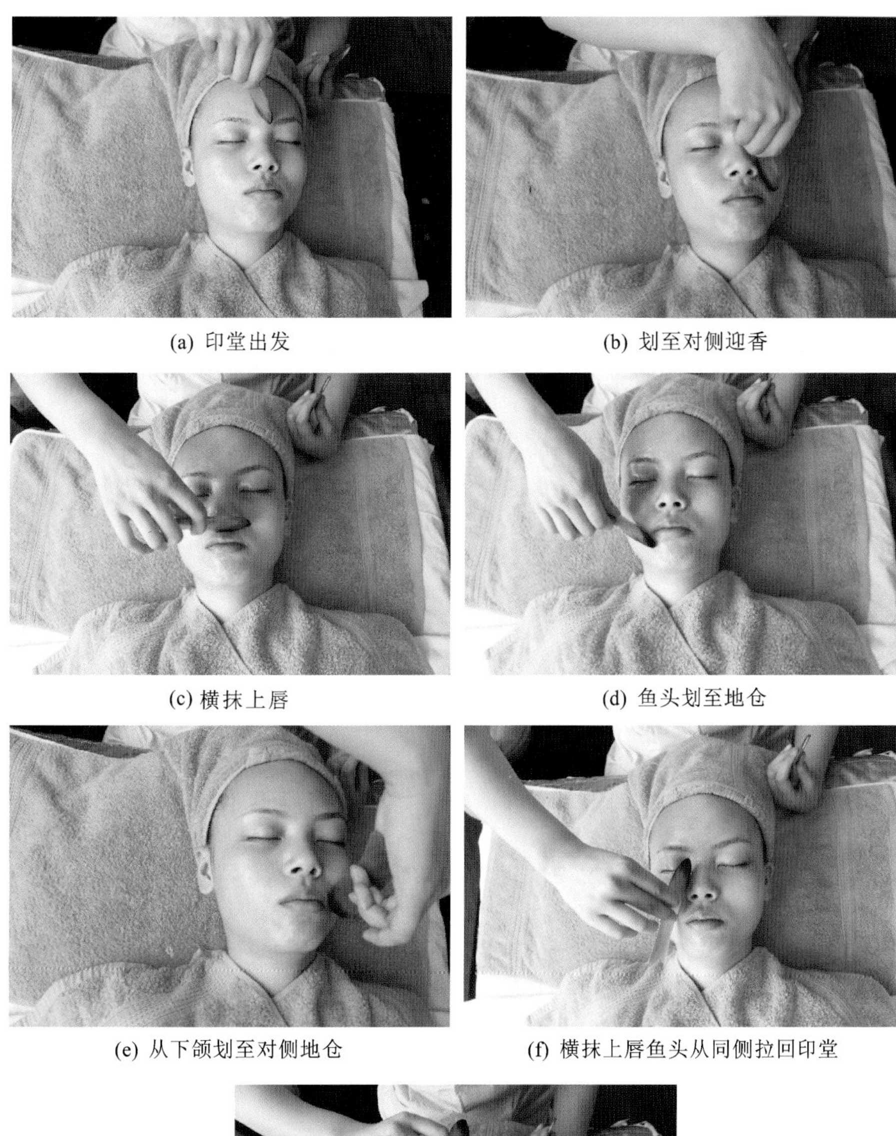

(a) 印堂出发
(b) 划至对侧迎香
(c) 横抹上唇
(d) 鱼头划至地仓
(e) 从下颌划至对侧地仓
(f) 横抹上唇鱼头从同侧拉回印堂
(g) 结束S路线

图 3-56 塑造面部轮廓

(4) 严重敏感性皮肤者慎用。

(5) 妊娠期慎用。

(6) 中、重度贫血者慎用。

(7) 过度饥饱、过度疲劳、醉酒者慎用。

5. 刮痧板 使用前后，都需严格消毒。

6. 面部刮痧 以皮肤发红、发热为度，不要求出痧。

三、面部芳香疗法

（一）面部芳香疗法基础知识

1. 面部芳香疗法的概念 芳香疗法是使用从芳香植物中萃取的高浓度芳香精华（精油），进行养生、保健和美容的疗法。芳香（aroma）意为芬芳、香味。疗法（therapy）意为对疾病的医疗。芳香疗法（aroma therapy）是传统的自然疗法之一，是一种辅助性的疗法，与正统医疗相似，但并不取代正统医疗。

"芳香疗法"一词是在20世纪中期才被正式提出，并在欧美流行，20世纪80年代传入香港，20世纪80年代末90年代初传入台湾，20世纪90年代末传入中国内地。目前越来越多的人逐渐熟悉精油和芳香疗法。但回顾香料发展的历史，我国早在五千年前已应用香料植物驱疫避秽。古巴比伦和亚述人在三千五百年前便懂得运用熏香治疗疾病。古埃及人在沐浴时已使用香油或者香膏，并且认为有益肌肤。古希腊和古罗马人也早就知道使用一些新鲜或干燥的芳香植物令人镇定、止痛或振奋精神。因此，芳香疗法是融合了几千年来的文明智慧及本世纪医学家和科学家的研究成果，提供了有效又愉悦的保健选择，使之能达到平衡身、心、灵的整体效果。芳香疗法表现的是精致的生活内涵和高品质的生活理念。

面部芳香疗法提倡运用天然的护肤品及精油，通过对面部皮肤香熏、涂抹、按摩等方法，达到面部皮肤的保养美容及人体身、心、灵的整体疗愈。

2. 基础油 基础油（base oil 或 carrier oil），也称之为媒介油或基底油，是从植物的种子、花朵、根茎或果实中萃取的非挥发性油脂，可以滋润、软化肌肤，使肌肤保持柔软和光泽，能直接用于肌肤按摩，也是稀释精油的最佳基底油，让纯精油能够安全地应用于皮肤。植物基础油本身就具有医疗效果，是营养和精力的良好来源，身体有了它就能产生热，它是蛋白质的绝佳来源。有好几百种植物的种子可以生产出油，其中只有少数几种油是用在商业的用途上的。

根据不同肤质选择合适的基础油来进行调配：

(1) 油性皮肤：甜杏仁油、杏桃仁油、荷荷巴油。

(2) 干性皮肤：鳄梨油、小麦胚芽油。

(3) 敏感性皮肤：甜杏仁油。

(4) 衰老性皮肤：小麦胚芽油。

(5) 皱纹皮肤：鳄梨油。

(6) 粉刺：荷荷巴油。

3. 针对常见的皮肤类型推荐精油及配方

(1) 中性皮肤：

①单方精油：薰衣草、迷迭香、柠檬、尤加利、天竺葵等。

②复方精油配方：1滴薰衣草＋1滴佛手柑＋10 mL 甜杏仁油。

(2) 干性皮肤：

①单方精油：薰衣草、迷迭香、檀香、天竺葵等。

②复方精油配方：1滴天竺葵＋1滴罗马洋甘菊＋8 mL 甜杏仁油＋2 mL 荷荷巴油。

(3) 油性皮肤：

①单方精油：薰衣草、丝柏、迷迭香、尤加利、柠檬、罗勒等。

②复方精油配方：1 滴丝柏+1 滴尤加利+8 mL 甜杏仁油+2 mL 荷荷巴油。

（二）面部芳香疗法的操作程序

1. 疗程指引　播放心灵音乐→点香熏灯→调整灯光→品花草茶→草本洁肤→花卉水净肤→美白去角质→美肤香熏法→美肤按摩法→草本修复面膜→花卉水净肤→涂芳香平衡保湿精华→涂隔离乳液→再品花草茶。

2. 疗程配方

（1）香熏灯：柠檬、桉树叶、薰衣草。

（2）花草茶：玫瑰、薰衣草、洋甘菊、柠檬、姜茶。

（3）洁肤：在清洁用品中加入一滴单方精油，打出泡沫彻底清洁皮肤。

（4）花卉水：玫瑰纯露、洋甘菊纯露、薰衣草纯露、金缕梅纯露。

（5）去角质：将精磨燕麦粉浸泡 2 h，加上 1 滴单方精油。在面部轻轻打圈，去除老化角质。

（6）香熏法：香熏精油为杜松、丝柏、马郁兰、松树、茶树精油调配而成。

（7）美肤按摩法：薰衣草、茶树、迷迭香、玫瑰精油。

（8）面膜：茶树、玫瑰、橙花、甘草精油。

3. 手法按摩　动作要轻、柔、慢，每个动作重复 4～6 遍，心神合一。

（1）前奏：①洁肤（面部、胸部）；②敷压（热、冷）（图 3-57）；③上精油（胸、颈、面）（图3-58）；④口述（语言引导），打圈从额往下每个部位放松，与口述同步进行。

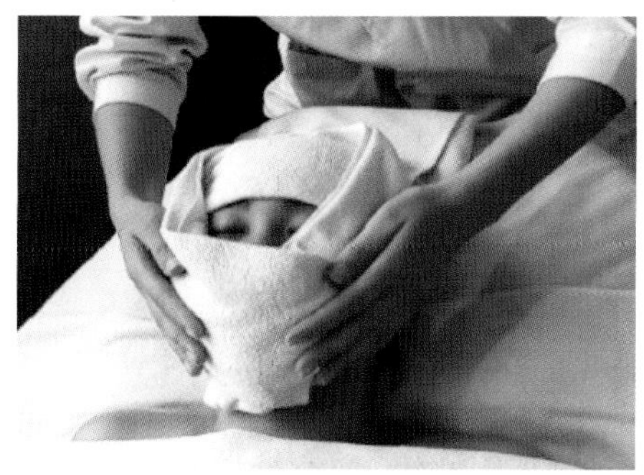

(a) 敷压面部

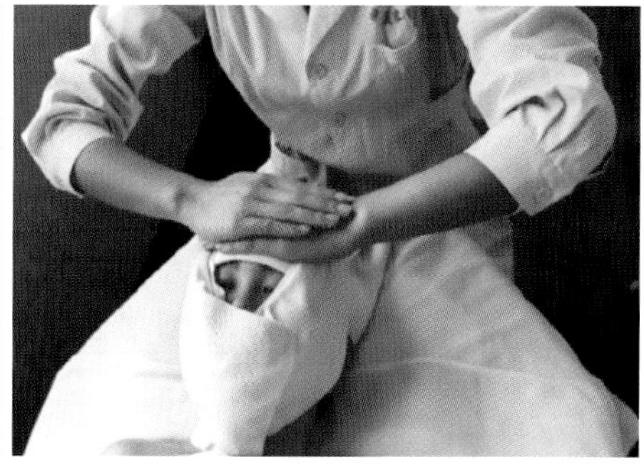

(b) 敷压额头

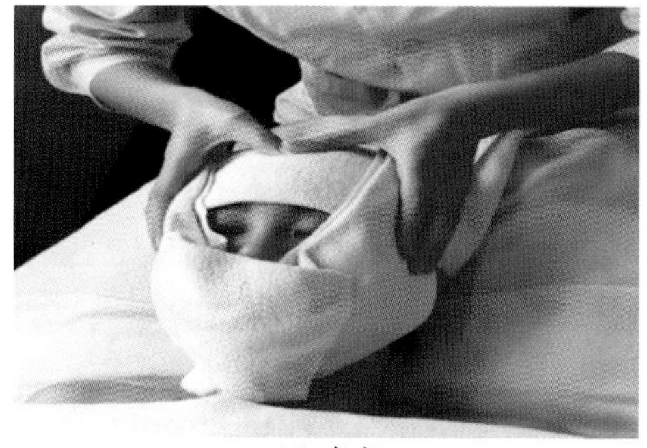

(c) 点穴

图 3-57　敷压

（2）胸部按摩：①横抹法：前胸至腋下，叠掌背，掌心抹（图 3-59）。②游泳式：前胸→肩峰→肘关节，双臂游，掌心掌指打圈（图 3-60）。

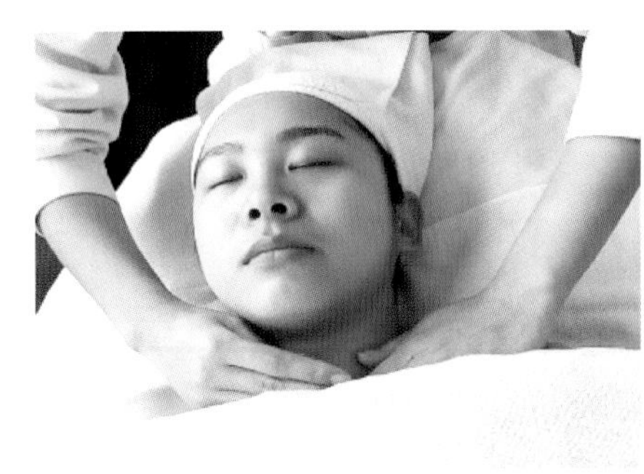

(a) 胸部上精油

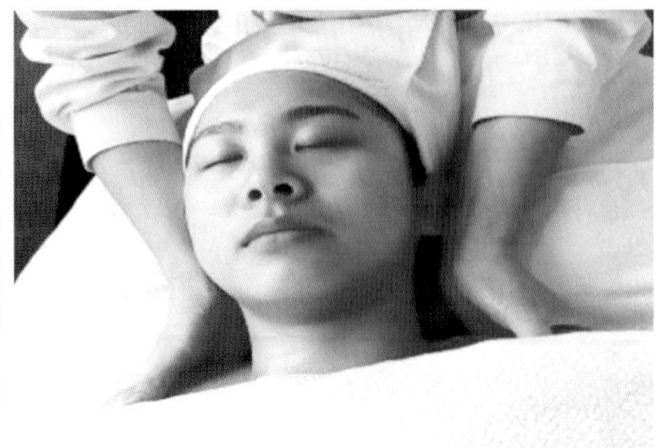

(b) 肩颈上精油

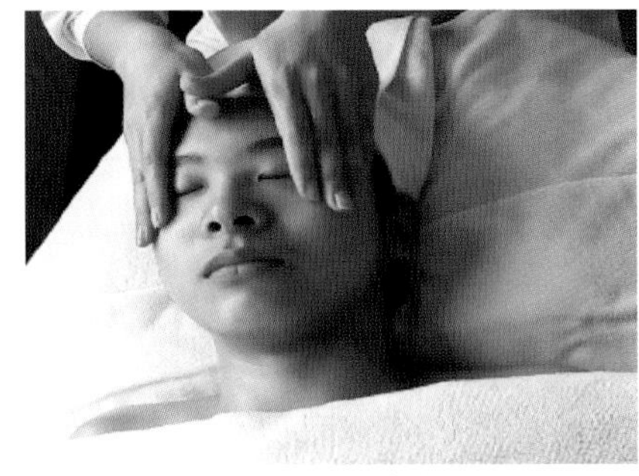

(c) 面部上精油

图 3-58 上精油

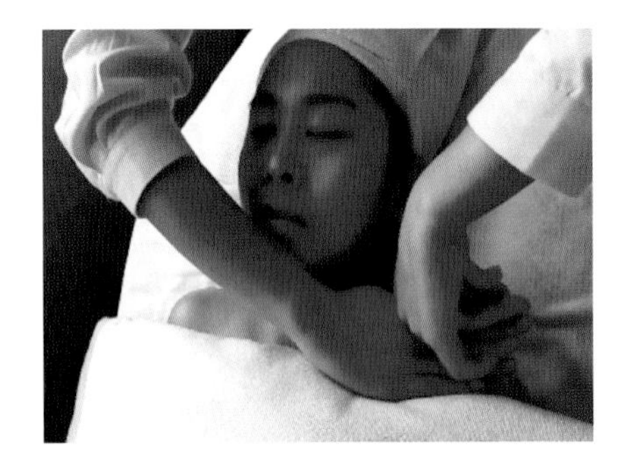

图 3-59 横抹法

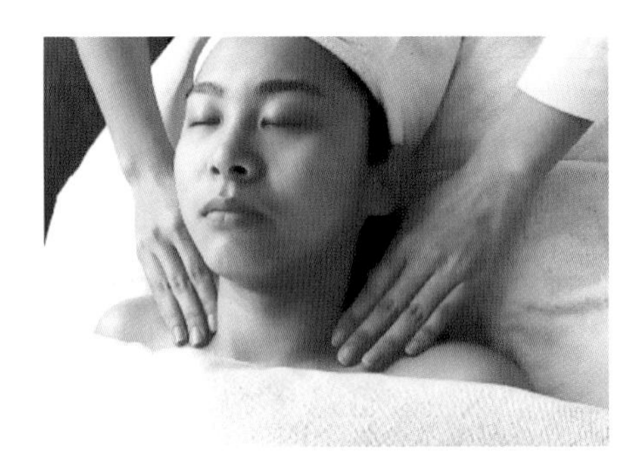

图 3-60 游泳式

(3) 颈部按摩:①揉颈部两侧,按压颈根部:由上至下打圈(图 3-61)。②横抹颈部:左右

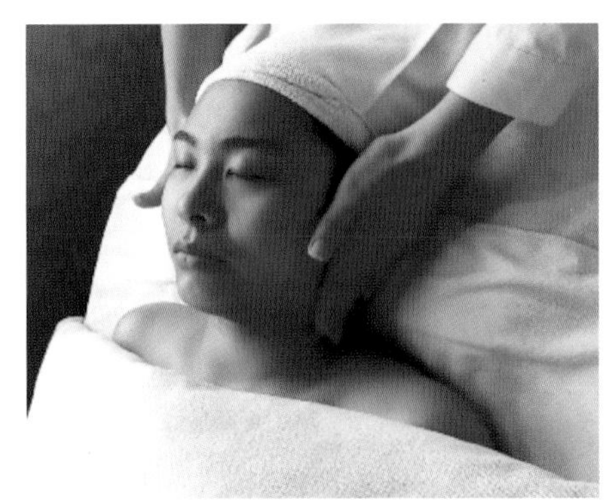

(a) 揉颈部

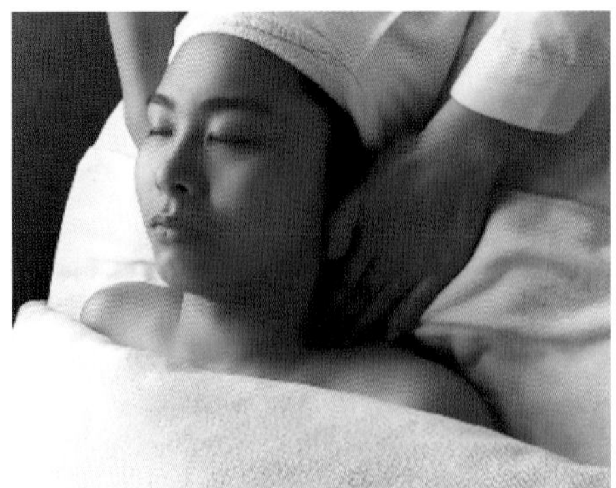

(b) 按压颈根

图 3-61 揉颈部两侧

交替横向拉抹前颈部。③推抹颈部两侧：左右交替推抹颈侧部（图 3-62）。

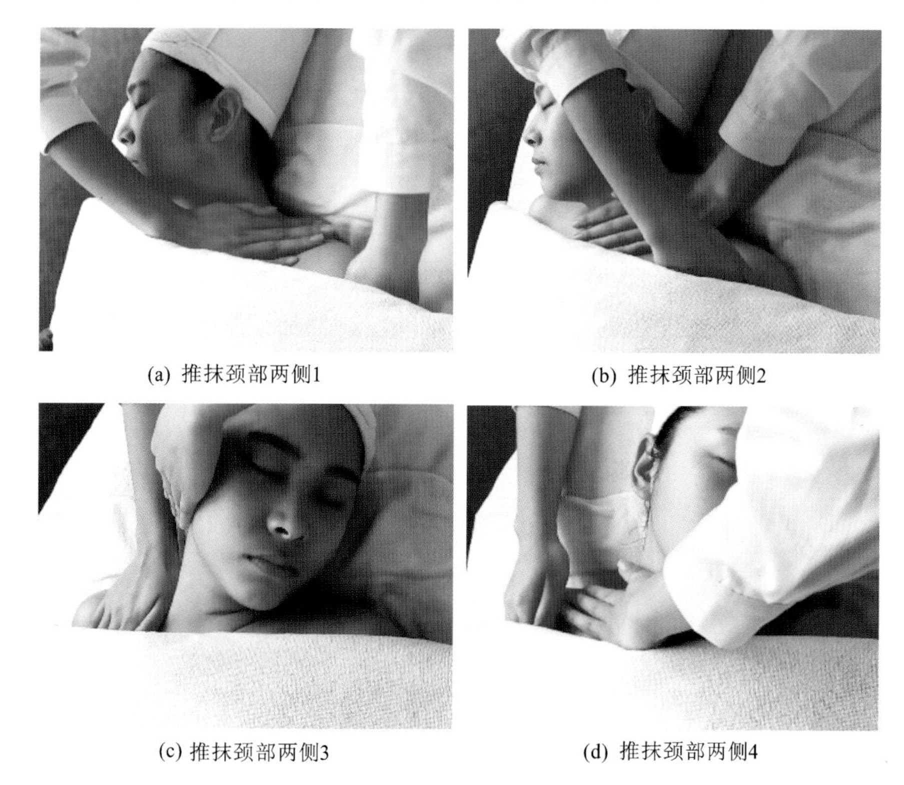

(a) 推抹颈部两侧1　(b) 推抹颈部两侧2

(c) 推抹颈部两侧3　(d) 推抹颈部两侧4

图 3-62　推抹颈部两侧示意

（4）面部按摩：①拉抹下颌（图 3-63(a)）；②提拉嘴角（图 3-63(b)）；③提拉面颊（图 3-63(c)）；④提拉眼角（图 3-63(d)）；⑤安抚额头（图 3-63(e)、(f)）。⑥点压面部穴位（图 3-64）。

（5）收势：由额部至前胸全面安抚（图 3-65）。

（三）面部芳香疗法的注意事项

1. 精油的保存方法

（1）存放：精油一定要存放在具有遮光效果的深色玻璃瓶中，可减少 90%的紫外线照射，禁用塑胶瓶存放精油，塑胶的化学成分会破坏精油的品质。

（2）密封：为了避免精油氧化及快速挥发，密封保存是非常重要的。精油的瓶盖一定要拧紧，减少开启次数，若经常开启瓶盖，精油很容易因接触空气而变质。所以最好选择小包装精油。避免接触阳光及强光。阳光是精油的头号杀手，需绝对避免阳光照射，强度较高的日光灯、灯泡也避免照射到精油。

（3）远离热气及高温：精油需置于干燥、阴凉的地方，不应离电器用品太近，更不能放于厨房或浴室。

（4）注意存放温度：精油适宜的存放温度为 18～30 ℃，最佳温度约为 25 ℃，精油不可存放在冰箱内，温差太大会加速精油品质变化。

（5）避免强烈的震动：震动有可能会使精油变质。

（6）避免与药物、茶叶、活性炭存放在一起。

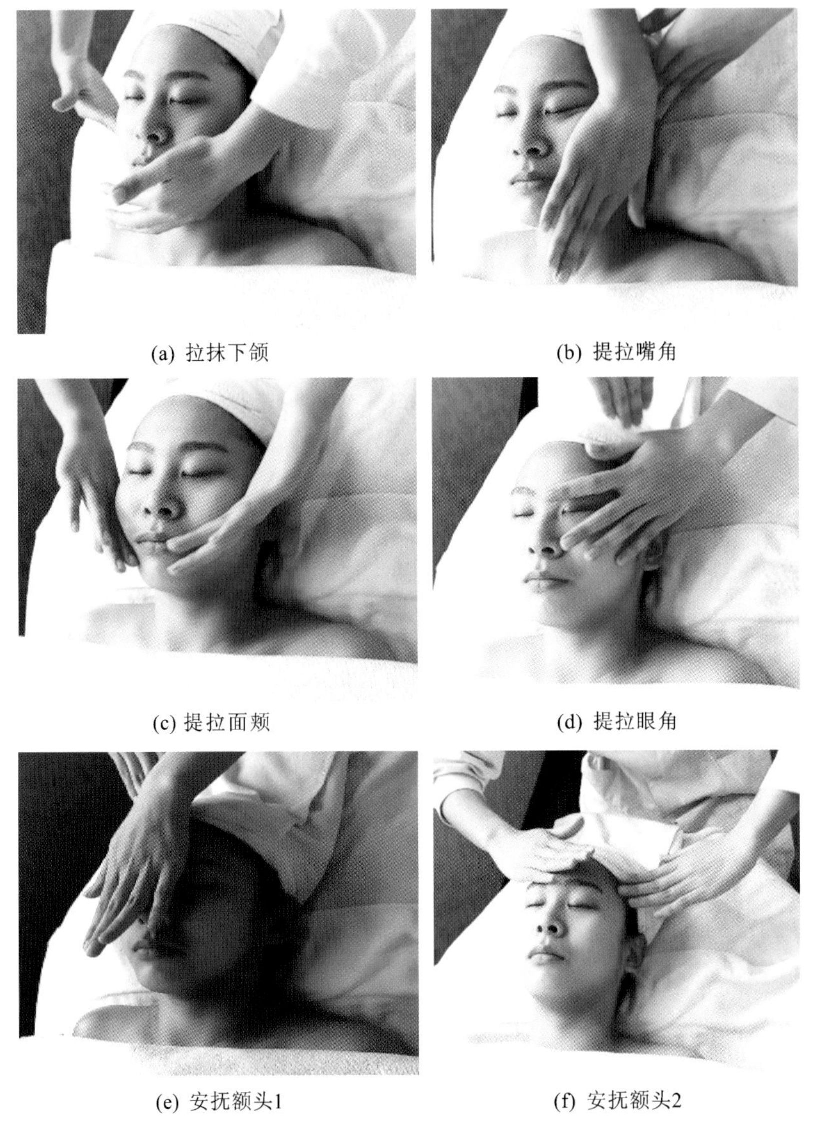

(a) 拉抹下颌　(b) 提拉嘴角

(c) 提拉面颊　(d) 提拉眼角

(e) 安抚额头1　(f) 安抚额头2

图 3-63　面部按摩

2. 使用精油的注意事项

(1) 未经稀释的纯精油不能直接使用。紧急情况下，个别精油可以直接涂抹。如烫伤，可直接将薰衣草精油滴在烫伤部位。

(2) 孕妇、婴幼儿、高血压患者、心脏病患者、癫痫患者不可使用酮、酚含量过高的精油，如鼠尾草、艾草、牛膝草、丁香、肉桂等。

(3) 孕妇不可用通经作用的精油，如鼠尾草、茉莉、玫瑰、丝柏、艾草、茴香、罗勒、薄荷、马郁兰等。

(4) 剂量是安全的关键，剂量越高效果不一定越好。如薰衣草有催眠效果，但是高剂量的薰衣草有提神作用，让人无法入睡。不安全的剂量会引发副作用及不良后果，因此任何情况下都要按照计算原则来调配精油。

(5) 同一种精油不要持续使用超过 3 周，要数种精油交替使用。

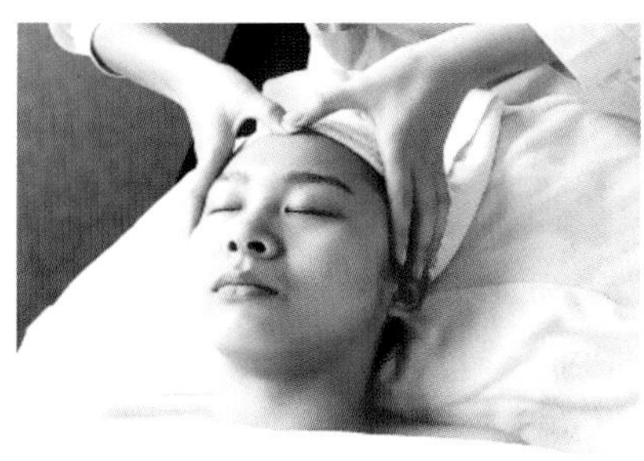

(a) 点压面部穴位1

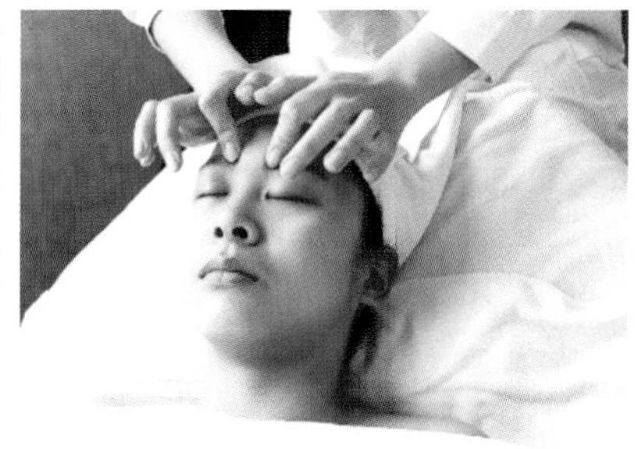
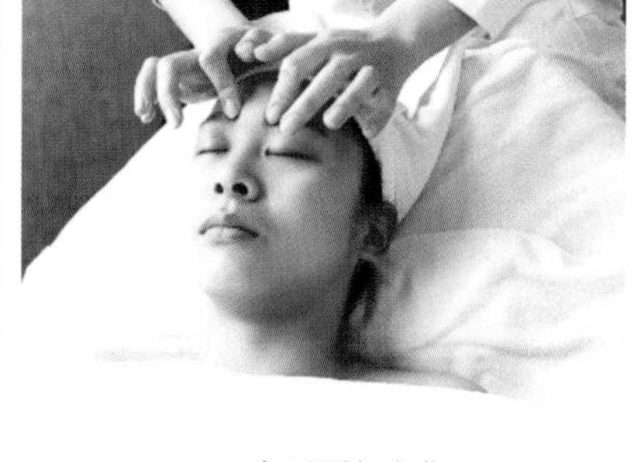

(b) 点压面部穴位2

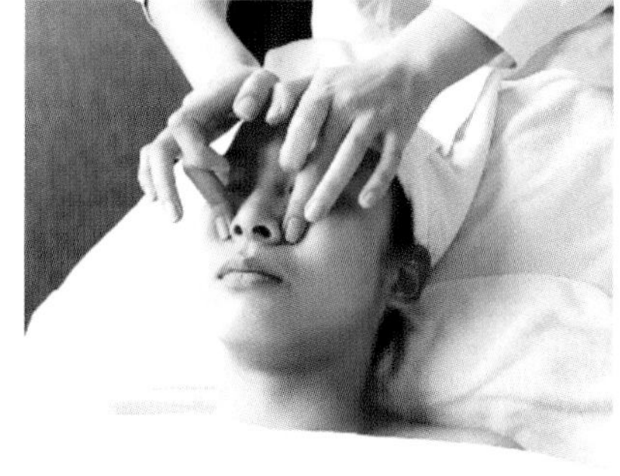

(c) 点压面部穴位3

(d) 点压面部穴位4

图 3-64 点压面部穴位

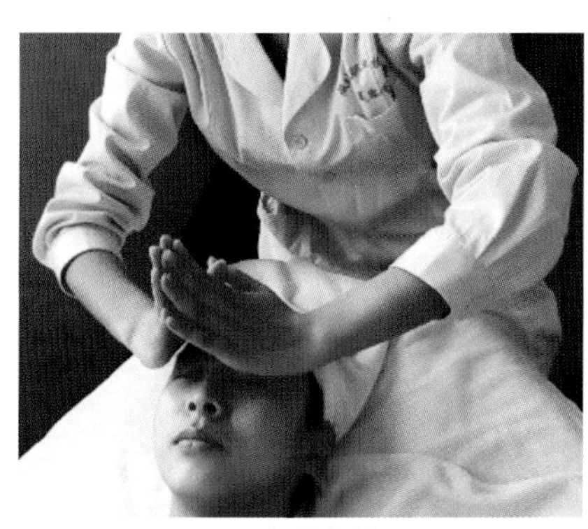

(a) 全面安抚1

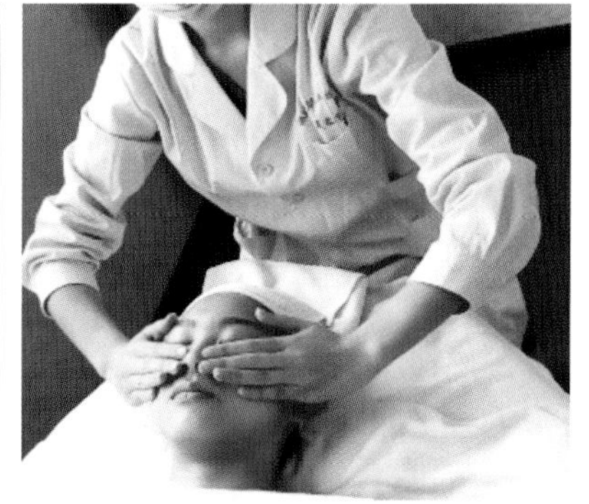

(b) 全面安抚2

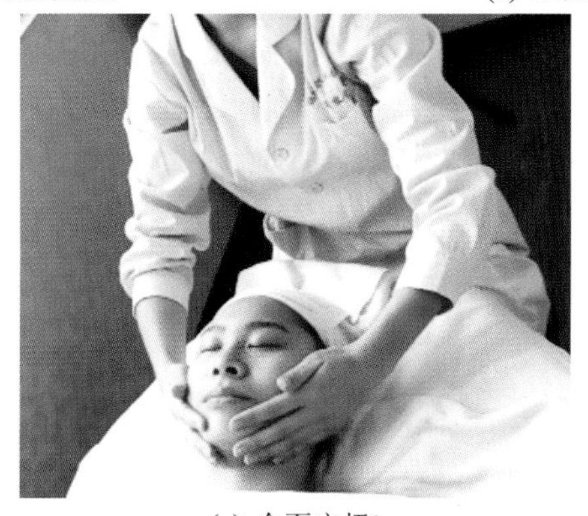

(c) 全面安抚3

图 3-65 收势

(6) 含呋喃香豆素的精油具有光敏性,使用后要避免阳光照射,以免引起色素沉着。

(7) 如果是敏感肌肤,首次使用要做皮肤测试。

(8) 精油不能取代药物。某些病症使用后如症状没有缓解,一定要及时就医。不可因使用芳香疗法而放弃治疗。

(9) 气喘患者不宜采取蒸汽吸入法使用精油。

本项目重点提示

(1) 面部护理程序包括准备工作、清洁皮肤、面部按摩、敷面膜、后续工作等五大部分。

(2) 需要掌握的操作技能有面部清洁、面部按摩、敷面膜、面部刮痧、面部芳香疗法。

(3) 在整个面部护理过程中,美容师需要随时注意顾客感受,及时调整,工作细致、认真,全心全意地为每一位顾客服务。

能力检测

一、选择题

1. 清洁皮肤最适宜的水温是(　　)。

A. 34～38 ℃　　B. 38 ℃以上　　C. 28 ℃以下　　D. 28～37 ℃

2. 冷喷适合于哪种类型的皮肤?(　　)

A. 中性　　B. 油性　　C. 干性　　D. 任何皮肤

3. 正确点穴后的感觉是(　　)。

A. 酸、麻、胀　　B. 痒、痛　　C. 刺痛　　D. 无感觉

4. 磨砂膏的操作时间为多少?(　　)

A. 2～3 min　　B. 3～4 min　　C. 4～5 min　　D. 5～6 min

二、填空题

1. 清洁皮肤包括________、________、________。

2. 面部按摩的手法要求包括________、________、________、________、________。

3. 面膜的作用包括________、________、________、________。

三、问答题

1. 皮肤清洁主要包括哪几项操作?冷喷的作用有哪些?

2. 面部按摩的作用是什么?

3. 使用面膜的注意事项有哪些?

4. 面部刮痧的作用原理是什么?

(李文娟　徐　静　熊　蕊)

项目四　面部常见损美性皮肤护理

学习目标

(1) 掌握面部常见损美性皮肤的分析诊断和护理方案。

(2) 熟悉面部常见损美性皮肤的概念、成因与表现。

(3) 了解面部常见损美性皮肤的类型与特点。

项目描述

本项目主要介绍面部常见衰老性皮肤、色斑性皮肤、痤疮性皮肤、敏感性皮肤、毛细血管扩张性皮肤和晒伤皮肤的概念与分类，成因与分析诊断及护理原则。学生通过本项目的学习，具备分析诊断面部常见损美性皮肤和设计护理方案的能力。

案例引导

张某，女，45岁。26岁怀孕期间面部颧骨区域出现色斑，生完孩子后未见好转。面部皮肤干燥、触之粗糙、肤色较晦暗，色斑对称分布在颧骨两侧，边界清晰，呈褐色。两颊可见毛细血管，易敏感，偶尔出现红肿、痒、灼热感。

问题：

1. 该顾客面部色斑属于哪种类型？形成的主要原因有哪些？
2. 请为她设计科学合理的护理方案。
3. 作为美容师或顾问，可给予哪些日常生活护理建议？

一、衰老性皮肤护理

人出生后皮肤组织日益发达，功能逐渐活跃，当到达一定年龄就开始慢慢退化。皮肤组织的成长期一般在25岁左右结束，此期称为“皮肤的弯角”。此后，皮肤组织生长与老化同时进行，皮肤弹力纤维渐渐变粗。40～50岁进入初老期，皮肤老化逐渐明显，老化程度因人而异。

(一) 衰老的概念

衰老是生物随着时间的推移，自发的必然过程，它是复杂的自然现象，表现为结构和功能衰退，适应性和抵抗力减退。人体皮肤老化是指皮肤在外源性或内源性因素的影响下引起的皮肤外部形态、内部结构和功能衰退。

（二）衰老性皮肤的成因与表现

1. 衰老性皮肤的成因

（1）皮肤衰老的外在因素：

①紫外线的伤害：紫外线损伤又称光老化，是造成皮肤老化的主要因素之一。

②重力的作用：由于重力的作用，使自然老化松弛的皮肤加速下垂。

③错误的保养：使用过热的水洗脸，过度的按摩和去角质，使用劣质的化妆品等，均会使皮脂含量减少，角质层受损，丧失对皮肤的保护和滋润作用，皮肤老化加快。

④不良生活习惯：过于丰富的面部表情，如挤眉弄眼、皱眉、眯眼等；不适当的快速减肥或缺乏锻炼；不当的饮食造成肥胖或消瘦，经常接触刺激性的食物，如酒、咖啡等；长期熬夜，过度疲劳。

⑤恶劣的生活环境：空气污染、汽车排放废气和化工厂排放刺激性气体影响皮肤的新陈代谢；噪声影响听力、伤害神经系统，也会造成衰老；吸烟的烟雾会损耗体内的维生素 C 而影响皮肤胶原纤维，致使皮肤松弛；空气干燥会使皮肤中的水分流失过快，导致皮肤粗糙、产生皱纹；寒风、强冷刺激也会导致皮肤血管收缩，皮脂、水分减少而导致皮肤提前老化。

（2）皮肤衰老的内在因素：

①年龄增加：随着青春期结束，皮肤的生理机能便开始降低。

②植物性神经功能紊乱：生活节奏加快，工作压力大，家庭纷争均可引起植物神经功能紊乱，导致内脏功能异常、失眠等，进而引起皮肤早衰。

③内脏机能病变：肝脏具有参与物质代谢、解毒、助消化等多种功能，若有病变将影响人体新陈代谢；肾是机体内重要的排泄器官，若肾脏病变，发生功能障碍，体内的有害物质不能及时排出，会妨碍机体新陈代谢；心脏功能不全，不能及时将氧气和营养物质通过循环系统带给身体各部位，会造成营养不足进而影响皮肤新陈代谢。这些有害因素均会导致皮肤老化、色素沉着。

④内分泌紊乱：内分泌系统是调节人体新陈代谢、生长繁殖的重要系统。内分泌腺通过分泌激素来调节代谢，如雄激素和肾上腺皮质激素能刺激皮脂腺生长增殖与分泌，使皮肤保持滋润与光滑；雌性激素则可使皮下脂肪丰厚，维持皮肤弹性等。当激素分泌减少，皮肤机能便逐渐衰退，肌肤萎缩，失去光泽，更年期妇女尤为明显。

2. 衰老性皮肤的表现

（1）皮肤组织衰退：

①皮肤变薄：皮肤厚度随着年龄的增加而有明显的改变。20 岁时人的表皮最厚，以后表皮的增殖能力减退，到老年期颗粒层可萎缩至消失，棘细胞生长周期缩短，表皮逐渐变薄。

②肤色变化：皮肤的色素调节会造成黑色素增加，脂褐质沉积而产生黑斑；由于黑色素细胞退化，而产生色素脱失，呈雨滴状白点；表皮细胞不正常的角化，会产生脂漏性角化症或俗称的“老人斑”。

③失去光泽：角质层细胞脱落减慢，产生不规则角化，已衰老死亡的细胞堆积于表皮角质层，使得皮肤表面粗糙不光滑。

④失去弹性：真皮结缔组织在 30 岁时最厚，以后逐渐变薄并伴有萎缩。皮下脂肪减少，弹性纤维与胶原纤维逐渐失去弹性和张力，导致皮肤松弛与皱纹的产生。

⑤失去血色：真皮层变薄，真皮乳头层的血管减少，血流量降低，皮肤缺乏红润色泽，出现萎黄。

知识拓展

皮肤皱纹的分类

1. 自然性皱纹　多呈横向弧形，与生理性皮肤纹理一致。自然性皱纹与皮下脂肪堆积有关，伴随年龄增大皱纹逐渐加深，纹间皮肤松垂。如颈部的皱纹，为了颈部能自由活动，此处的皮肤会较为充裕，自然形成一些皱纹，甚至刚出生就有。早期的自然性皱纹不表示老化，只有逐渐加深、加重的皱纹才是皮肤老化的象征。

2. 动力性皱纹　面部表情肌与皮肤相附着，表情肌收缩，皮肤与表情肌垂直的方向上会形成皱纹，即动力性皱纹。动力性皱纹是由于表情肌的长期收缩所致。早期只有表情肌收缩，皱纹才出现，后期表情肌不收缩，动力性皱纹亦不减少。如长期额肌收缩产生前额横纹，在青年即可出现；而鱼尾纹是由于眼轮匝肌的收缩作用所致，笑时尤甚，也称"笑纹"。

3. 重力性皱纹　重力性皱纹是在皮肤及深面软组织松弛的基础上，外加重力的作用而形成皱襞和皱纹，重力性皱纹多分布在眶周、颧弓、下颌区和颈部。

4. 混合性皱纹　由多种原因引起，机制较复杂，如鼻唇沟、口周的皱纹。

从皱纹的形态上可以分为假性皱纹和定性皱纹。

(1) 假性皱纹：面部出现的不稳定的、可自行消退的皱纹。

(2) 定性皱纹：已经形成的具有稳定性的皱纹，是由于皮肤的胶原蛋白和弹性纤维性能下降造成。

(2) 生理功能低下：

①皮脂腺、汗腺功能衰退，皮脂与汗液排出减少，皮肤逐渐失去昔日光泽而变得干燥。

②血液循环功能减退，不足以补充皮肤必要的营养，因此老年人皮肤伤口愈合难。

总之，衰老性皮肤是由于表皮、真皮交界处、真皮及附属器发生退行性改变，导致皮肤形态、弹性、色泽等方面的改变，外观特征主要表现为皮肤干燥、粗糙、无光泽、皱纹增加、松弛下垂伴随黑斑、老年斑、毛细血管扩张、血管瘤等。

（三）衰老性皮肤护理方案

1. 护理原则

(1) 为肌肤补充充足的养分和水分。

(2) 清洁皮肤要彻底，防止残留污物侵害皮肤。清洁产品性质要温和，避免造成皮肤的天然油分流失。

(3) 调整好个人生活规律，解除精神上的压力，减少不良的刺激因素。注意防晒，防止皮肤老化。

(4) 注意合理饮食，定期养护。

(5) 如有内脏病变，应先积极进行治疗，再配合皮肤养护。

2. 护理程序

(1) 清洁面部：使用干性皮肤适用的滋润轻柔的清洁霜或洗面奶进行面部清洁，避免使用泡沫型的洁面膏，防止过度的清洁而产生皮肤脱水现象。

(2) 爽肤：使用保湿滋润型的柔肤水爽肤2～3遍。

(3) 观察、分析皮肤。

(4) 喷雾：选择热喷3～5 min，可以达到补充水分和舒展皱纹的作用。禁止使用奥桑喷雾仪。

(5) 去角质：去除角质层老死细胞。衰老性皮肤建议2个月做一次去角质。由于衰老性皮肤的角质层含水量下降，皮肤的新陈代谢减慢，因此需选择去角质啫喱进行去角质，在去角质的同时可以给皮肤补充水分。

(6) 面部按摩：选择滋润度高的按摩膏，按摩10～15 min，提高皮肤的温度，促进血液循环，为皮肤补充氧气和养分。

(7) 仪器护理：利用超声波美容仪导入具有补水去皱、淡化色素、抗衰老等作用的精华素，时间为5～8 min；射频美容仪可以拉紧皮下深层组织和收紧皮肤，使下垂或松弛的面部达到提升的效果。

(8) 面膜：使用具有抗衰、去皱、补水、滋养功能的面膜，可选择海藻、人参、鹿茸、珍珠等补水去皱的面膜。敷膜时间为15～20 min。

(9) 爽肤：使用保湿滋润型的柔肤水爽肤2～3遍。

(10) 基本保养：使用可以滋养皮肤的精华液、乳液、面霜、防晒和隔离霜。

(11) 后续工作：告知顾客居家保养方法，预约下次到店时间，书写护理记录，归还用物，环境整理。做好跟踪服务。

3. 预防

(1) 衰老性皮肤日常护理洗脸时，可冷热水交替进行，以增加皮肤血液循环。

(2) 避免外界因素(风、霜、雪、紫外线等)对皮肤的直接伤害。

(3) 合理、正确选用化妆品和护肤品。

(4) 饮食营养均衡，干性皮肤应增加蛋白质、维生素等物质的摄入。

(5) 生活规律，保持充足的睡眠，并适当进行体育锻炼，劳逸结合，不抽烟酗酒。

(6) 定期到美容院进行皮肤护理。

(7) 改变大笑、皱鼻、皱眉、眯眼等不良习惯。

(8) 面部按摩动作要轻柔、缓慢、服帖，不得过度牵拉皮肤。

二、色斑性皮肤护理

(一) 色斑的概念

色斑是由于多种因素影响所致的皮肤黏膜色素代谢失常，色素沉着，是生活中常见的面部损美性皮肤问题。包括雀斑、黄褐斑、黑斑、老年斑、色素痣、炎症后色素沉着，属色素障碍性皮肤病。

(二) 色斑性皮肤的成因、分类与表现

1. 色斑性皮肤的成因

(1) 遗传因素：常染色体遗传是色斑的主要成因。淡褐色至黄褐色针尖到米粒大小的斑点，对称分布在面部，特别是鼻部。

(2) 紫外线照射：日光中的紫外线照射是色斑形成的重要原因。当皮肤接受过多日光照

射时，表皮就会产生更多的黑色素颗粒，引起色素沉着。

(3) 内分泌失调：经期和妊娠期体内的性激素水平变化，可以影响黑色素的产生。另外，内分泌不稳定时通常引起情绪不稳定，也会间接引起色斑形成。

(4) 不良生活习惯：压力大、偏食、睡眠不足等不良生活习惯也会令黑色素增加。

(5) 化妆品使用不当：使用含有重金属铅、汞、砷类化妆品。

2. 色斑性皮肤的分类与表现

(1) 雀斑：又称为夏日斑，雀斑是极为常见的，发生在日光暴晒部位，如面部或颈、手背、鼻、臂、胸及四肢等部位，常见于鼻和两颊(图 4-1)。一般是浅咖啡色、棕色、褐色或褐黑色斑，如针头、绿豆大小，直径一般在 0.5 mm 以下。呈圆形或椭圆形，常对称或分散分布，且受气温影响。夏天颜色深，数目增多；冬天颜色变浅，数目减少。与遗传有关。

(2) 黄褐斑：面部常见的局限性淡褐色或黄褐色色素沉着斑。成年女性多见，好发于育龄期妇女。呈对称性分布，在面颊部形成蝴蝶状，遍及前额、颧部、颊部，偶见于颏和上唇部(图 4-2)。边缘清楚，与邻近斑块趋向融合，无任何主观上的不适感，在中医上也被称为"黧黑斑""肝斑""妊娠斑""蝴蝶斑"。

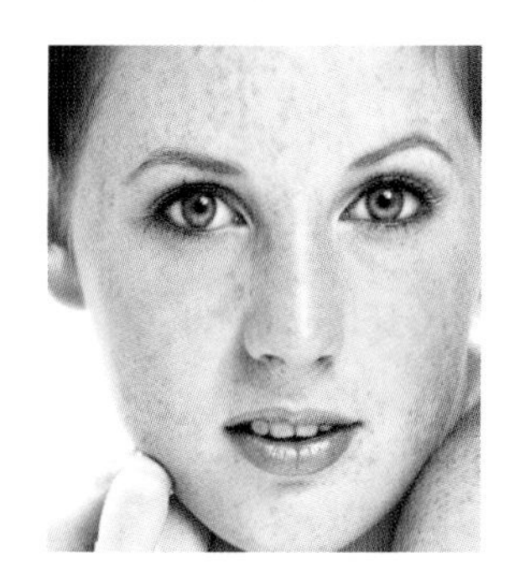

图 4-1 雀斑

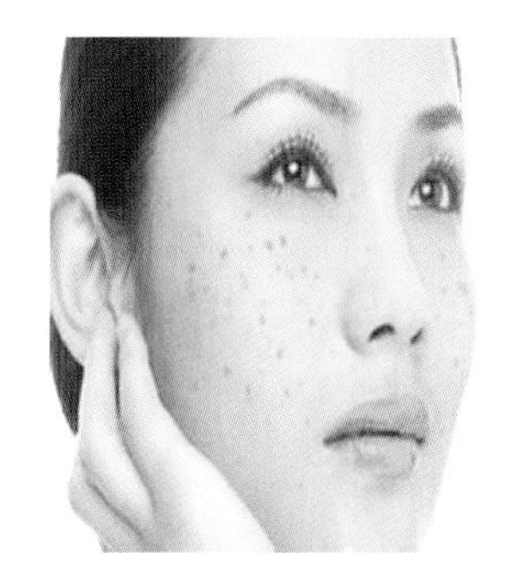

图 4-2 黄褐斑

(3) 黑斑：黑斑多见于面颊、前额、颈、手背、前臂、脐等处(图 4-3)。如针尖、米粒大小，呈点状、网状、片状的黑斑，较褐斑色重而浓，在中医上被称为"面尘"。

(4) 老年斑：呈黄褐色，斑点较雀斑稍大，常发生在太阳照射的部位，尤其是手背、手臂、双颊和前额(图 4-4)。

图 4-3 黑斑

图 4-4 老年斑

(5) 色素痣：色素痣又称痦子，很常见。主要表现为局限性淡黑色、暗黑色或黑色斑疹，大小不等，形状不一，有些痣上有黑色短毛，常常是幼年开始出现，可长在身体的任何部位(图 4-5)。

(6) 炎症后色素沉着:化妆品使用不当、纹饰术或炎症消退等因素导致的局部皮肤色素沉着,浅褐色或深褐色,散状或片状分布,表面平滑(图 4-6)。

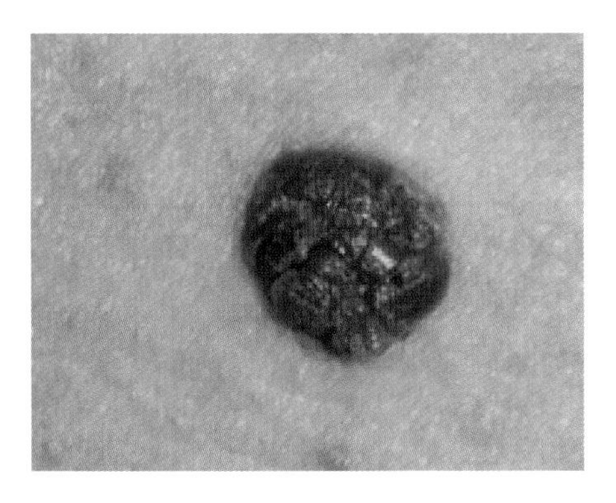

图 4-5 色素痣

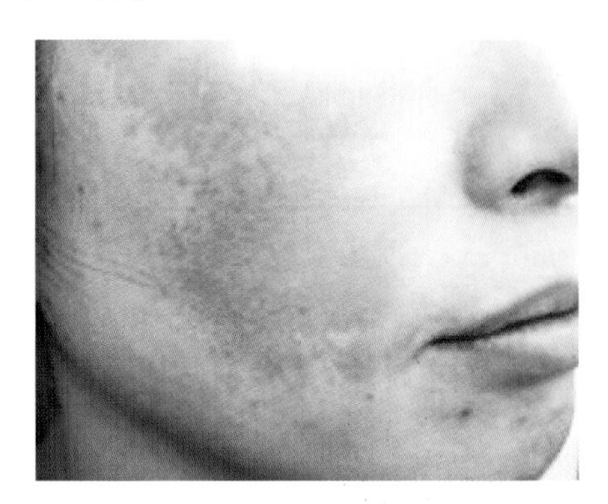

图 4-6 炎症后色素沉着

知识拓展

色斑形成的原理

色斑是黑色素在皮肤浅表层的沉淀,由于内分泌失调,皮肤代谢不畅导致黑色素不能有效排除而形成的。

黑色素是人体内的一种蛋白质,存在于皮肤基底层的细胞中间,不是真正意义上的黑色素,而是一种黑色素原生物质,也被称为"色素母细胞"。色素母细胞分泌麦拉宁色素,当紫外线(B 波、A 波)照射到皮肤上(B 波即 UVB 作用于皮肤基底层,A 波作用于皮肤的真皮层),肌肤就会处于"自我防护"的状态,紫外线刺激麦拉宁色素,激活酪氨酸酶的活性,来保护皮肤细胞。酪氨酸酶与血液中的酪氨酸发生反应,生成一种叫"多巴"的物质。多巴是黑色素的前身,经酪氨酸氧化而成,释放出黑色素。黑色素又经由细胞代谢,层层移动,到了皮肤表皮层形成色素沉着,即为色斑。

(三) 色斑性皮肤护理方案

1. 护理原则

(1) 根据色斑形成的原理尽量减少黑色素形成,阻断黑色素形成的各个阶段,防止黑色素过量产生。

(2) 加强按摩,促进新陈代谢,增加血液循环,淡化色斑。

(3) 防止紫外线的照射,为皮肤补充充足的油分和水分,谨慎选择化妆品。

(4) 积极治疗身体内部疾病,保持良好的精神状态,饮食均衡,多吃含维生素 C、B 族维生素的食物,正确使用护肤品。

(5) 保证睡眠,调节情绪,保持心情舒畅。

2. 护理程序

(1) 清洁面部:使用含有美白成分的洁面乳洁面。

(2) 爽肤:使用美白保湿水爽肤 2～3 遍。

(3) 观察、分析皮肤。

(4) 喷雾:慎用热喷,热喷时间为 5～8 min。禁止使用奥桑喷雾仪,以免再造成色素沉着。

(5) 去角质:选择去角质霜或去角质啫喱进行全脸深层清洁,可加速黑色素的代谢速度,促进祛斑精华素的吸收。

(6) 面部按摩:选择滋润度高的按摩膏或用美白精华素按摩 10～15 min。按摩可以促进皮脂腺的分泌,色斑部位采用震颤法可激活维生素 C,加速其淡化色斑的效果。

(7) 仪器护理:利用超声波美容仪导入美白祛斑精华素,采用低挡位。导入时间为 5～8 min。光子嫩肤仪祛斑效果更为明显。

(8) 面膜:使用美白祛斑面膜,配合热膜效果更佳。敷膜时间为 15～20 min。

(9) 爽肤:使用美白保湿水爽肤 2～3 遍。

(10) 基本保养:使用可以美白、滋润、祛斑、营养皮肤的营养霜和防晒霜。

(11) 后续工作:告知居家保养方法,预约下次到店时间,书写护理记录,归还用物,环境整理。做好跟踪服务。

知识拓展

色斑的位置与成因

(1) 额头的斑点:内分泌失调。

(2) 眼皮的斑点:内分泌失调,流产次数多。

(3) 太阳穴、眼尾的斑点:甲状腺功能减退、妊娠、更年期;神经质的人;长期心理压抑。

(4) 鼻下的斑点:妇科疾病、处女斑。

(5) 眼睛皱纹的斑点:妇科疾病,流产次数多,情绪不稳定。

(6) 两颊的斑点:肝肾功能失调,日晒,更年期,老年斑。

(7) 下颚的斑点:妇科疾病,白带多,对化妆品过敏。

3. 预防

(1) 避免电离辐射,电离辐射比日光照射对皮肤的损伤还要大。

(2) 做好防晒,皱纹和斑点大部分都是由于光老化引起。

(3) 色斑与疾病有关系,若身体有疾病应及时医治,尤其是妇科病,如乳腺增生、月经不调等。对于各种皮肤创伤,一定要谨慎治疗。

(4) 保持营养均衡,注意各种维生素的均衡摄入。

(5) 对于激素类产品要慎用,不要使用含汞、铅等有害物质的祛斑产品。

(6) 保持心情舒畅,精神愉快,避免忧思恼怒,并保持充足的睡眠。

(7) 改善微循环,调节内分泌,激活细胞活力,促进新陈代谢,增强营养物质吸收,疏通经络,平衡阴阳,调和气血。

三、痤疮性皮肤护理

(一) 痤疮的概念

痤疮是一种常见的皮肤炎症疾病,以粉刺、丘疹、脓疱、结节、囊肿及瘢痕为其特征,常伴皮脂溢出。青春期是痤疮的高发年龄段,特别是油性肤质的人。

(二)痤疮的成因、分类与表现

1. 痤疮的成因

(1)雄激素过多。

(2)皮脂分泌调节失控。

(3)毛囊、皮脂腺过度角化与皮脂腺腺体增生。

(4)微生物增多与聚集。

(5)炎性介质启动炎症。

(6)肠胃疾病导致的内分泌失调。

(7)遗传、环境和药物因素。

2. 痤疮的分类与表现

(1)粉刺型痤疮:粉刺是痤疮的最初的皮疹,粉刺周围由于炎症反应及微生物的作用,可演变为丘疹、脓疱、囊肿及瘢痕。分为白头粉刺和黑头粉刺。

①白头粉刺:又称为闭合性粉刺。堵塞时间短,为灰白色小丘疹,不易见到毛囊口,表面无黑点,挤压出来的是白色或微黄色的脂肪颗粒。

②黑头粉刺:又称开放性粉刺。表现为圆顶状丘疹伴显著扩张的毛囊开口。毛囊开口脂栓氧化变成黑色,呈点状黑色,可挤出脂栓。

(2)丘疹型痤疮:炎症继续发展扩大并深入,表现为炎性丘疹和黑头粉刺。

(3)脓包型痤疮:表现为脓包和炎性丘疹。当白头粉刺和黑头粉刺未及时清除时,由于细菌的大量繁殖,表面皮肤不洁净,造成感染面积扩大,局部出现脓性分泌物。此时,有肿胀和疼痛。粉刺在毛囊顶部形成破溃,可挤出脓血。若处理不及时可因反复感染形成囊肿。

(4)囊肿型痤疮:表现为大小不等的皮脂腺囊肿,内含有带血的黏稠脓液,破溃后可形成窦道及瘢痕。脓包若清除不及时、不彻底,残留的部位易形成反复感染或炎症。严重时造成皮脂腺囊肿,化脓破溃会形成瘢痕。平时皮肤表面皮疹不明显,用手可触及皮下有囊性物,深部肿痛,按之有移动感。当发生炎症时,可形成红色大痤疮囊性丘疹。此时感染不易愈合,病程长,且反复发作。

(5)结节型痤疮:炎症已深入到毛囊根部,脓肿造成毛囊壁的破溃,造成毛囊内容物及痤疮杆菌、脓性分泌物流入真皮层,造成真皮层的感染,出现凹陷状萎缩性瘢痕。脓包型痤疮漏治、误治以后,可以发展成壁厚、大小不等的结节。位于皮下或高于皮肤表面,呈淡红色或暗红色,质地较硬,称为结节型痤疮,又称硬结型痤疮。

(6)萎缩型痤疮:丘疹或脓包型痤疮破坏腺体而形成凹坑状萎缩性瘢痕者,称为萎缩型痤疮。

(7)聚合型痤疮:数个痤疮结节在深部聚集融合,有红肿,颜色青紫,称为融合型痤疮或聚合型痤疮(图4-7)。

(8)恶病质型青春痘:超重型青春痘,虽极少见,但却相当严重。表现为小米至黄豆大的紫红色丘疹、脓包或结节,黑头粉刺不多,经久不愈。多并发于贫血、结核病或其他全身性疾病。

(三)痤疮性皮肤护理方案

1. 护理原则

(1)抑制皮脂腺过度分泌。

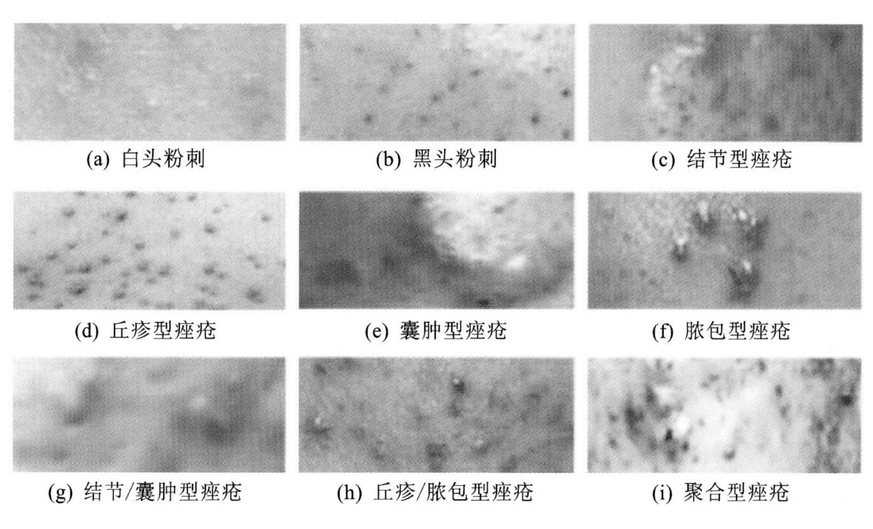

(a) 白头粉刺　(b) 黑头粉刺　(c) 结节型痤疮

(d) 丘疹型痤疮　(e) 囊肿型痤疮　(f) 脓包型痤疮

(g) 结节/囊肿型痤疮　(h) 丘疹/脓包型痤疮　(i) 聚合型痤疮

图 4-7　痤疮

（2）清洁、消炎、杀菌。

2. 护理程序

（1）清洁面部：使用清爽型、控油祛痘型、收缩毛孔型洁面乳洁面。

（2）爽肤：使用清爽型爽肤水 2～3 遍。

（3）观察、分析皮肤。

（4）喷雾：使用奥桑喷雾仪喷 5～8 min，起到消炎杀菌的作用；亦可冷喷 20 min，达到镇静皮肤，防止感染的目的。

（5）去角质：使用去角质霜去角质，避开痤疮部位，严重者不去角质。

（6）针清：采用针清方式清除白头粉刺和黑头粉刺，涂抹消炎膏。

（7）仪器治疗：清除痤疮后，用红蓝光仪器照射。

（8）面部按摩：使用滋润清爽、控油祛痘型按摩膏或痤疮膏按摩 5～10 min。应避开炎症期、脓包期、囊肿期痤疮部位。痤疮较多者不做按摩。

（9）面膜：使用消粉刺软膜、痤疮面膜或冷膜。痤疮部位也可用甲硝唑涂敷打底后涂冷膜，或痤疮膏打底后涂冷膜。调理期每周 2～3 次，巩固期可每周 1 次。

（10）爽肤：使用消炎、镇静或平衡油脂的爽肤水或者痤疮消炎水爽肤 2～3 遍，可暂时收敛毛孔，平衡油脂。

（11）基本保养：痤疮部位涂痤疮膏，其他部位涂平衡调理霜。

（12）后续工作：告知居家保养方法，预约下次到店时间，书写护理记录，归还用物，环境整理。做好跟踪服务。

3. 仪器护理和特殊养护

（1）针清：

①用物准备：暗疮针、酒精棉球、无菌持物钳、无菌敷料罐、碗、盘。

②适应证：适用于粉刺型痤疮，在粉刺顶端有白色分泌物时，可以用暗疮针进行挑刮，将脓性分泌物排出，以防止感染向深层发展。

③操作方法：

a. 用 75%酒精消毒局部皮肤。

b. 选择在粉刺顶端最白最薄、已化脓破溃的地方进针。

c. 暗疮针应与皮肤呈平行状态进针，迅速刺破，迅速拔出，不必穿透整个粉刺。切忌与皮肤方向呈直角进针，否则，无法将粉刺内容物彻底刮出。

d. 将暗疮针的另一端用酒精消毒，然后轻轻压住进针部位的对侧皮肤，向针眼处平移暗疮针，使粉刺内容物顺针眼被挤压出，以达到清除脓血的目的。

④注意事项：

a. 凡是红肿部位切不可挑刮，有白色化脓时才可以进针。

b. 进针前做好消毒工作，针清前后对局部皮肤消毒，暗疮针每挑一个部位也要进行消毒，避免交叉感染。

c. 压刮的时候不可过于用力，以免造成出血；挑、刮前后用酒精消毒；出血部位可以用共鸣火花作短时间电疗，以消炎、杀菌、止血。

d. 挑、刮后的创面按摩时要避开；爽肤前，面部应用酒精再消毒一次。

(2) 红蓝光治疗：详见项目七。

(3) 超声波导入：

①适应证：粉刺型痤疮、脓包型痤疮、结节型痤疮等创面的恢复期。

②操作方法：

a. 清洁面部皮肤。

b. 清除化脓部位的脓液，并用3%双氧水彻底杀菌。

c. 开启电源，调节时间与强度。

d. 将治疗痤疮的药物均匀涂抹至全脸。

e. 用超声头导入20 min。

f. 治疗结束后，关闭机器。可直接在面部敷冷膜或洗净后敷暗疮膜。若不敷膜，将面部残留的暗疮膏洗净后涂抹爽肤水，整个疗程结束。

③注意事项：

a. 对化脓部位处理后，视感染轻重，创面大小，涂擦3%双氧水彻底杀菌，以免超声导入时，尚残留有脓性分泌物。

b. 治疗痤疮的药物应以膏剂、霜剂为宜，水剂痤疮药物更适合离子导入仪导入。

c. 全脸导入应不超过20 min，时间过长，反而会导致皮肤组织疲劳。

d. 导入时，超声头应避开眼球及眼睑。

e. 疗程设置，一般10次为一个疗程，隔日1次或每周2次。特别严重者，也可以每天1次，两个疗程之间应间隔7～10天。

4. 预防

(1) 保持面部清洁，选用弱碱性洁面乳，每天洁面2～3次，常用温水洗脸，选择清爽型乳液。

(2) 禁用油性护肤品和含有粉质的化妆品，避免堵塞毛孔，导致细菌感染，从而加重炎症。

(3) 不可挤压痤疮，以免炎症扩散，愈后遗留瘢痕。

(4) 多食蔬菜水果，保持大便通畅。忌食高脂、高糖、刺激性食物。

(5) 保持心情舒畅，保证充足睡眠，使面部肌肉得到有效的放松与自我修复。

四、敏感性皮肤护理

(一) 敏感性皮肤的概念

敏感性皮肤是指皮肤较薄,面颊及上眼睑处可见微细的毛细血管,对花粉、灰尘、化妆品中的某些成分的刺激,产生不同程度的瘙痒、灼热、疼痛、红斑、丘疹、水疱甚至水肿、糜烂或渗出等症状。

(二) 敏感性皮肤的成因与表现

1. 敏感性皮肤的成因 皮肤细胞受损而使皮肤的免疫力下降,角质层变薄导致皮肤滋润度不够,最终导致肌肤的屏障功能过于薄弱,无法抵御外界刺激。皮肤的神经纤维由于经常受到外界刺激过于亢奋,从而产生泛红、发热、瘙痒、刺痛和红疹等不适现象。具体原因如下。

(1) 环境因素:过冷、过热刺激,温度的迅速改变,季节的变化,空气污染,日光等。

(2) 年龄:年轻健康的皮肤表面有一层弱酸性的皮脂膜锁住水分,以保护皮肤不受到外界的侵害。随着年龄的增长,皮肤分泌功能减退,以致敏感物质容易侵入皮肤。

(3) 遗传因素:敏感皮肤个体大部分有家族史。

(4) 生理因素:内分泌失调、月经周期等会影响皮肤的敏感性。

(5) 化学因素:如化妆品、肥皂、清洁剂等。

(6) 生活方式:辛辣刺激饮食、酒精可加重皮肤敏感反应。

(7) 心理因素:压力大、情绪波动激发或加剧皮肤敏感反应。

(8) 疾病:某些皮肤病可使皮肤敏感性增高,如特应性皮炎、脂溢性皮炎、鱼鳞病等。

(9) 护肤不当:使用劣质护肤品和碱性较强的洗涤用品、换肤术后等。

2. 敏感性皮肤的表现

(1) 敏感性皮肤多表现为瘙痒、刺痛感、针刺感、烧灼感、紧绷感。

(2) 皮肤较薄,红血丝明显。

(3) 当气温变化(过冷或过热)时,皮肤出现泛红、发热。

(三) 敏感性皮肤护理方案

1. 护理原则

(1) 使用温和、无刺激成分的面部清洁和保湿产品,镇静安抚皮肤。

(2) 补充水分,保持皮肤角质层健康。

(3) 补充皮肤油脂,加固皮肤屏障,增加皮肤的抵抗力。

2. 护理程序

(1) 清洁面部:选择温和无泡沫的洁面乳洁面,避免使用碱性洁面产品,易产生皮肤脱水或敏感现象。

(2) 爽肤:选择温和无刺激保湿的柔肤水爽肤 2～3 遍。

(3) 观察、分析皮肤。

(4) 去角质:T 区去角质,避开敏感部位。

(5) 喷雾:冷喷 20 min,镇静舒缓皮肤,为皮肤补充一定的水分。

(6) 面部按摩:选择温和无刺激、滋润度高的按摩膏按摩 5～10 min,按摩时间不宜过长,力度不能过重。按摩时避开敏感部位,选择大面积安抚为主,在按摩的过程中避免过度牵拉。

(7) 导入：利用超声波美容仪导入防敏、补水、舒缓的精华素或胶原蛋白，采用低挡位，时间为 5～8 min。

(8) 面膜：选择防敏、保湿补水面膜，时间为 15～20 min。敷片状面膜，可以用棉片遮盖双眼，同步冷喷。

(9) 爽肤：选择温和无刺激保湿的柔肤水爽肤 2～3 遍。

(10) 基本保养：选择性质温和、滋润、营养皮肤的乳液、防敏面霜、防晒霜。

(11) 后续工作：告知居家保养方法，预约下次到店时间，书写护理记录，归还用物，环境整理。做好跟踪服务。

3. 预防

(1) 远离过敏原，忌蒸桑拿、热喷。

(2) 选择合适的护肤品，使用弱酸性的洁面乳。

(3) 敏感性皮肤在卸妆、洁面的时候基本要做到“快速、温和”，卸妆一定要彻底。

(4) 为肌肤补充水分和油分，做好防晒工作。

(5) 护肤方法宜简单，不要随意更换护肤品。

(6) 合理饮食，保持充足的睡眠。

(7) 减少去角质的次数，少按摩，切记不要用温度过高的水洗脸。

五、毛细血管扩张性皮肤护理

（一）毛细血管扩张性皮肤的概念

毛细血管扩张症，俗称“红血丝”，是指皮肤或黏膜表面的细静脉、毛细血管和细动脉呈持久的扩张状态，形成红色或紫色斑块、点状、血丝状或星网状损害，压之褪色，可长期不变或缓慢发展，多无自觉症状。毛细血管扩张症任何年龄均可发生，是一种发生在面部或躯干部位的皮肤问题。

（二）毛细血管扩张性皮肤的成因与表现

1. 毛细血管扩张性皮肤的成因

(1) 物理因素刺激：如温度变化的刺激，使毛细血管的耐受性超过了正常范围，引起毛细血管扩张甚至破裂。过度的日晒，也会引起慢性光线性皮炎，造成皮肤干燥等。

(2) 气候环境因素：长期生活在较为恶劣的生活环境中，如高原空气稀薄，皮肤缺氧，导致红细胞数量增多及血管代偿性扩张，久而久之血管收缩功能产生障碍，引起永久性毛细血管扩张。

(3) 激素依赖性：不恰当的糖皮质激素治疗后遗症，面部滥用外用药物，如皮炎平、皮康霜等。

(4) 护肤不当：劣质化妆品或长期“皮肤包月护理”及换肤不当引起后遗症。换肤产品的酸性成分严重破坏了皮肤角质层的保护作用和毛细血管的弹性，使毛细血管扩张或破裂。

(5) 局部或全身并发症：患有某种皮肤病（如酒渣鼻），或患有皮肤病但原因不明的人群，面部容易发红，在发热、情绪激动、剧烈运动或饮酒时，颜色不但会加深，整个面部还会潮红，即“红脸”“关公脸”，很难消退，影响美观。

2. 毛细血管扩张性皮肤的表现　面部毛细血管扩张是影响面部美容的主要原因，多发于女性，临床表现为面部的丝状、点状、星芒状或片状红斑。仔细看皮肤上有许多红色血管，象“红线头”。

（三）毛细血管扩张性皮肤护理方案

1. 护理原则

（1）标本兼治，缓解伴随的不适症状。可采用激光治疗，外用去红血丝产品。

（2）日常保养注意防晒，忌去角质，坚持用冷水洗脸以增加毛细血管弹性。

2. 护理程序

（1）清洁面部：用温和无泡沫的洁面乳洁面，洁面时间不超过 1 min。

（2）爽肤：用温和无刺激、保湿的柔肤水爽肤 2～3 遍。

（3）观察、分析皮肤。

（4）面部按摩：选择温和无刺激的按摩膏按摩 5～10 min，按摩时间不宜过长，力度不能过重。按摩时避开红血丝部位，在按摩过程中避免过度牵拉。

（5）导入：利用超声波美容仪导入去红血丝产品，采用低挡位，时间为 5～8 min。

（6）面膜与冷喷：选择去红血丝和保湿补水面膜，时间为 15～20 min。敷面膜的同时做冷喷护理。

（7）爽肤：选择温和无刺激、保湿的柔肤水爽肤 2～3 遍。

（8）基本保养：涂擦去红血丝保养品。

（9）后续工作：告知居家保养方法，预约下次到店时间，书写护理记录，归还用物，环境整理。做好跟踪服务。

3. 预防

（1）慎重选择化妆品，避免使用含有重金属的劣质化妆品，否则，劣质化妆品毒素残留在面部皮肤内，会使面部红血丝加重和色素沉积。

（2）饮食合理，戒掉过度辛辣、烟酒等不良饮食习惯。

（3）保持充足的睡眠，常用冷水洗脸，增加皮肤的耐受力。

（4）经常轻轻按摩红血丝部位，促进血液流动，有助于增强毛细血管弹性，以促进血液循环。动作宜轻柔。

六、晒伤皮肤护理

（一）晒伤皮肤的概念

皮肤晒伤又称日光性皮炎，是由于日光的中波紫外线过度照射后引起皮肤发生的光毒反应，即皮肤过度暴露于 UVB 射线（280～315 nm）所致，强烈日光照射后引起局部急性红斑，水肿性皮肤炎症。晒伤后的皮肤容易出现红肿、刺痛甚至出现水疱、脱皮等现象，皮肤易过早老化。

（二）晒伤皮肤的成因与表现

1. 晒伤皮肤的成因 日光是连续的电磁辐射波，波长以纳米（nm）为单位。波长越短，能量越大；波长越长，穿透力越强。波长由短到长依次为 R 射线、X 射线、紫外线、可见光、红外线、微波及无线电波等。与皮肤有关的主要是紫外线，在太阳的辐射量中，紫外线占 6%左右，是光线波长中最短的一种。紫外线能将皮肤中的脱氧胆甾醇转变为维生素 D，并能促进全身的新陈代谢，还具有杀菌、消毒的作用。如果紫外线辐射累积量过大，并有外源性光敏物质参与时，则可引起一系列的生物学效应或诱发和加重某些皮肤病。现代社会由于环境污染严重，导致地球上空的臭氧层遭到破坏，使地球表面的紫外线强度日益增加，紫外线对人们的皮肤损伤也日益加剧。

2. 晒伤皮肤的表现 皮肤具有反射、散射和吸收紫外线的能力。长时间受紫外线照射，UVB 波段透射达真皮乳头层，使部分毛细血管发生障碍，造成发炎和毛细血管扩张，表皮发红、水肿、疼痛，严重的会产生红斑和水疱，并伴有脱皮现象。红斑一般在阳光下直接照射 2～3 h 开始产生，在 12 h 内达到高峰，以后逐渐减退，4～7 天后消失，随之皮肤开始黑化。皮肤角质层的厚薄直接影响到对紫外线的敏感程度，未成年人的皮肤角质层比成年人薄，故儿童皮肤晒伤对成年后皮肤癌的发生影响很大。嘴唇由黏膜组成，比表皮角质层薄很多，并且它不会自身产生黑色素来保护自己，故极易晒伤。

（三）晒伤皮肤护理方案

1. 护理原则

（1）心理护理：顾客皮肤晒伤可能产生暴躁、恐慌，应稳定顾客的情绪使其能较好地配合护理和治疗。

（2）局部护理：弥漫性红斑给予炉甘石洗剂外涂，冰水湿敷或气雾剂冷喷。水疱者加强无菌换药，防止感染。

（3）预防护理：做好预防和防晒工作。

（4）健康指导：加强皮肤营养，平时多食富含维生素 C 的新鲜果蔬，适量吃点脂肪，以保证皮肤的足够弹性，增强皮肤的抗皱活力。维生素 C 可使皮肤黑色素沉着减少，从而减少黑斑和雀斑，使皮肤白皙。夏季应多食含多种维生素的食品。

知识拓展

认识紫外线

紫外线是电磁波谱中波长 10～400 nm 辐射的总称，不能引起人们的视觉。分为 A 射线、B 射线和 C 射线（简称 UVA、UVB 和 UVC），波长范围分别为 315～400 nm、280～315 nm、190～280 nm。

UVA 又称晒损伤性紫外线（长波紫外线），虽然不会引起皮肤的急性反应，但可引起真皮层细胞功能的改变，对衣物、玻璃、水等物质的穿透力极强，因此对皮肤的损害很大，而且其作用缓慢、持久，呈累积性，可导致皮肤晒黑、老化甚至癌变。其约有 52%可到达真皮层。

UVB 又称晒斑紫外线（中波紫外线），主要引起表皮细胞的功能改变，使皮肤产生以红斑为主的急性反应，玻璃可阻止其穿透。约有 10%通过真皮层，其余的 90%只能透过表皮层。

UVC 又称杀伤性紫外线（中波紫外线）。对生命细胞的杀伤能力最强，但一般均被臭氧层吸收，到达地面量少，不会对人体产生大的危害。

2. 护理程序

（1）清洁面部：晒伤后，一般不要马上清洗皮肤，等皮肤温度降至正常后再彻底清洁皮肤。尽量使用冷水或温水，避免用热水。停用护肤品。

（2）分期护理：

①红、烫期：采用冷敷、冷喷等方式给皮肤降温。

②肿痛期：可适当使用皮肤修复液，禁用碱性洁面乳。多补水，让肿痛自行消退。

③水疱：应及时到医院把水疱中液体吸出，疱膜留在皮肤上起保护作用。不可自己挤破，以免发生感染。

④色素沉着期：一般正常的色素沉着会在半年左右消退。多补水，可以淡化色素沉着。

⑤毛孔粗大、皮肤松弛：一般采用激光来处理，安全无副作用，效果较好。

(3) 仪器护理：红蓝光交替照射 15 min，修复晒伤皮肤。

(4) 喷雾：选择冷喷 15～20 min。

(5) 面膜：敷晒后修复面膜 15～20 min。

(6) 保湿：使用晒后修复、高效保湿保养品。

3. 预防

(1) 外出穿长袖上衣和长裤，戴太阳帽，搽防晒霜剂等，以避免太阳对皮肤的直接照射。

(2) 外出尽量避免日光中紫外线最为强烈的时间(6—8 月份 10:00—14:00)。

(3) 外出前避免食用光敏性食物，如菠菜、油菜、紫云英、马兰头、苋菜、马齿苋、茄子、土豆等，以免引起植物性日光性皮炎。

(4) 治疗期间应避免食用海鲜、辛辣食物等。

本项目重点提示

(1) 面部常见损美性皮肤包括衰老性皮肤、色斑性皮肤、痤疮性皮肤、敏感性皮肤、毛细血管扩张性皮肤和晒伤皮肤。

(2) 色斑分为雀斑、黄褐斑、黑斑、老年斑、色素痣和炎症后色素沉着。

(3) 痤疮分为粉刺型痤疮、丘疹型痤疮、脓包型痤疮、囊肿型痤疮、结节型痤疮、萎缩型痤疮、聚合型痤疮和恶病质型青春痘。

(4) 敏感性皮肤表现为面部泛红、发热、瘙痒、刺痛感、针刺感、烧灼感和紧绷感。

(5) 面部损美性皮肤的护理原则、护理程序和预防措施。

能力检测

一、选择题

1. 下列哪项不是衰老性皮肤形成的外因？(　　)

A. 紫外线的伤害　B. 地心引力的作用　C. 错误的保养　D. 内分泌紊乱

2. 下列哪种痤疮适合针清？(　　)

A. 白头粉刺　B. 丘疹型痤疮　C. 脓包型痤疮　D. 结节型痤疮

3. 痤疮红肿期应选择下列哪项护理项目？(　　)

A. 针清　B. 高频电疗仪　C. 超声波导入　D. E 光祛痘

4. 敏感性皮肤禁忌做哪项护理？(　　)

A. 冷喷　B. 敷冰膜　C. 超声波导入　D. 热喷

5. 下列哪项是毛细血管扩张性皮肤的不当保养方式？(　　)

A. 防晒　B. 去角质

C. 坚持用冷水洗脸增加毛细血管弹性　D. 使用去红血丝产品

6. 一日中紫外线最强烈的时间是(　　)。

A. 08:00—12:00　B. 09:00—13:00　C. 10:00—14:00　D. 11:00—15:00

二、填空题

1. 色斑分为________、________、________、________、________和________。

2. 痤疮分为________、________、________、________、________、________、________和________。

3. 白头粉刺可采用________处理,痤疮的红肿期可采用________治疗。

4. 毛细血管扩张性皮肤的成因包括________、________、________、________和________。

三、问答题

1. 色斑的形成原因有哪些?如何正确护理?

2. 哪种类型的痤疮适合做针清?针清的适应证和注意事项有哪些?

3. 敏感性皮肤的预防措施有哪些?

4. 如何正确护理衰老性皮肤?

5. 色斑性皮肤的成因有哪些?如何正确护理?

(张　颖　高　琼)

项目五　眼部护理

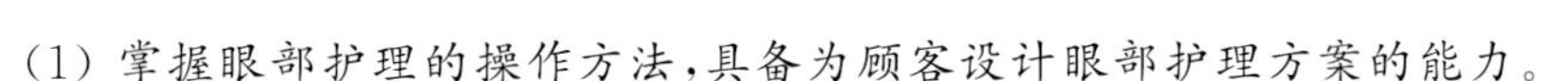

学习目标

（1）掌握眼部护理的操作方法，具备为顾客设计眼部护理方案的能力。

（2）熟悉常见的眼部问题、保养方法及护理的注意事项。

（3）了解眼部皮肤的特点。

项目描述

本项目主要介绍眼部皮肤的特点、眼部护理的目的、常见眼部问题和保养方法以及眼部护理操作方法和注意事项。学生通过本项目的学习，能了解如何保养眼部，预防眼部问题的产生，掌握眼部护理的方法，具备为顾客设计眼部护理方案的能力。

案例引导

张某，女，35岁，白领，因经常加班，劳累、生活不规律，并且有睡前喝水的习惯，眼睛浮肿并伴有黑眼圈，眼角有鱼尾纹，眼袋明显。影响了个人形象，想改善眼部问题。

问题：

1. 作为美容师，你认为应该给这位顾客做什么护理项目？有哪些注意事项？
2. 请你给顾客介绍其眼部问题形成的原因和眼部保养的方法。

一、眼部皮肤的特点

眼部皮肤不同于身体其他部位的皮肤。其特点为：①眼部皮肤是人体皮肤最薄的部位，表皮与真皮的厚度总和约0.55 mm，与面部表皮与真皮的综合厚度2 mm相比，要薄1/5～1/3，更易受到外界的伤害；②眼部皮下几乎没有皮脂腺和汗腺分布，眼周皮肤干燥、缺少水分；③眼部周围的肌肤有丰富的微血管和淋巴及神经组织，对外界刺激敏感，微血管极为细小，易出现血液循环不良；④眼部皮下疏松结缔组织丰富，柔软而富有弹性，疏松结缔组织的纤维结构中，分布着极为丰富的毛细血管和神经末梢，毛细血管的管壁非常薄，有一定的渗透性，疏松结缔组织容易充血、积水，形成血肿或水肿；⑤眼部肌肤的工作负荷重，每天眨眼(1.2～2.4)万次，易造成眼部疲劳，出现皱纹。

知识拓展

眼部保养小贴士

(1) 把一小杯茶放入冰箱中冷冻约 15 min,然后用一小块化妆棉浸在茶水中,再把它敷在眼皮上,能减轻眼袋浮肿。

(2) 常吃些含有胶体、优质蛋白、动物肝脏及番茄、土豆之类的食物。注意膳食平衡,可对此部位组织细胞的新生提供必要的营养物质,对消除下眼袋亦有裨益。

二、常见的眼部美容问题

(一) 眼袋

1. 形成的原因 眼袋系下睑皮肤、皮下组织、肌肉及眶隔松弛,眶后脂肪肥大,突出形成袋状突起(图 5-1)。不论男女均可发生,它是人体开始老化的早期表现之一。眼袋的形成受年龄、家族遗传、劳累和疾病等因素的影响。

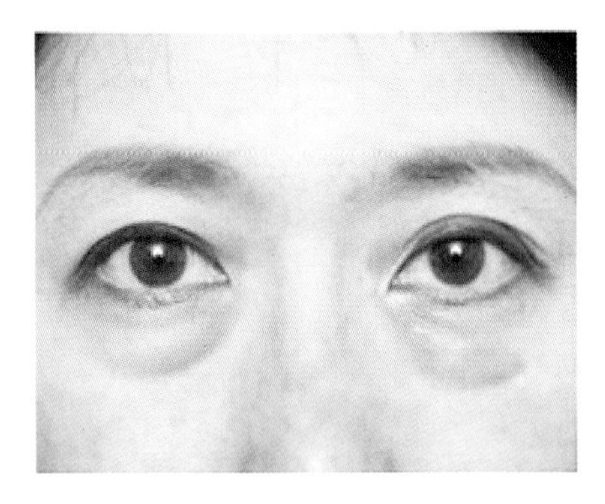

图 5-1 眼袋

(1) 年龄:常见于 40 岁及以上的中老年人,随着年龄的增长,出现皮肤松弛和肌肉松弛,导致眼袋的形成。也有女性在 25～30 岁之间就会出现眼袋。

(2) 遗传:眼袋的形成与遗传有密切关系,很多人的原发性眼袋往往有家族遗传史,多见于年轻人,其主要原因多为眶内脂肪过多。

(3) 劳累:现代生活方式下压力过大,休息睡眠不好,容易引起眼袋。此外,长期在电脑前工作,以及长期的脑力劳动,容易产生眼袋。

(4) 疾病:眼部神经疲乏或老化,使得肌肉衰弱,累积脂肪,造成眼皮组织松弛。当心理压力大、紧张、苦恼、悲伤或职业倦怠达到一定程度时,脾胃功能减弱,水湿运化不畅,亦可形成眼袋。急性肾炎,晨起上、下眼睑肿,代表脾虚、脾热。

2. 眼袋的类型 根据形成原因不同,眼袋分为:

(1) 单纯眼轮匝肌肥厚型眼袋:由于遗传性因素,年轻时就有下睑眼袋。其突出特点为靠近下睑缘,呈弧形连续分布,皮肤并不松弛,多见于 20～30 岁年轻人。

(2) 单纯皮肤松弛型:此种情况为下睑及外眦皮肤松弛,但无眶隔松弛,故无眶隔脂肪突出,眼周出现细小皱纹,多见于 33～45 岁的中年人。

(3) 下睑轻中度膨隆型:主要是眶隔脂肪的先天过度发育,多见于 23～36 岁的中青年人。

(4) 下睑中重度膨隆型:同时伴有下睑的皮肤松弛,主要是皮肤、眼轮匝肌及眶隔松弛,造成眶隔脂肪由于重力作用脱垂,严重者外眦韧带松弛,睑板外翻,睑球分离,常常出现流泪,多见于 45～68 岁的中老年人。

(5) 眼皮水肿:感觉眼睛睁不开,而且眼皮严重水肿,多是因为睡眠不足。

知识拓展

眼膜消眼袋的最佳时机

敷眼膜也要讲究时间，选对时机敷眼膜，可以起到事半功倍的护眼效果。一般来说，敷眼膜的最佳时间有三个：一个是女性生理期后一周，此时人体内雌性激素分泌旺盛，是敷眼膜最有效的时间；其次就是在泡澡时，借着热水蒸气，可以促进眼部循环；最后便是在睡觉前，在眼部运动最少时敷眼膜，发挥的效果也是很好的。

（二）眼周脂肪粒

脂肪粒是一种长在皮肤上的白色小疙瘩，约针头般大小，看起来像是小个白芝麻，一般在脸上，特别是女性的眼周。脂肪粒的起因是皮肤上有微小伤口，而在皮肤自行修复的过程中，生成了一个白色小囊肿（图 5-2）。也有可能是由于皮脂被角质所覆盖，不能正常排至表皮，从而堆积于皮肤内形成的白色颗粒。

（三）黑眼圈

1. 形成原因 黑眼圈是由于经常熬夜，情绪不稳定，眼部疲劳、衰老，静脉血管血流速度过于缓慢，眼部皮肤红细胞供氧不足，静脉血管中二氧化碳及代谢废物积累过多，形成慢性缺氧，血液较暗并形成滞留以及造成眼部色素沉着，也就是我们常说的“熊猫眼”（图 5-3）。常见于月经不调的女性，如功能性子宫出血、原发性痛经、经期长、经量大，均会出现黑眼圈。

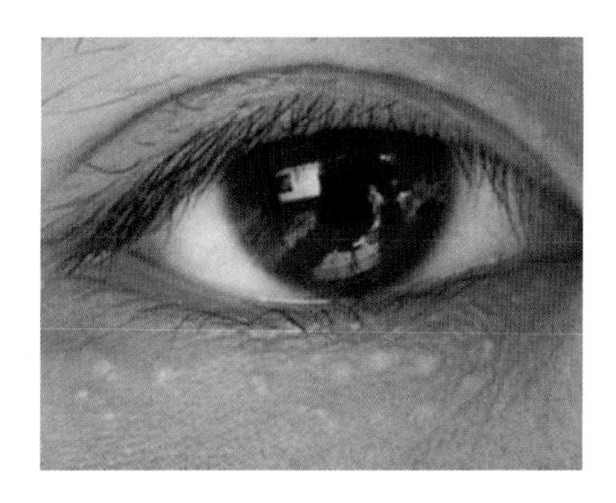

图 5-2 眼周脂肪粒

图 5-3 黑眼圈

2. 缓解方法 一是热敷法，即橄榄油加温，用棉签蘸取，下眼睑处轻轻擦拭，指腹拍打吸收；二是冷敷法，即冷冻铁汤勺熨眼袋；三是精油收敛法。

（四）眼部细纹

出现在眼角和鬓角之间的皱纹，其纹路与鱼尾巴上的纹路很相似，故被形象地称为鱼尾纹。上、下眼睑皮肤也易出现细纹。

1. 形成原因 是由于神经内分泌功能减退，蛋白质合成率下降，真皮层的纤维细胞活性减退或丧失，胶原纤维减少、断裂，以及日晒、干燥、寒冷、洗脸水温过高、表情丰富、吸烟等导致纤维组织弹性减退，眼部细纹增多。眼部细纹是由于眼轮匝肌长期收缩引起的，呈放射状（图 5-4）。

2. 缓解方法 多喝水是自然疗法，正确的涂抹眼霜可润泽干燥的眼部肌肤。

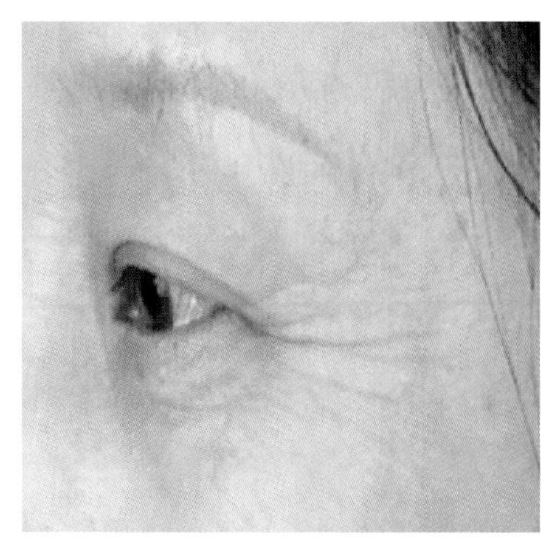

图 5-4 眼部细纹

三、眼部护理的目的

眼睛是最容易老化并产生美容问题的地方。一般自 25 岁以后眼周肌肤就开始老化。眼部护理可以缓解眼部疲劳，尽可能地缓解眼袋、黑眼圈、鱼尾纹等眼部损美性问题的发生。

四、眼部护理操作程序

（一）准备工作

1. 美容师准备 化淡妆、着工作服、穿工作鞋、戴口罩、去首饰、修剪指甲、洗手、消毒双手。

2. 用物准备

（1）床位准备：床单 2 条、毛巾 4 条，铺好美容床。

（2）护理产品准备：洁面乳、爽肤水、面盆、面巾纸、乳液、面霜、眼部卸妆液、卸妆棉、棉签、眼霜、眼部按摩精油、眼部精华液（蛋白霜）、眼膜（图 5-5）。

图 5-5 眼部护理用物

3. 环境准备 环境整洁卫生，温度、湿度适宜，光线柔和，备好香熏，播放舒缓的音乐等。

4. 顾客准备 协助顾客更衣，安置顾客。

（二）眼部卸妆和清洁

详见项目三。

（三）眼部按摩

1. 上按摩膏 由内眼角到外眼角均匀涂抹按摩膏（图 5-6）。

2. 点压下眼眶 左手呈虎口状固定左眼外眼角，右手中指指腹从内眼角到外眼角分 5～6 点按压下眼眶（图 5-7）。

3. 一手推，一手拉 左手中指拉，右手中指推，顺着眼袋朝斜向上方推到发际线，重复 3～5遍（图 5-8）。

4. 眼周排毒 右手剪刀指，左手提拉外眼角，重复 5～8 遍（图 5-9）。

5. 提拉上眼睑 双手四指指腹上提上眼皮，重复 5～8 遍（图 5-10）。

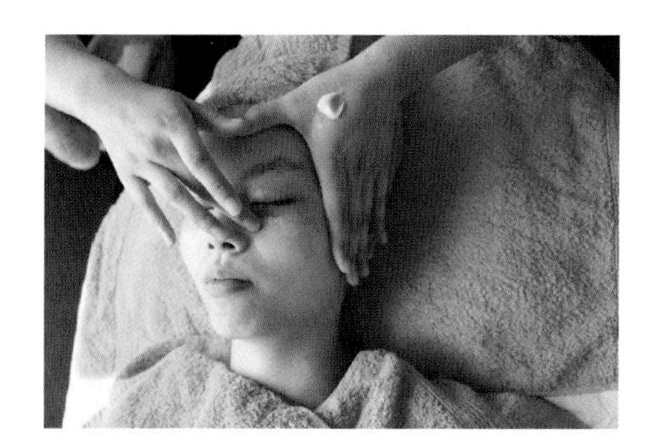

图 5-6 上按摩膏

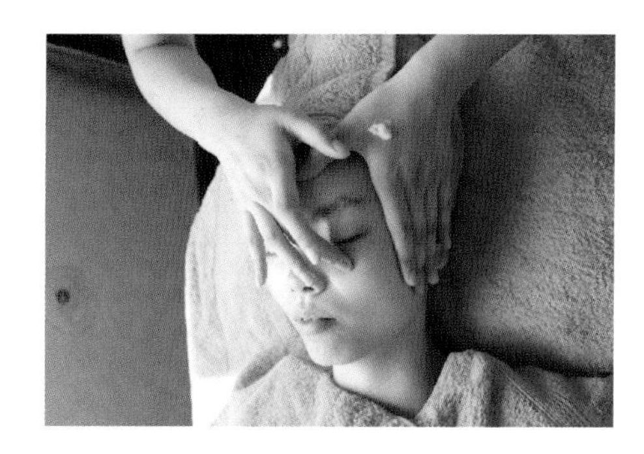

图 5-7 点压下眼眶

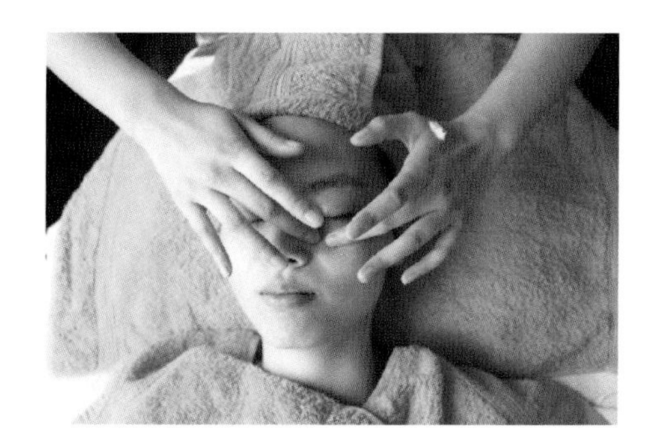

图 5-8 一手推一手拉

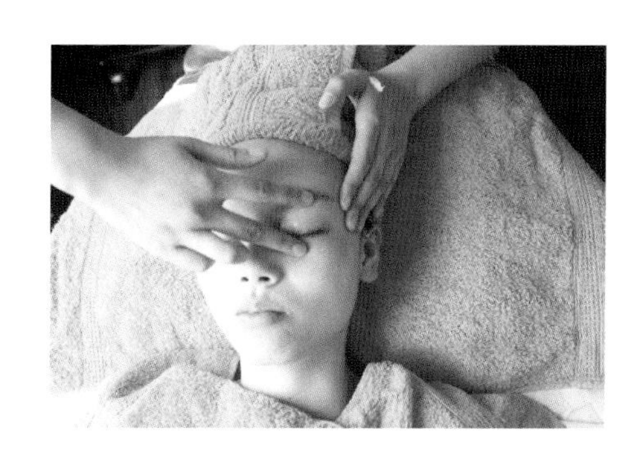

图 5-9 提拉外眼角

6. 提拉外眼角 虎口卡住眉骨，提拉眉骨至发际线，重复 5～8 遍(图 5-11)。

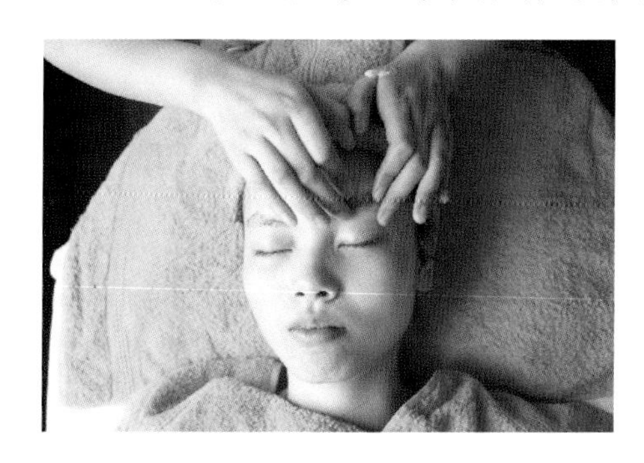

图 5-10 提拉上眼睑

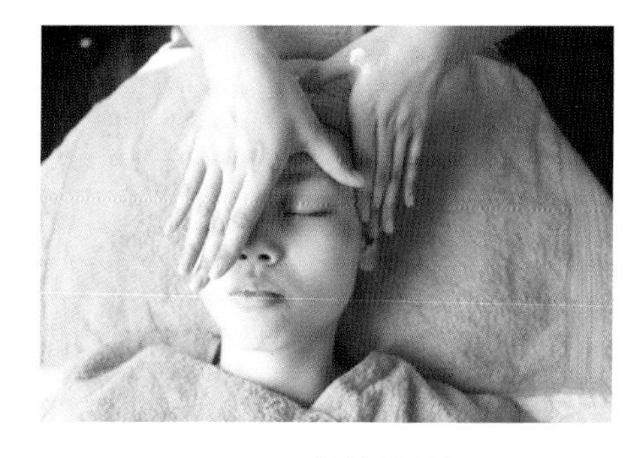

图 5-11 提拉外眼角

7. 同法按摩 同法按摩另一侧眼部。

8. 点压眼周穴位 依次点压睛明、攒竹、鱼腰、丝竹空、太阳、瞳子髎、球后、承泣，提拉睛明，安抚额头(图 5-12)。

9. 安抚眼部 安抚眼部，结束操作(图 5-13)。

(四) 仪器导入

1. 激光仪 目前最成熟的激光祛眼袋技术因其时间短、恢复快、无出血、痛苦小等优势，受到众多求美者的青睐。它采用新型的高能超脉冲激光，从睑板内侧结膜处切一约 0.3 mm 的切口，然后从切口处依次取出内、中、外三块脂肪团，再使用激光气化收缩真皮组织，整个手术过程只需 10 min 左右。此外，激光祛眼袋对皮肤表面无损伤，适应范围广，不易复发。

2. 射频仪 利用射频将眶隔脂肪液化，通过仪器产生的离子负压使汗腺导管扩张将眶

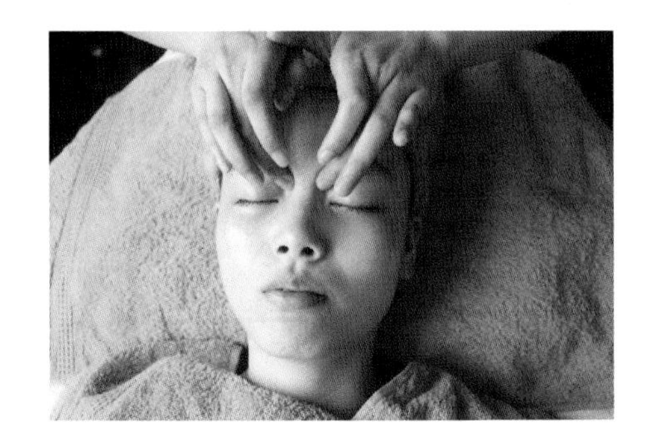
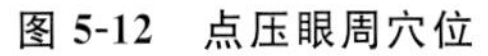
图 5-12　点压眼周穴位

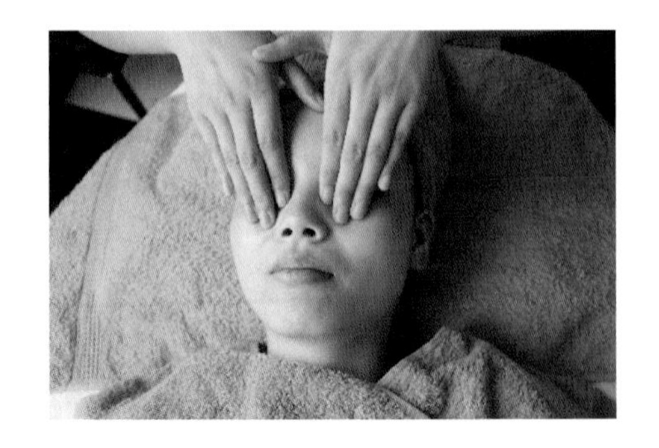
图 5-13　安抚眼部

隔脂肪导出。同时用微型组织定位器重新组织、定位收紧皮肤。优点:采用先进的射频仪使手术中无出血、无痛苦、恢复快,且能促进皮肤弹性恢复,对减少眼部细纹有独特的效果。

3. 超声波导入仪　详见项目七。

（五）敷眼膜

眼膜能够迅速深层滋润眼周围因缺水而产生细纹的肌肤,通过阻断空气,将养分大量导入眼周肌肤底层,并通过血液循环来促进吸收。通常情况下,敷 10～15 min 即可,顾客可利用这段时间闭目养神。

（六）基本保养

取下眼膜,美容指以打圈的方式促进眼膜精华液的吸收,全脸爽肤。取绿豆粒大小的眼霜,用双手无名指以打圈的方式均匀地涂在眼睛周围。涂擦面部保养。

（七）后续工作

引导顾客买单,告知居家保养方法,预约下次到店时间,书写护理记录,归还用物,环境整理。

五、注意事项

(1) 眼部皮肤娇嫩,按摩动作要轻柔,不易用力牵拉皮肤。

(2) 眼霜或精华膏使用量不可过大,否则容易引起脂肪粒。

(3) 眼霜、眼膜、精华膏等眼部保养品短期使用达不到好的效果,需要持之以恒长期护理方可延缓眼部衰老,也可保护视力。

(4) 睡眠充足,不熬夜,睡前避免饮水。

(5) 情绪乐观,避免阳光直射,改变眯眼、眨眼、挤眼睛等不良习惯。

(6) 注意眼部卸妆,合理选择并坚持使用眼霜。

(7) 经常按摩眼周穴位,促进眼周血液循环。

本项目重点提示

(1) 常见的眼部问题包括眼袋、眼周脂肪粒、黑眼圈、眼部细纹。

(2) 眼部护理操作流程包括准备工作、眼部卸妆和清洁、眼部按摩、仪器导入、敷眼膜、基本保养、后续工作。

(3) 眼部护理注意事项。

能力检测

一、选择题

1. 下列哪项不是眼部皮肤的特点？(　　)

A. 皮肤薄　　B. 几乎没有皮脂腺和汗腺的分布

C. 对外界刺激敏感　　D. 血液循环良好

2. 关于眼周脂肪粒，下列哪项描述不正确？(　　)

A. 针头般大小　　B. 眼霜使用不当　　C. 熬夜　　D. 不易改善

二、填空题

1. 常见的眼部问题包括________、________、________和________。

2. 眼部卸妆的顺序是________、________和________。

3. 眼袋的类型包括________、________、________、________和________。

三、问答题

1. 常见的眼部美容问题有哪些？

2. 简述眼袋的形成原因。

3. 试从眼部皮肤结构特点分析眼部为什么容易出现眼袋、细纹和黑眼圈等损美性问题。

(韩　慧　张　颖)

项目六 唇部护理

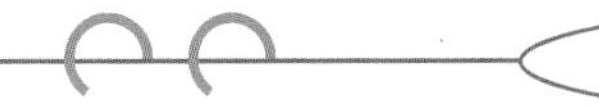

学习目标

(1) 掌握唇部护理的操作方法，具备为顾客设计唇部护理方案的能力。

(2) 熟悉常见的唇部问题、保养方法及注意事项。

(3) 了解唇部皮肤的特点。

项目描述

本项目主要介绍唇部皮肤的特点、唇部护理的目的、常见唇部问题和保养方法，唇部护理的操作方法和注意事项。学生通过本项目的学习，能了解如何保养唇部，预防唇部问题产生，掌握唇部护理的方法，具备为顾客设计唇部护理方案的能力。

案例引导

李某，女，30岁，IT行业培训讲师。经常在北方城市出差，唇部色泽暗沉，脱皮，唇纹明显，嘴角干裂并伴有疼痛感，为此十分苦恼，想尽快改善唇部问题。

问题：

1. 该如何帮助李某解决其唇部问题？请为她制订科学合理的护理计划。
2. 请告知顾客唇部问题产生的原因和唇部日常保养方法。

唇部护理就是对嘴唇进行养护。双唇是整个身体肌肤中最薄弱的部位，是人体比较容易"衰老"的部位。唇部皮肤一直裸露在外，易受环境的侵害。因此，根据唇部皮肤的特点定期做唇部护理，可以有效预防和改善唇部常见的美容问题。

一、唇部皮肤的特点

唇部皮肤具有独特性，其特点有：①缺乏皮脂腺和汗腺，缺乏天然保护膜，易失去水分且不具备锁水功能，容易发生嘴唇干燥甚至干裂；②缺乏黑色素细胞保护，对于阳光中的紫外线缺乏抵御能力，唇部皮肤颜色比其他部位的肌肤颜色深，更容易受到紫外线伤害，因此无论什么季节，都要选用有防晒成分的唇膏；③唇部皮肤比眼部肌肤还要薄，自身保护力弱，随着年龄增长，唇部皮肤角质层中的胶原质数量会不断减少，弹性变弱，会直接导致皮肤松弛，皱纹增多，甚至蔓延到唇线以外。

知识拓展

性感美唇的标准

1. 无痕　没有皱纹是娇嫩美唇的标志。
2. 水润　即古人说的“娇艳欲滴”。水润柔软的红唇是性感美唇的前提。
3. 色泽红润　美丽而健康的唇部，应该是唇红齿白。

二、常见的唇部问题

1. 嘴唇干裂　缺水是嘴唇干裂的主要原因。

2. 干燥脱皮　嘴唇皮肤比较薄，很容易被紫外线灼伤而引起脱皮。

3. 嘴唇颜色黯哑　暴晒容易导致唇色黯淡无光。经常化妆者要注意彻底卸妆，卸妆不彻底会导致唇色暗沉和干燥，严重的还可能染上“口红病”。

4. 唇纹　唇纹是指因嘴唇干燥和老化而形成的纹路。唇纹一般有两个方面的原因：遗传或是天生体质较干燥，缺水是形成皱纹的主要原因。另外，由于气候或体质造成干燥脱皮时，如若不保养呵护，唇纹会越来越深。唇部皮肤原本就很脆弱，加上一直裸露在外，所以极易受环境的侵害而变得缺乏生气。

5. 嘴角开裂　嘴角开裂常出现在秋冬季节，由天气干燥、身体内缺乏水分和维生素等原因引起。

三、唇部护理的目的

唇部是面部最活跃的部分，一个美丽的唇形，对五官的美化起着至关重要的作用。唇部一直以来都是女性化妆、美容的重点和亮点之一。有效预防和改善唇部问题，改掉影响嘴唇健康的不良习惯，会给自己带来一个健康完美的嘴唇。

（一）预防和改善嘴唇干裂

(1) 摄取 B 族维生素，多食新鲜蔬菜等。

(2) 补充水分，充足的饮水量对于人体机能的均衡有极大的帮助。

(3) 使用护唇膏呵护双唇。当唇部出现干裂时，可先用热毛巾敷唇 3～5 min，再用柔软的刷子轻轻刷掉唇上的死皮，然后抹上润唇膏，注意不要立即抹口红，会伤害唇部柔嫩的皮肤。如唇部皮肤干裂严重，则要进行唇部的特别护理。通常可以选择睡觉前，在双唇上涂抹含有金盏草及甘菊精华成分的润唇膏，这两种成分能舒缓干裂的双唇。

（二）预防和改善唇部脱皮

(1) 随身携带优质的润唇膏，特别是含有维生素 E 等滋润成分的润唇膏最为理想，能随时滋润唇部以防止双唇干燥脱皮。

(2) 要擦去口红补妆时，先在唇上加入少许凡士林，再用纸巾同时抹去残存的唇膏。因为口红中的石蜡、色素都具有带走水分的作用，长期化妆者容易出现嘴唇脱皮现象，所以一个星期最好有两天不化妆，只涂抹润唇膏。

（三）改善嘴唇颜色黯哑

（1）选用具有隔离与防晒功能的唇膏，多喝水，食用含有丰富维生素的蔬菜和水果。

（2）唇部护理必须使用护唇油，然后使用口红，最好是无色的。

（四）注意防晒、润唇，淡化唇纹

（1）给双唇涂抹蜂蜜，可以达到很强的保湿嫩肤效果。

（2）临睡前在唇部涂上一层橄榄油，防止水分蒸发，滋润效果非常好。

（3）外出注意做好防晒工作。

（五）预防和改善嘴角开裂

（1）嘴角开始干裂是缺少维生素的现象，应多补充维生素，多吃水果、青菜，避免偏食。

（2）缺少水分、说话太多都会导致唇部干燥，使嘴唇或嘴角处堆积废皮。因此，要多饮水，保持嘴唇湿润。避免舔嘴唇、撕皮等陋习。

知识拓展

橄榄油蜂蜜唇膜——水润防干唇

1. 美唇问题　干裂脱皮。

2. 材料　维生素 E 2 滴、橄榄油 2 滴、蜂蜜 1 茶匙。

3. 做法

（1）取适量橄榄油和蜂蜜倒入容器中。

（2）滴入维生素 E 2 滴（从胶囊里面取），混合调匀。

（3）用刷子在双唇上涂抹厚厚一层，敷约 15 min。

4. 效果　能使嘴唇保持湿润。此法特别适合干燥的季节和大风天气，可有效防止干燥季节嘴唇干裂。

四、唇部护理操作程序

（一）准备工作

1. 美容师准备　化淡妆、着工作服、穿工作鞋、戴口罩、去首饰、修剪指甲、洗手、消毒双手。

2. 用物准备

（1）床位准备：床单 2 条、毛巾 4 条，铺好美容床。

（2）护理产品准备：卸妆液、洁面乳、爽肤水、面盆、面巾纸、乳液、面霜、眼霜、唇部去角质产品、唇部精华素或营养油、唇部按摩膏或唇部保养液、唇膜、热棉片、保鲜膜、棉签（图 6-1）。

3. 环境准备　环境整洁卫生，温度、湿度适宜，光线柔和，备好香熏，播放舒缓音乐等。

4. 顾客准备　协助顾客更衣，安置顾客。

（二）唇部清洁

首先进行面部清洁，唇部卸妆要彻底。

（1）取一块棉片滴上适量的唇部专用卸妆液。

（2）先将充分蘸湿卸妆液的棉片按压在唇上 5 s，再将唇部分为 4 个区，分区清理唇部妆容。分别从中间向两边擦拭，直至擦拭干净为止(图 6-2)。唇部褶纹里的残妆，可用棉签蘸取卸妆液清除。

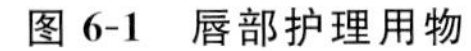

图 6-1 唇部护理用物

图 6-2 唇部清洁

(3) 取一块干净的湿棉片擦拭唇部残留的卸妆液，直至擦拭干净为止。

(三) 热敷唇部

用热棉片敷唇 3～5 min，软化唇部角质。

(四) 唇部去角质

选用合适的唇部去角质产品，每个月做 1 次，如果唇部已经受损就不可再去角质。唇部去角质可有效去除死皮细胞，加速新陈代谢，令双唇更加润泽。只有彻底清除干燥翘起的唇皮，双唇才会恢复细腻光滑的感觉。

(1) 用棉签蘸取“黄豆粒”大小去角质产品均匀地涂抹在唇部(图 6-3)。

(2) 停留 3～5 min 后，用棉签来回轻推，清除软化的小皮屑。

(3) 取一块湿棉片擦拭干净。

(五) 唇部按摩

(1) 选择合适的唇部按摩膏或唇部保养液，用棉签均匀地涂抹在唇部。

(2) 用食指和大拇指捏住上唇，大拇指不动，食指以画圈的方式按摩上唇，注意动作轻柔。再用食指和大拇指捏住下唇，食指不动，轻动大拇指按摩下唇(图 6-4)。然后，反方向有节奏地按摩上、下唇，反复数次，这样可以消除或减少唇横向皱纹。

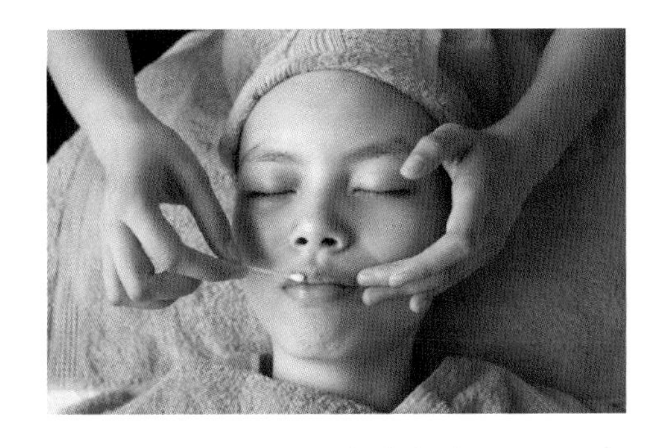

图 6-3 唇部去角质

图 6-4 唇部按摩

(3) 美容指轻拍嘴角部位数次，可减少嘴角皱纹。

(六) 敷唇膜

唇膜被称作唇部的“面膜”，唇膜中含有去角质酶和滋润唇部的功能性成分。使用唇膜的目的是滋润软化、去除老化角质，为双唇提供营养，淡化唇部色素沉着（图 6-5）。敷上专用唇膜，加盖热棉片，10 min 后取下。每周可做 1～2 次。

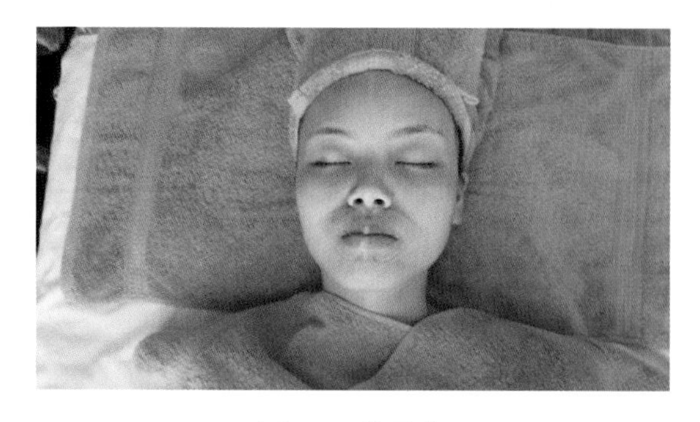

图 6-5　敷唇膜

（七）清洗唇部

用湿棉片擦拭干净。

（八）基本保养

唇部涂上保湿精华素或营养油，供给唇部营养，使唇部更加健康、柔润。涂擦面部保养品。

（九）后续工作

引导顾客买单，告知居家保养方法，预约下次到店时间，书写护理记录，归还用物，整理环境。

知识拓展

唇部健美操

1. 上唇运动　有平滑上唇线之功效，每个星期可练习 3 次。

（1）将拇指置于上排牙齿与牙床之间。

（2）将上唇轻轻伸展八次。

（3）维持伸展动作 5 s，然后让上唇放松。

2. 下唇运动　有助于强化下唇，使皮肤更有光泽，每个星期可练习 3 次。

（1）将口张开，上下排牙齿距离一寸左右。

（2）将食指扣在下唇内。

（3）将下唇伸展 8 次，然后放松 5 s 左右。

五、注意事项

（1）唇部卸妆要彻底，使用无酒精、成分温和的专用唇部卸妆液，避免清洁力过强而带走唇部油脂，对嘴唇造成刺激。

（2）避免舔唇、咬唇或抽烟等不良习惯，唇上起皮时，千万不要用手硬撕。

（3）选择优质护唇产品，劣质唇膏基本上都含有过量的蜡质，非但不能滋润唇部皮肤，还会影响正常新陈代谢。

（4）按摩动作轻柔，注意护理产品的用量和用法，护理产品不得流入顾客口腔内。

（5）注意做好唇部晨、晚间护理及周期护理，日常注意涂抹具有滋润功能的唇膏，防止唇部问题产生。

本项目重点提示

（1）常见的唇部问题包括嘴唇干裂、干燥脱皮、嘴唇颜色黯哑、唇纹、嘴角开裂。

（2）唇部护理操作流程包括准备工作、唇部清洁、热敷唇部、唇部去角质、唇部按摩、敷唇膜、清洗唇部、基本保养和后续工作。

（3）唇部皮肤特点：不具备锁水功能，易干裂；缺乏黑色素细胞保护，易受到紫外线伤害；唇部皮肤薄，自身保护力弱，易衰老。

能力检测

一、选择题

1. 下列哪项是唇部干裂的主要原因？（　　）

A. 缺水　　B. 日晒　　C. 唇部皮肤薄　　D. 体内缺乏维生素

2. 下列哪项不属于唇部护理的目的？（　　）

A. 防止干裂脱皮　　B. 预防唇纹

C. 改善嘴唇颜色黯哑　　D. 敷唇膜

二、填空题

1. 常见的唇部问题包括________、________、________、________和________。

2. 唇部皮肤特点包括________、________和________。

三、问答题

1. 常见的唇部问题有哪些？请说明原因和预防措施。

2. 请简述唇部日常保养方法和注意事项。

（张　颖　熊　蕊）

项目七　面部护理常用美容仪器

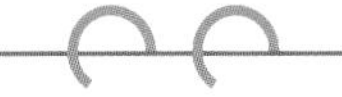

学习目标

（1）掌握面部护理常用仪器的作用及操作方法。

（2）熟悉面部护理常用仪器的工作原理及注意事项。

（3）了解面部护理常用仪器的日常养护。

项目描述

本项目主要介绍了面部护理常用仪器的作用、操作方法、注意事项以及仪器的日常保养。学生通过本项目的学习，能掌握面部护理常用仪器的正确操作方法，了解面部护理常用仪器的日常养护，具备根据皮肤检测的结果，正确选择美容仪器进行养护的能力。

案例引导

李某，女，48岁，白领，脸上有较多黄褐斑，眼角鱼尾纹也很明显，为此十分困惑，想改善面部肌肤问题。

问题：

1. 作为美容师，应该给这位顾客做什么仪器项目？有哪些注意事项？
2. 还需要向顾客介绍哪些面部肌肤的护理常识？

一、分析与检测仪器

（一）美容放大镜

美容放大镜有手持式、落地式、台灯式三种（图7-1）。

1. 作用

（1）提供放大及不刺眼的照明光线，以便进行肉眼观察，详细检视皮肤的微小瑕疵。

（2）增加皮肤治疗的专业性，借助美容放大镜，可有效地清除面部黑头粉刺、白头粉刺等。

2. 操作方法

（1）清洁面部，待皮肤紧绷感消失后，请被测试者闭眼，再用清洁纱布块盖住双眼，以免双眼被放大镜折射的光线刺伤。

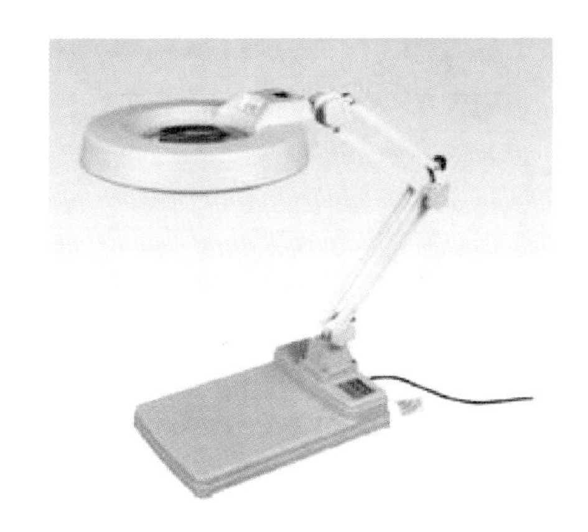

图 7-1 美容放大镜

(2) 将放大镜对准被测试者皮肤，操作者俯身近距离观察皮肤纹理、毛孔等情况。

3. 结果判断 镜下可观察到不同类型皮肤的特点(表 7-1)。

4. 注意事项

(1) 观察前，顾客必须彻底清洁面部皮肤。

(2) 顾客的皮肤可能会受到季节、环境、气候以及本人的休息、健康状况等诸多因素的影响，观察时应以当时的皮肤状态为基准。

表 7-1 美容放大镜下不同类型皮肤的特点

皮肤类型	镜下特点
干性缺水性皮肤	①肤色一般较白皙； ②皮肤干燥，松弛，缺乏弹性，不润滑，无光泽； ③表皮纹路较细，毛孔小，皮肤毛细血管和皱纹均明显； ④常有粉状皮屑自行脱落
干性缺油性皮肤	①皮肤干燥，但与干性缺水性皮肤比较，略有滋润感； ②皮肤缺乏弹性，松弛，缺乏光泽； ③表皮纹路细致，毛孔细小不明显，有皱纹，皮肤粗糙； ④常见微小皮屑
中性皮肤	①面色红润而富有弹性，皮肤滋润光滑，既不干燥，也不油腻； ②皮肤细嫩，无松弛老化迹象； ③表皮部位纹理清晰，肌理不粗不细，毛孔较细，无粗糙及黏滑感； ④无粉刺
油性皮肤	①皮肤油腻光亮，颜色黄； ②毛孔明显，皮肤纹理较粗，但不易发现皱纹； ③皮脂分泌过多堵塞毛孔，形成白头粉刺； ④皮脂被空气氧化可形成黑头粉刺，若被感染，则可形成痤疮
混合性皮肤	在面部 T 区(额、鼻、口周、下颌)呈油性皮肤特点，其余部分呈干性皮肤特点
敏感性皮肤	①皮肤毛孔紧闭细腻，表面干燥缺水； ②皮肤薄，粗糙，有皮屑； ③自觉红肿及瘙痒，多能看到丘疹，毛细血管表浅，可见不均匀潮红

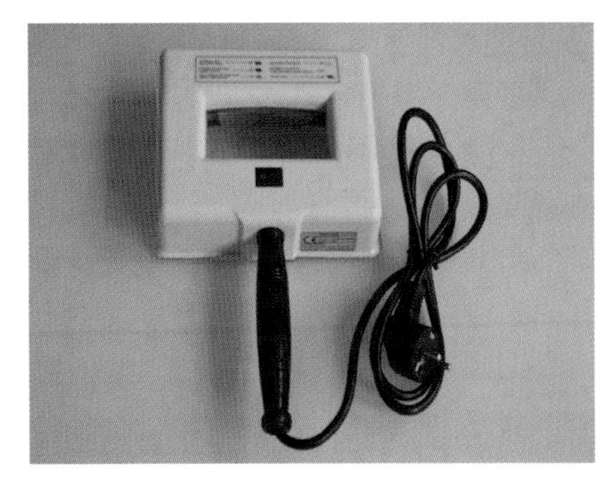
图 7-2 伍德灯

（二）美容透视灯

1. 工作原理 美容透视灯又称滤过紫外线灯，是由美国物理学家罗伯特·威廉姆斯·伍德(Robert Williams Wood)发明的，故称之为伍德灯(图 7-2)。它是由普通紫外线通过含镍的玻璃滤光器制成，由于不同的物质在它的深紫色光线照射下，会发出不同颜色的光，由此判断皮肤情况。

2. 作用

(1) 紫外线灯射出的光线能够穿透皮肤，帮助美容工作者仔细检查顾客皮肤的表面及深层组织情况，判定皮肤类型。

(2) 根据观察结果，便于制订和采取适宜的养护方案及措施。

3. 操作方法

(1) 清洁皮肤后，用清洁棉片盖住顾客眼睛。

(2) 关闭观察室窗帘及灯源，打开透视灯开关，使灯源距离顾客面部 15～20 cm，开始观察。

(3) 据观察所得资料进行分析判断(表 7-2、表 7-3)。

表 7-2 美容透视灯下皮肤色泽情况

皮肤状况	美容透视灯下显示
正常皮肤	蓝白色荧光
皮肤角质层及坏死细胞	白色斑点
厚角质层	白色荧光
水分充足的皮肤	很亮的荧光
较薄的、水分不足的皮肤	紫色荧光
缺乏水分的皮肤	淡紫色
皮肤上的深色斑点	棕色
痤疮及油性部位	橙色、黄色或粉红色

表 7-3 黄褐斑在肉眼观察和美容透视灯下的色泽对比

皮肤类型	肉眼观察	美容透视灯下观察
表皮型	灰褐色	色泽加深
真皮型	蓝灰色	不加深
混合型	深褐色	斑点加深

4. 注意事项

(1) 检测前，应清洁皮肤，不可涂任何药物或护肤品。

(2) 美容透视灯应在暗室内使用。

(3) 透视灯使用时间不能过长，以免仪器过热，缩短使用寿命。

(4) 透视灯不能直接接触皮肤及眼睛。

（三）皮肤检测仪

1. 工作原理 皮肤检测仪主要用于检测皮肤的性质，以便为皮肤病的治疗或美容护肤提供依据。皮肤检测仪由紫外线光管和放大镜两个部分组成(图 7-3)。它是基于不同物质对光的吸收、反射的差异原理以及光的特点工作的：不同性质的皮肤在吸收紫外光后，会反映出各不相同的颜色，此时再用放大镜加以扩放，就能清晰鉴别出不同性质的皮肤。

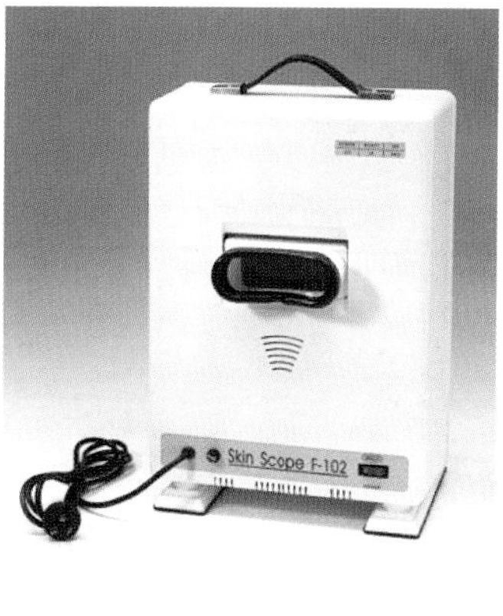

2. 作用 通过观察皮肤的颜色，可测试皮肤的性质，并根据其性质制订相应的治疗和护肤计划。

图 7-3 皮肤检测仪

3. 操作方法

(1) 清洁皮肤后，请被测试者闭上双眼，再用湿棉片覆盖被测试者的眼部。

(2) 美容工作者坐在被测试者对面，手持皮肤检测仪，灯管朝向被测试者，水平面置于被测试者面部，检测仪与面部间距为 15～20 cm，打开紫外光进行观察，测试时间不超过 2 min。

(3) 仔细观察皮肤颜色特征，以便区别皮肤类型，检测完毕及时关闭开关，移开湿棉片后，再请被测试者睁开眼睛。

(4) 根据颜色进行结果判断(表 7-4)。

表 7-4 皮肤检测判断标准

颜色	皮肤类型
青白色	健康中性皮肤
青黄色	油性皮肤
青紫色	干性皮肤
深紫色	超干性皮肤
橙黄色	粉刺皮脂部位
淡黄色	粉刺化脓部位
褐色、暗褐色	色素沉着
紫色	敏感性皮肤
悬浮的白色	表面角质老化
亮点	灰尘或化妆品的痕迹

4. 注意事项

(1) 测试前必须请被测试者闭上双眼，并用湿棉片覆盖其眼部，以防视觉疲劳。

(2) 测试时间最多不能超过 2 min，避免出现色斑。

(3) 面部有色斑者不宜使用检测仪，以免促使原有色斑加重。

(4) 严格掌握检测仪与被测试者面部之间的距离，不能少于 15 cm，以免引起光敏性皮炎。

5. 皮肤检测仪的日常养护

(1) 使用时注意轻拿轻放，以免紫外线光管被损坏。

(2) 不要使用刺激的清洁剂或有机溶剂清洁仪器。

图 7-4 皮肤、毛发显微成像检测仪

(3) 避免测试镜头接触油、蒸气和灰尘。

(4) 不能直接用水清洗,每天用干布擦拭仪器,放置于常温通风处,防止受潮。

(四) 皮肤、毛发显微成像检测仪

1. 工作原理 该仪器是利用光纤显微技术,采用新式的冷光设计,再放大足够的倍数,通过彩色银幕,直接观察局部皮肤基底层的细微情况,微观放大,及时成像,顾客可以目睹自身皮肤与毛发的受损情况,因此,它又被喻为皮肤的"CT"(图 7-4)。

2. 作用 通过观察皮肤的颜色,可测试皮肤的性质,并根据其性质制订相应的治疗和护肤计划。

3. 操作方法

(1) 接通电源,调整好镜头,用酒精棉球消毒镜头。

(2) 将镜头接近顾客受检部位,轻轻接触皮肤,显示屏即出现高清晰图像。

(3) 如需留资料,可启动彩色影像印制机,使之印成相片。

4. 注意事项

(1) 检测时皮肤应保持干燥,以免损伤镜头。

(2) 受检部位皮肤不得涂抹任何化妆品。

(3) 该仪器是光纤显微成像的精密检测仪,价格昂贵,需谨慎操作,轻拿轻放,避免碰撞使仪器受损。

(五) 专业皮肤检测分析系统

1. 工作原理 随着科学技术水平的提高,相继出现一系列高科技美容检测设备。专业皮肤检测分析就是利用专用皮肤电子数字水分计、皮脂测试仪、pH 值检测仪、色素测试仪、弹性分析仪及电子显微镜表面成像系统等,通过直接接触皮肤或将图像及相关参数输入电脑进行分析,准确而量化地诊断出皮肤的水分含量、油脂含量、皮脂膜的酸碱值、皮肤的色素含量、弹性强弱程度及皱纹、粗糙度等皮肤的综合状况,帮助美容工作者及时发现顾客皮肤的各种问题,从而选择正确的处理方法。

2. 作用

(1) 检测皮肤油分水分:了解皮肤表面水分和皮脂分泌的状况,正确判断顾客皮肤的类型,判断皮脂腺分泌是否正常。

(2) 检测皮肤酸碱度:人体表面的皮脂膜属于弱酸性。通过该测试仪所提供的皮肤 pH 值的资料和数据,可以帮助选择适合皮肤 pH 值的护肤品,制订合适的护肤疗程。

(3) 检测皮肤黑色素及血红素:可准确测出这两种色素的含量,有助于美容工作者观察肤色、色斑及色素沉着的形成和变化,以便评定养护效果,进而找到有效的养护方法。

(4) 检测皮肤水分流失情况:可以定量检测皮肤表面水分流失情况,以便确定保湿化妆品的效果,使皮肤处于最佳状态。

(5) 检测皮肤弹性状况:该皮肤测试仪可正确分析顾客皮肤的弹性情况,也可间接检测出各种增强皮肤弹性的方法是否有效。

(6) 检测皮肤衰老状况:该皮肤检测仪可以通过分析皮肤表面的图像,提供皮肤皱纹、粗糙度等参数，从而分析皮肤衰老状况，为延缓衰老的美容护肤品及肌肤养护方法的功效评定提供科学依据。

3. 操作方法

（1）在测试点上作一标记。

（2）将双面胶圈粘在探头上，掀去覆盖物。

（3）将平面测试探头垂直压在皮肤上，选择测试模式。注意探头与皮肤的接触适当，不能压得过紧，否则皮肤压入探头时可能被透镜擦伤或探头被皮肤的油脂污染。压得过紧也会影响皮肤血液循环，从而导致测量结果出现误差。如果需要在皮肤上多毛的部位进行测试，则需剃掉测试区域的毛发，防止玻璃透镜被毛发或其附着物擦伤。

（4）测试完毕会直接出现数据或有一个结果曲线出现在相应的显示器上，利用相关软件即可分析该曲线。

（5）在探头使用完毕及时盖上原来的保护盖。

4. 注意事项

（1）测试前，避免使用酸性或碱性洁肤用品，以免影响测试结果。

（2）电极探头只能用来检测未受伤的皮肤。

（3）测量在相同的室内条件下进行，即温度和湿度要保持恒定。只有这样才能对测试结果作比较。较为理想的室内温度为 20 ℃左右，湿度为 40%～60%。

（4）被测试者需要经过约 10 min 的自我调节，以便让活动后的血压恢复到正常水平，情绪激动会引起出汗。过高的血压或出汗都会给测量结果带来误差。

5. 仪器保养

（1）仪器探头不能受震动或碰撞，以防玻璃透镜被损坏。

（2）使用探头要十分小心，探头内部要保持清洁，任何物品与玻璃透镜的接触都将导致它的损坏，探头内部不干净将引起测量结果的不准确。

二、超声波美容仪

（一）超声波导入仪

物体在进行机械性振动时，空气中产生疏密的弹性波，其中，振动频率为 20～20000 Hz 机械振动波到达耳内能引起正常人的听觉，形成声音，称为声波。频率高于 20000 Hz 的机械振动波不能引起正常人的听觉，被称为超声波。

1. 工作原理 该仪器由高频振荡发生器和超声波发射器组成(图 7-5)。其工作原理是：由高频振荡发生器提供高频交流电，超声波发射器中的晶体薄片能随着交流变电场频率迅速而准确地改变体积（周期性的压缩与伸展），由此形成机械振动，此振动向周围介质传播而产生疏密交替的波形，即为超声波。当其作用于人体后被机体吸收，声能转变为热能，加上本身的机械振荡作用，使超声波具有如下作用。

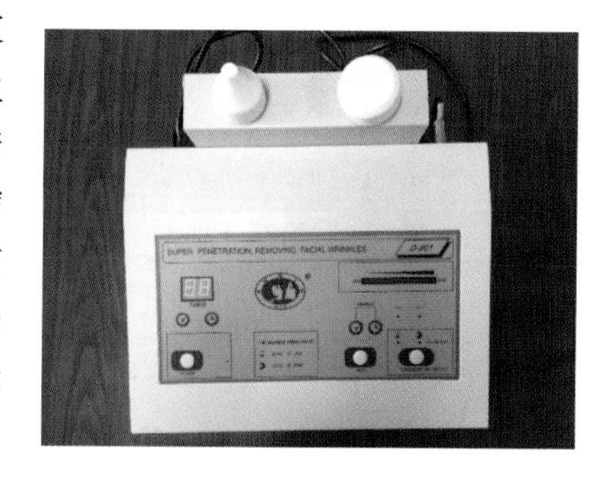

图 7-5 超声波导入仪

（1）温热作用：超声波传入皮肤后，引起组织细胞间的摩擦而产生热能，同时声能被吸收的部分也转化为热能，促进血液与淋巴循环，新陈代谢加强，使细胞吞噬功能也增强，从而提高机体防御能力，促进炎症吸收。

（2）化学作用：超声波的化学作用主要表现为聚合反应和解聚反应。聚合反应是将许多相同或相似的小分子合成一个大分子的过程，小剂量超声波作用于机体时，实质就是促进损伤组织的再生能力；解聚反应是使大分子黏度下降，相对分子质量降低，用超声波时，利于药物透入和吸收，增强药物疗效。

（3）机械作用：超声波具有比一般声波强大的能量，频率越高，振动速度就越快，提供的动能也就越大。当超声波作用于人体时，可引起组织中的细胞随之波动，组织得到微细而迅速的按摩，从而增强细胞膜的通透性，加强细胞新陈代谢，提高组织的再生能力，使皮肤富有光泽和弹性。它还可使坚硬的结缔组织延长、变软，使细胞内部结构发生改变，引起细胞功能的变化。

目前所用超声波发射功率一般在 25～30 W，其振动频率为 800～1000 kHz。

2. 作用

（1）减轻或消除皮肤色素沉着：一方面超声波美容仪的声波冲击能破坏色素细胞内膜，干扰色素细胞的繁殖；另一方面利用其化学解聚作用帮助祛斑精华素渗透于肌肤，从而化解色素，使色斑变浅变小。常用于化学性皮肤剥脱术后、磨削术后、激光术后、外伤、冷冻、炎症及痤疮愈后遗留的皮肤色素沉着、黄褐斑和晒斑等。

（2）消除眼袋和黑眼圈：超声波加上机械按摩产生的能量，可加速血液和淋巴循环，促使皮下脂肪溶解，增加皮下吸收，或使积聚过多的水分和脂肪消散，眼袋也随之减轻或消失，通过加快静脉血液循环，使血液流通正常，达到消退黑眼圈的目的。

（3）防皱除皱，散血去淤：超声波本身具有机械按摩作用，可调节皮下细胞膜的通透性，使药物抗皱霜迅速渗透到皮肤内，促进血液循环，增强新陈代谢，使皮肤缺水缺氧的情况得到改善，细小皱纹日渐消失，延缓衰老。机械按摩还可起到活血化瘀的作用，促使组织更快吸收，使淤斑消退。

（4）软化血栓，消除“红脸”：利用超声波的机械作用按摩扭曲变形的血管，再配合使用活血化瘀的药膏，从而软化血栓，扩张血管，促进血液回流，矫正变形的毛细血管，使之恢复正常，从而达到消除“红脸”的作用。

（5）治疗炎性硬结痤疮及其愈后瘢痕：超声波加痤疮消炎膏，再配合适当的按摩（可以轻轻按摩痤疮表面，待皮肤适应后再稍加压力），促进局部血液和淋巴循环；利用药物导入，使炎性痤疮的充血现象得到改善，皮下硬结逐渐软化，同时也避免了硬结的形成。

（6）治疗皮肤粗糙：利用超声波的机械按摩和温热作用，加上液状石蜡、甘油或润肤霜，能有效改善粗糙的皮肤状态。

（7）治疗螨虫感染：超声波可将药物渗透到螨虫感染部位，治疗被螨虫感染的皮肤。

（8）其他：除了以上作用外，超声波美容仪还具有祛除妊娠纹、减消双下巴、减肥，以及镇静、镇痛的作用。

3. 操作方法 一般采用直接接触辐射法，即超声头与治疗部位的皮肤直接接触，然后超声头在治疗部位作均匀缓慢的直线往返式“之”字形移动或作均匀缓慢圆圈式螺旋形移动，移动速度以 0.5～2 cm/s 为宜。

（1）接电源线与仪器，根据治疗面积的大小选择合适的超声头，一般面积小的部位皮肤有凹凸、狭窄处选择 1 cm 超声头，面积大且平坦的部位选择 2 cm 超声头，插入输出端，接通电源。

(2) 将仪器工作旋钮调至预热位置,时间为 3～5 min。

(3) 清洁顾客面部皮肤、蒸气喷面清除黑头粉刺。

(4) 选择适量的药膏或精华素、油剂、水剂或霜膏等均匀地涂擦在面部和超声头上,以超声头操作时能灵活转动为准。

(5) 根据顾客的肤质、年龄和个人感受调节超声波强度,一般皮肤较薄的部位超声波强度调为 0.5～0.75 W/cm^2,皮肤较厚的部位超声波强度调为 0.75～1.25 W/cm^2。

(6) 设定治疗时间,一般为每次 5～10 min。将工作按钮由预热调至工作位"连续"或"脉冲",即开始工作。

(7) 美容工作者手持超声头,力度均匀地呈"之"字形或螺旋形缓慢移动。

(8) 操作完毕,超声头离开皮肤,及时关掉电源。药物、精华素在皮肤上保留 5～8 min,使其充分渗透。

(9) 取下超声头进行清洗、消毒,擦干后保存,以防交叉感染。

4. 注意事项

(1) 超声波美容仪用前先要清洁面部,并涂上足够的面霜或药物后再使用,以防皮肤受损。使用的药物最好有一定黏度,黏度较好的介质可将超声头与皮肤较好地耦合起来,防止出现空隙,造成声能反射现象而不利于声能吸收。

(2) 治疗时间不超过 15 min,时间加长不会增加效果;隔日治疗一次,10 天为一个疗程,两个疗程之间间隔 7 天。

(3) 如果局部面积小,可用小探头做,但声波输出要减至 0.5～0.75 W/cm^2,时间为 8～10 min。如果顾客皮肤敏感,则最初强度要低,力度要轻,逐渐调整声波强度,并询问有没有灼热感和刺痛感,正常皮肤和敏感性皮肤有温热感已足够。超声头热度不代表声波输出功率,调得太高易灼伤面部皮肤。

(4) 严禁将处于工作状态的超声头置于顾客眼部,以免伤害眼球。

5. 仪器养护

(1) 超声头用后消毒擦干,保持洁净干燥;仪器及配件置于干燥环境,避免与酸、碱性物质接触。

(2) 应用干布擦拭超声头,轻拿轻放,用后放回原位。

(二) 超声刀美容仪

超声刀美容技术利用超声波聚集于单一一个点,以产生高能量,直达皮肤的肌肉腱膜系统产生收缩,立即产生提位效果,是目前风靡全球的一种美容技术(图 7-6)。

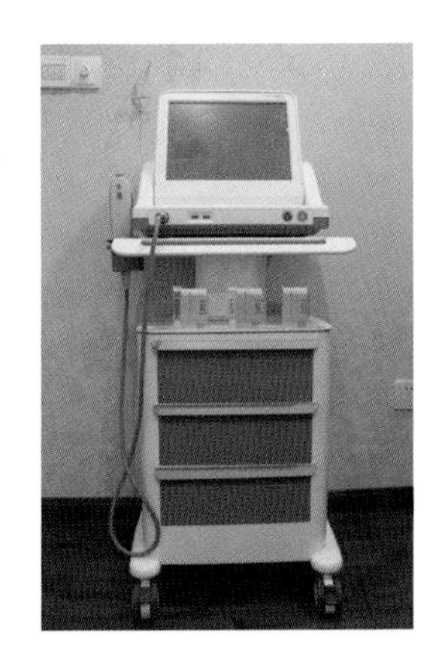

图 7-6 超声刀美容仪

1. 治疗原理 超声刀能释放一种高强度聚能超声波,它是以超声波为基础发展而来的技术,利用超声波在生物组织内的方向性、穿透性和可聚焦性等物理特性,将体外能量超声波穿透正常组织,精准聚焦于体内美疗区域,肌肤真皮层、基底层、SMAS 筋膜层相应位置点,其周围组织由于不在焦点上而不受影响,在 0.5 s 内的瞬间产生 65～70 ℃的热凝结点,使目标区域组织细胞即刻产生蛋白凝固反应,以凝固点为中心向四周产生拉力,即刻显现紧缩提拉效果。

2. 工作原理 超声刀聚焦以其独特的高能聚焦超声波直达 SMAS 层，提升 SMAS 筋膜悬吊，全面解决面部的下垂及松弛问题，它将超声波的能量精准地定位于皮下的 4.5 mm 筋膜层，使筋膜层肌肉生长牵拉，达到塑型、提拉、紧致的最佳效果。作用于皮下 3 mm 的胶原蛋白层，使胶原蛋白重组新生，达到使皮肤恢复弹性、美白、去皱、收细毛孔等美容抗衰效果。由于能量是掠过表皮，不用担心表皮受伤。使用该仪器后，可达到紧致皮肤、塑造面部轮廓、快速抚平皱纹的功效。

3. 作用

(1) 紧致皮肤，塑造面部轮廓，快速抚平皱纹。

(2) 改善肤质和皮肤弹性，让皮肤细致、有光泽。

4. 禁忌人群

(1) 发热、患传染病、急性疾病的人。

(2) 有心脏疾病、配置心脏起搏器的人。

(3) 严重高血压、恶性肿瘤、哮喘、深度静脉血栓、静脉曲张、甲状腺肿、癫痫等患者。

(4) 有出血性疾病、外伤、血管破裂、皮肤发炎、皮肤病患者。

(5) 孕妇。

(6) 月经期勿使用在腹部。

(7) 医学整形的部位、体内有金属物部位。

(8) 免疫系统功能异常者。

5. 操作方法

(1) 确认治疗部位，如面部、颈部或全方位。

(2) 与手术者签署治疗同意书，让手术者了解术前及术后的注意事项。

(3) 清洁手术部位以保证仪器治疗的效果。

(4) 拍摄治疗前的照片并存档，以便于后期对比。

(5) 根据所需治疗部位换上相应的工作头。

(6) 仔细观察手术者皮肤情况，规划最佳治疗线路，涂抹上专用凝胶(图 7-7)。

(7) 术前准备完毕后，给手术者按照线路进行超声刀的全面操作。选择相对应的工作头，调节适合于手术者的能量。每次在手术者变动体位后，确保工作头与手术者皮肤接触良好。治疗过程中确保探头四角完全接触皮肤表面，并有足够的介质(图 7-8)。

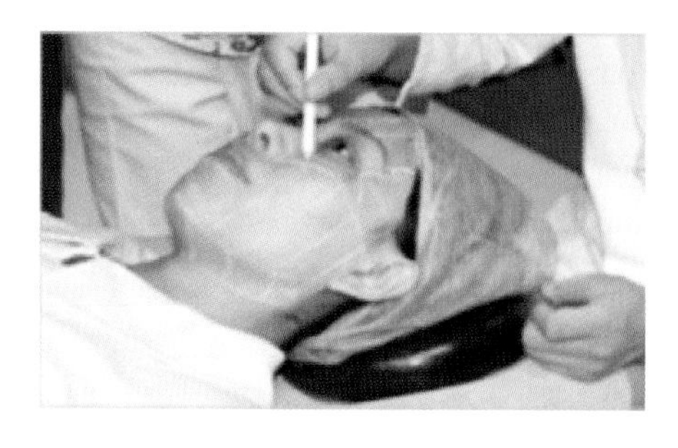

图 7-7 面部画线

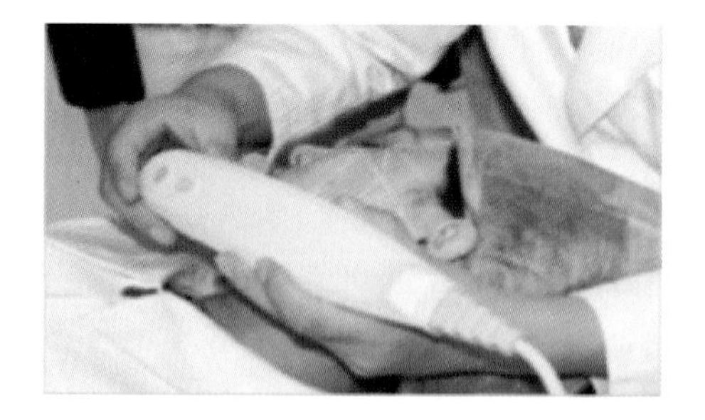

图 7-8 超声刀实操

(8) 操作完成后，给手术者用温水清洁面部，并在 1 h 后为手术者敷上医用的修复面膜。

6. 注意事项

(1) 高能量聚焦超声刀治疗后的轻微泛红水肿乃正常现象，通常可于治疗后数小时内缓解。

(2) 治疗后使用冷水与温和清洁用品清洁治疗区域。若肌肤仍处泛红状态,应避免碰触热水,直到泛红情况缓解。

(3) 治疗后至少1周内,勿于治疗区域进行去角质操作,如出现敏感现象也勿去角质,直至该处敏感缓解。

(4) 超声刀拉皮治疗后可立即上妆,以矿物质彩妆为优。

(5) 治疗后可使用舒缓、无刺激性的乳霜或保养品,推荐使用皮肤修复套组,配合超声刀拉皮使用,可获得理想抗衰效果。

(6) 如需外出接触阳光,使用SPF30及以上的防晒乳以避免日照伤害。

7. 仪器养护

(1) 使用时避免周围放置医疗设备或家用电器。

(2) 禁止在高温潮湿、灰尘多、油烟多、水蒸气多的环境中使用。

(3) 使用中如发生故障,应立即关闭电源,必须由专业人员拆卸、维修,或送到经销商、厂家维修,不得自行修理。

(4) 使用时请勿靠近墙壁,保持仪器四周有30 cm的空间,以便散热。

三、热疗美容仪器

(一) 射频美容仪

射频(radio frequency,RF)美容技术是一种非手术、准医学的全新美容方法,可以拉紧皮下深层组织和收紧皮肤,达到使下垂或松弛的面部重新提升的效果(图7-9)。

1. 工作原理 射频美容仪利用每秒600万次的高速射频技术作用于皮肤,皮肤内的电荷粒子在同样的频率下会变换方向,随着射频高速运动后产生热能,真皮层胶原蛋白在60～70 ℃时,会立即收缩,让松弛的肌肤马上得到向上提拉、紧实的拉皮效果,促使皮肤快速恢复到年轻健康的状态。同时皮肤组织在吸收大量热能后也会源源不断地合成新的胶原蛋白,使真皮层的厚度和密度增加,皱纹得以抚平,达到消除皱纹、收紧皮肤、延缓皮肤衰老的美容效果。

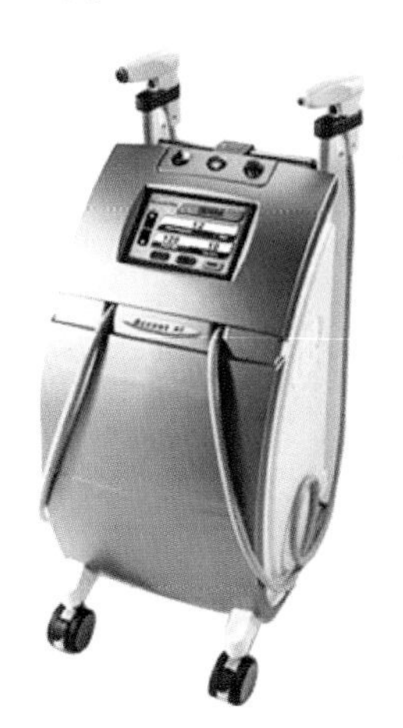

图7-9 射频美容仪

2. 作用

(1) 收紧皮肤,提升面部皮肤。

(2) 改善肌肤的新陈代谢,光嫩皮肤。

(3) 去除皱纹,修复妊娠纹。

3. 操作方法

(1) 用适合顾客皮肤的洗面奶初步清洁皮肤。

(2) 接通电源,向仪器插进IC卡,仪器处于待机状态。

(3) 在顾客面部涂抹一层冷凝胶。

(4) 连接射频探头和紧肤电流棒,设置工作时间,一般为20～40 min。

(5) 美容工作者分别用射频探头和紧肤电流棒在面部皮肤上轻轻滑动,操作手法由内向外,由下向上,与皱纹方向垂直,与肌肉走向相一致,重点集中在眼角、嘴角的表情纹和其他有皱纹的部位,每个部位养护时间约为15 min。

(6) 养护完毕,清洗凝胶,涂抹营养霜。

4. 注意事项

(1) 安装心脏起搏器、有金属植入、发热、晚期病证、出血性疾病、治疗区有严重皮肤病者,以及有注射皮下填充物者和孕妇禁止使用。

(2) 通常情况下,射频美容养护需要 20～40 min。如果顾客对疼痛或者热度敏感,可以在治疗部位涂抹一层具有镇静或者缓解疼痛作用的冷凝胶或喷雾剂。

(3) 少数顾客在养护后皮肤有微红现象,不必处理,可在几小时后自行恢复正常。

(4) 加强皮肤保湿和防晒养护。

(5) 一周内勿用热水洗脸、泡温泉及桑拿浴。

知识拓展

RF、e 光与 IPL

RF(radio frequency),即射频,是一种高频交流变化电磁波的简称,可以辐射到空间的电磁频率为 300 kHz～300 GHz,主要利用其射频能量进行祛皱、美白等。

IPL(intense pulsed light),即强脉冲光,是一种很柔和、有良好光热作用的光源。基于光的选择性吸收和强热量原理,照射皮肤后会产生生物刺激作用和光热解作用,而被用于治疗痤疮、老年斑、色斑以及改善皮肤肤质等。

e 光的核心技术主要是射频＋光能＋表皮冷却,是射频能量与强光优势互补结合进行治疗的技术,在光能强度较低的情况下强化靶组织对射频能的吸收,极大地消除了光能过强的热作用可能引起的副作用和不适,广泛用于祛斑、脱毛、祛除红血丝、除痣等。

(二) 光子嫩肤仪

1. 工作原理 光子具有良好的光热分离性,即对组织具有选择性,只对目标组织产生作用,而不会对周围正常组织产生影响。光子嫩肤就是采用宽光谱强脉冲的光解热能原理及生物激化作用,使特定波长的脉冲强光能量到达皮肤深层,有选择性地作用于皮肤组织中的病变色素,色素细胞分解吸收而不影响正常组织;作用于血管,使扩张的毛细血管闭合,祛除红血丝;同时还可以促使胶原组织增厚、弹性增加、皱纹减少(图 7-10)。

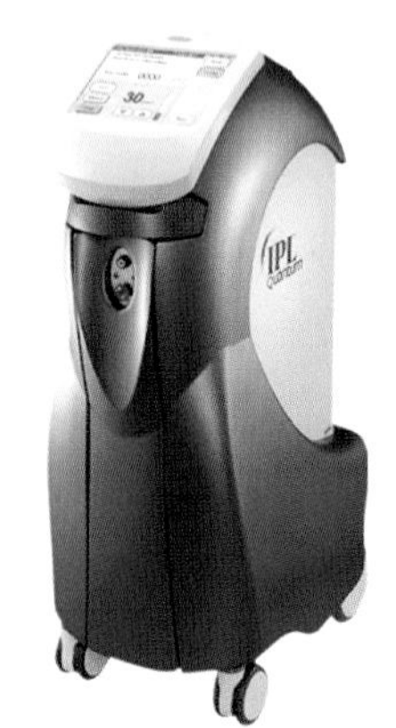

图 7-10 光子嫩肤仪

2. 作用

(1) 通过分解皮下色素而淡化雀斑、黄褐斑、日晒斑以及痤疮印。

(2) 闭合面部扩张的毛细血管,使皮肤发红以及毛孔粗大、细小皱纹、黑眼圈、晦暗皮肤和酒渣鼻引起的红鼻头等情况得到改善。

3. 操作方法

(1) 开机预热,观察顾客皮肤状态,根据其皮肤问题,确定治疗方案。

(2) 应用专用洗面奶洁面,彻底清除面部的污垢和死皮,提高治疗效果,同时为肌肤设置一层保护膜,避免强光刺激。

(3) 打开控制面板,根据治疗需求及顾客耐受度设置脉宽、脉冲数和能量等各项参数,并将它们调节到最佳组合状态。

(4) 让顾客戴上光子嫩肤专用护目镜,防止眼睛遭受强光刺激。

(5) 用专业工具将冷凝胶敷于需治疗部位的肌肤上,以防止强光灼伤皮肤,减轻顾客疼痛，同时也起光导入作用。

(6) 操作者自己也需戴上光子美容专用眼镜。

(7) 将冷凝胶涂于仪器的光头上，从面部耳旁皮肤开始用光点击治疗，并均匀地向周围扩散。因为耳旁皮肤比其他部位更敏感，如出现过敏反应，可及时调整。

(8) 治疗结束后，再次清洁肌肤。

(9) 导入精华素。

(10) 最后涂抹无刺激的眼霜、润肤霜和防晒霜。一方面皮肤可以充分吸收营养，达到理想的效果，另一方面也避免皮肤因日光照射引起过敏反应。

4. 注意事项

(1) 治疗前应询问过敏史，避免服用引起过敏和抗凝的药物。

(2) 治疗期间尽量不化妆，即使上妆，也尽量不用粉底，应用性质温和的护肤品。如果在治疗区域出现裂口或结痂，应立即停止化妆并到医院就诊。

(3) 配合内服一些维生素 C、维生素 E，帮助色素减退。

(4) 一个月内建议顾客外出时做好防晒工作，每天使用无刺激性的防晒品。

(5) 夜间用冷水柔和地清洗皮肤，可少量使用无刺激性保湿护肤品。

(三) 红蓝光治疗仪

1. 工作原理 红蓝光治疗仪主要运用光动力疗法原理。光动力反应的基本机制是：生物组织中的内源性或外源性光敏性物质受到相应波长的光，(可见光、近红外线或紫外线)照射时，吸收光子能量，由基态变成激发态，产生大量活性氧，其中最主要的是单线态氧，活性氧能与多种生物大分子相互作用，产生细胞毒性作用，导致细胞受损甚至死亡，从而产生治疗作用(图 7-11)。

图 7-11 红蓝光治疗仪

蓝光治疗仪的治疗原理：痤疮丙酸杆菌可产生卟啉，它主要吸收 415 nm 波长的可见光，蓝光的波长正好在这一波段，照射后产生光动力学反应，导致痤疮丙酸杆菌死亡，减轻或治愈痤疮。

红光治疗仪对卟啉的光动力效应弱，但能更深地穿透组织。在红光的照射下，巨噬细胞会释放一系列细胞因子，刺激成纤维细胞增殖和生长因子合成，细胞的新陈代谢加强，促使细胞的新生，同时也增加了白细胞的吞噬作用，提高了机体免疫功能，因而使炎症愈合和组织修复更快。

光动力治疗仪除了有红蓝光头，还有黄光头、绿光头等。

2. 各种光的临床应用

(1) 红光：波长为 635 nm 的红光具有纯度高、光源强、能量密度均匀的特点，在皮肤护理、保健治疗中效果显著，被称为生物活性光。红光能让细胞的活性提高，促进细胞的新陈代谢，使皮肤大量分泌胶原蛋白与纤维组织来自身填充。加速血液循环，增加肌肤弹性，改善皮肤萎黄等状况，从而达到抗衰老、抗氧化、修复的功效，有着传统护肤无法达到的效果。主要功效包括美白淡斑、嫩肤祛皱、修复受损皮肤、抚平细小皱纹、缩小毛孔、增生胶原蛋白。

(2) 蓝光：波长为 415 nm 的蓝光具有快速抑制炎症的功效，在痤疮的形成过程中，主要是痤疮丙酸杆菌在起作用，而蓝光可以在对皮肤组织毫无损伤的情况下，高效地破坏这种细菌，最大限度减少痤疮的形成，并且在很短时间内使炎症期的痤疮明显减少至愈合。

(3) 紫光:红光和绿光的双频光,其结合了两种光的功效,尤其在治疗痤疮和祛痤疮印方面有着特别好的效果和修复作用。

(4) 黄光:波长为 590 nm 的黄光,对于敏感性皮肤及处于过敏期的皮肤有良好的缓解和治疗作用。

(5) 绿光:波长为 560 nm,自然而柔和的光色,有安定神经的功效。可改善焦虑或抑郁,调节皮肤腺体功能,有效疏通淋巴及去水肿,改善油性皮肤等。

3. 操作方法

(1) 彻底清洁皮肤,消毒,清理痤疮及粉刺。

(2) 根据治疗要求,选择治疗光头,置于顾客治疗部位上方,光板距离皮肤表面 1～4 cm,每次照射 20 min,每周 2 次,光照间隔至少 48 h,8 次为一个疗程。

(3) 痤疮性皮肤以红蓝光交替治疗为主,炎性皮损较明显者先予以蓝光照射,炎症后期或炎症不明显者给予红光照射。

4. 注意事项

(1) 禁忌证:卟啉症患者、孕妇、光过敏者等。

(2) 注意照射时间,照射局部可出现轻微疼痛,照射后可出现持续数小时头痛。

(四) 眼袋冲击仪

1. 工作原理 眼袋冲击仪采用了高频磁振和恒温技术,通过舒适的电极振动和适宜的恒定温度,促进眼部的血液及淋巴循环,从而增强眼部肌肉运动,活跃皮下组织,使下眼睑堆积的脂肪分解,改善眼袋的下垂感,增加眼部皮肤弹性,防止皮肤老化及松弛,达到预防眼袋发生的作用。

2. 作用

(1) 促进血液及淋巴循环,延缓皮肤衰老,经常使用可消除眼角皱纹。

(2) 解除眼球疲劳,帮助消散眼袋淤血及黑眼圈。

(3) 减轻下眼睑的脂肪堆积,避免眼袋因脂肪增加而下垂和臃肿。

(4) 补充营养,增加皮肤弹性,活跃皮下组织,防止眼周皮肤松弛。

3. 操作方法

(1) 清洁皮肤,保证皮肤无杂质。

(2) 眼袋部位涂适量专用眼袋霜。两个眼袋接触片用滋润液浸湿,然后将接触片分别置于两侧眼袋处。

(3) 开电源开关,主体开始振动,调节振动强度,由弱到强,一般以有轻微振动为宜。同时开始加热,约 1 min 后达到适合温度,即可使用。

(4) 按摩走向应与肌肉和血管方向一致,慢慢移动。

(5) 治疗时间约为 10 min。结束时,先将振动强度调至零,再关电源开关。每天按摩 1～2次,每次 5～10 min。

(6) 取下眼袋接触片洗净、晾干,妥为保存。

4. 注意事项

(1) 使用前,皮肤应保持清洁,无水分、油分等杂质。

(2) 养护过程中,眼袋接触片不可接触眼球,以免伤害眼球。

(3) 眼周皮肤细嫩,振动强度要适中,避免强度过大而伤及皮肤。

(4) 眼袋接触片用后一定要清洗干净,保持干燥,以免接触片被腐蚀。

(五) 奥桑喷雾仪

奥桑喷雾仪又称为离子喷雾机,可以产生普通蒸汽和奥桑蒸汽(图 7-12)。

1. 仪器结构 奥桑喷雾仪由支架、滚轮、牵拉环、喷孔、注水孔、奥桑开关、喷雾开关、奥桑指示灯、烧杯和电热元件组成。

2. 工作原理 奥桑是 ozone 的谐音,奥桑喷雾仪工作时有臭氧产生。普通喷雾仪喷雾时,盛在水瓶中的蒸馏水或去离子水经电热器加热至沸腾产生大量蒸汽,然后由喷气管均匀柔和地喷出。而当奥桑开关打开时,喷气管内的紫外线灯开始工作,空气中的臭氧经照射分解产生负氧离子,负氧离子与氧气结合产生臭氧。臭氧及负氧离子随蒸汽到达皮肤。

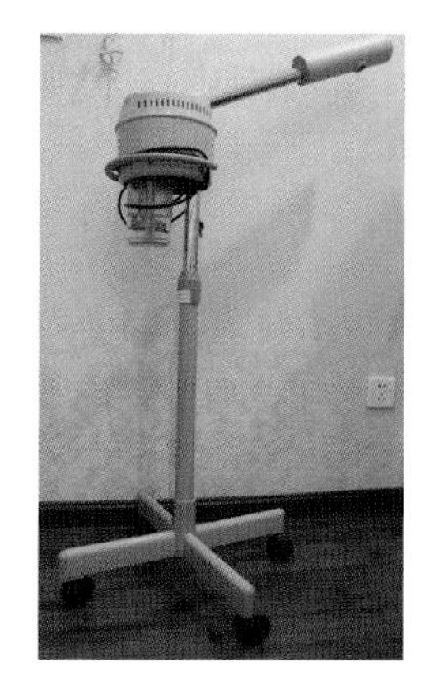

图 7-12 奥桑喷雾仪

3. 作用

(1) 软化表皮,便于清除老化的死皮细胞。

(2) 给皮肤补充水分。

(3) 扩张毛孔,便于清除毛孔内的污垢。

(4) 促进新陈代谢,利于皮肤排泄废物。

(5) 杀菌消炎,增强皮肤的免疫功能。

4. 使用方法

(1) 注入蒸馏水:注水时不宜超过烧杯容积的 4/5 或最高水位线,最低水位线不得低于发热元件。

(2) 预热:按下开关,仪器通电后预热 5～6 min,当蒸汽由喷口喷出后才能开臭氧灯,关闭时先关臭氧灯。

(3) 用湿棉片盖住眼睛。

(4) 喷雾使用时间和距离见表 7-5,喷射过程中可以点拍面部,使面部皮肤放松。

(5) 喷雾结束后,先将喷口从顾客面部移开,再关闭奥桑喷雾仪的开关。

表 7-5 喷雾使用时间和距离

皮肤类型及皮肤问题	喷口与面部的距离	应用时间
中性皮肤	25 mm 左右	3～5 min
油性皮肤	25 mm 左右	5～8 min(臭氧)
混合性皮肤	25 mm 左右	5～8 min
干性皮肤	35 mm 左右	3～5 min
痤疮性皮肤	25 mm 左右	5～8 min(臭氧),或者 20 min(冷喷)
敏感性皮肤	25 mm 左右	20 min(冷喷)
色斑性皮肤	35 mm 左右	5～8 min
衰老性皮肤	35 mm 左右	3～5 min
毛细血管扩张性皮肤	25 mm 左右	20 min(冷喷)
微血管破裂	35 mm	20 min(冷喷)

5. 注意事项

(1) 根据顾客皮肤类型调节好喷口与面部的角度和距离,避免喷出的蒸汽直射鼻孔而产生呼吸不畅。

(2) 喷雾时间不宜过长，最好不超过 15 min，以免皮肤出现脱水现象。

(3) 皮肤色斑、敏感及毛细血管扩张性皮肤不宜做奥桑喷雾。敏感性皮肤使用热喷时，需用湿棉片盖住敏感部位。

(4) 烧杯中的水位不得超过烧杯上面的红色标线或烧杯容积的 4/5。仪器在使用时要注意观察水位不得低于发热元件，避免干烧而损坏仪器。

(5) 奥桑喷雾仪必须使用蒸馏水，不能使用自来水。

(6) 当发现喷射的雾气不均匀或有水滴喷出时，必须马上将喷口从顾客面部移开，并关闭仪器，以免烫伤顾客。

(7) 注水前检查仪器是否处于完好备用状态，烧杯是否有裂缝。

6. 仪器养护

(1) 每周清洗 2 次玻璃杯，每天换水。

(2) 喷口产生喷水现象，可能是由于水中杂质将喷口堵塞。可用 6∶4 的白醋和水浸泡 24 h 后再刷洗。

(3) 蒸汽四散不集中时，可能是烧杯口上的胶垫老化，杯口不密封所致。

(4) 用完后先关闭开关，后切段电源，将仪器归位。

四、激光美容仪器

点阵激光技术是近两年美国乃至全球皮肤界最受关注的皮肤美容新技术，是介于有创和无创之间的一种微创治疗。点阵激光治疗理论称为点阵式光热分解作用理论，它由美国哈佛大学的激光医学专家 Dr. Rox Anderson 于 2004 年首先发表，立即得到世界各地专家认同并迅速应用于临床治疗(图 7-13)。

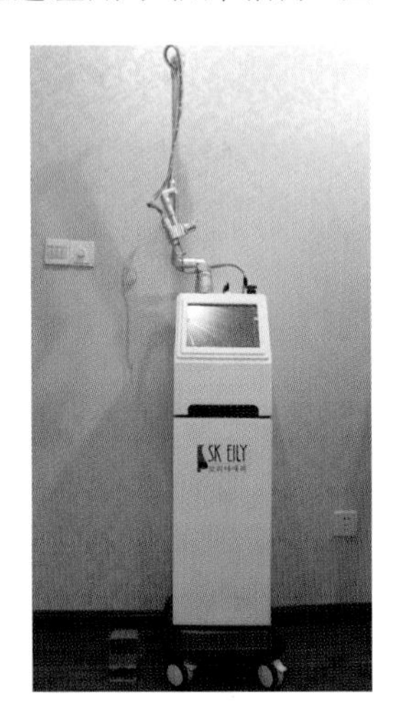

图 7-13 二氧化碳点阵仪

1. 治疗原理 点阵只是一种激光发射的模式，使用特殊的图像发生器(CPG)改变光的发射模式，点阵激光可透过高聚焦镜发射出 50～80 μm 的焦斑，并将这些焦斑扫描出多达 6 种图形(圆形、正方形、长方形、菱形、三角形、线形)，分别适用于不同部位和不同肤质的治疗。图像发生器把原本聚集的光斑分散成数十到数百个更微小的焦斑，即微量的热损伤被分隔，这样热损伤之间的正常组织不受影响，这部分皮肤可以作为热扩散区域，避免可能出现的热损伤等副作用，同时可以促进皮肤的愈合过程。减少一次性治疗对皮肤的热损伤，又能保证治疗的有效性，还可以减轻患者的疼痛感，使患者在更短的时间内恢复正常。

2. 仪器工作原理 CO_2激光的波长在 10.6 μm，几乎都能被人体生物组织吸收。利用激光的热效应和电磁效应能使手术的切割、灼烧、气化和微创治疗等过程少出血或不出血。因此，此类治疗几乎都采用 CO_2激光。CO_2激光治疗仪采用封离型 CO_2激光器，其输出的激光经导光关节臂传输，特点是方向性好，能量密度高。美容仪输出的激光是不可见光，为方便操作，加入了红色半导体激光作为指示光。美容仪输出的激光束经聚焦镜输出时，聚焦点产生的高温能将靶组织气化，可用于切割、烧灼，未经聚焦的激光束能量密度较低，可对靶组织做凝固手术。

3. 适应证

(1) 各型皱纹、妊娠纹，各类瘢痕包括创伤性瘢痕、烧伤性瘢痕、术后瘢痕。

(2) 皮肤松弛下坠、变薄，毛孔粗大、色素沉着。

(3) 痣、疣、小肿瘤。

(4) 光化性唇炎、痤疮、酒渣鼻、良性增生。

(5) 表皮色素病变，如雀斑、老年斑。

(6) 脆性增加、皮肤粗糙、血管瘤等。

4. 禁忌人群

(1) 发热，患传染病、急性疾病的人。

(2) 有心脏疾病、配置心脏起搏器的人。

(3) 严重高血压、恶性肿瘤、哮喘、深度静脉血栓、静脉曲张、甲状腺肿、癫痫等患者。

(4) 有出血性疾病、外伤、血管破裂、皮肤发炎、皮肤病患者。

(5) 孕妇。

(6) 月经期勿使用在腹部。

(7) 医学整形的部位、体内有金属物的部位。

(8) 免疫系统功能异常者。

5. 操作方法

(1) 点阵模式

①仪器治疗时需根据治疗项目切换相应的工作头，此时为点阵模式(图 7-14)。

②为手术者清洁手术部位，以保证仪器治疗的效果。

③操作者和手术者戴上防护镜，以保证手术过程中的安全性(图 7-15)。

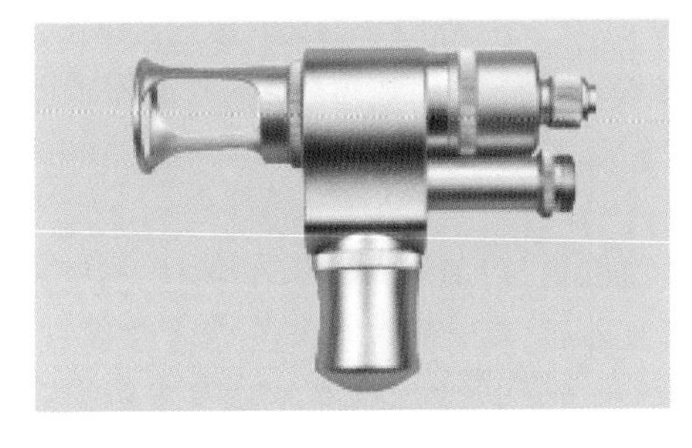

图 7-14 切换为扫描头

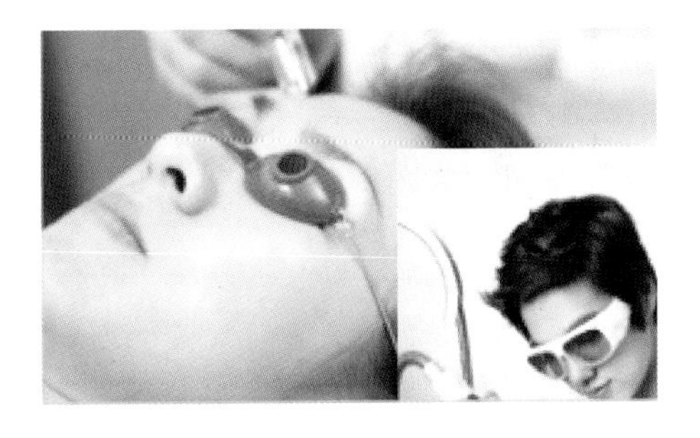

图 7-15 戴上防护镜

④准备完毕后，手术者进行操作(图 7-16)。

⑤手术完毕，给手术部位涂抹修复因子以加快手术部位的恢复(图 7-17)。

(2) 脉冲模式

①仪器治疗时需根据治疗项目切换相应的工作头(图 7-18)。

②清洁手术部位，以保证仪器治疗的效果。

③操作者和手术者戴上防护镜，以保证手术过程中的安全性。

④准备完毕后，手术者进行操作(图 7-19)。

⑤手术完毕，给手术部位涂抹修复因子，以加快手术部位的恢复。

6. 注意事项

(1) 禁止激光直接照射人的眼睛或健康的皮肤。

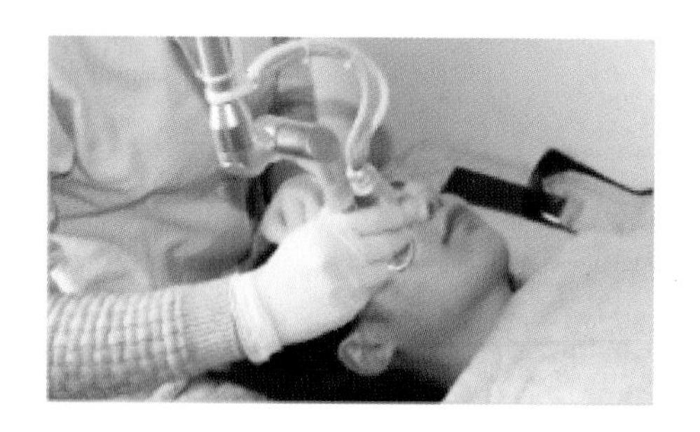

图 7-16 实施操作

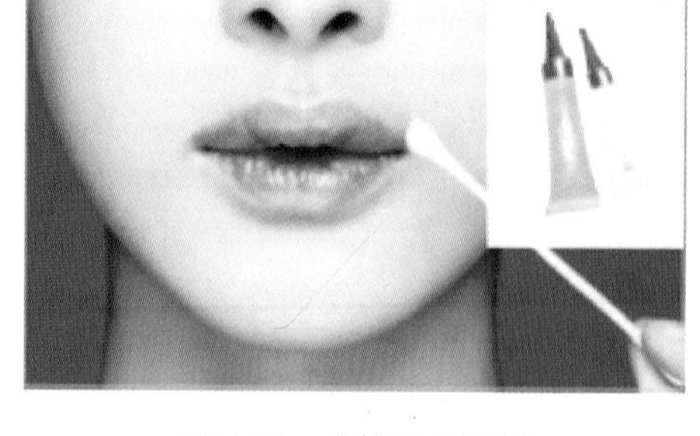

图 7-17 涂抹修复因子

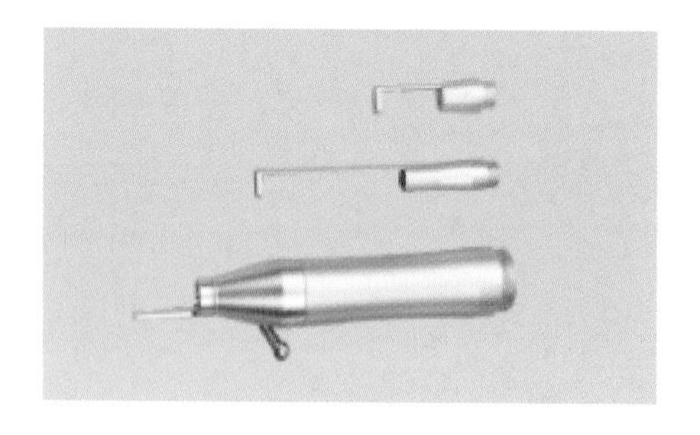

图 7-18 切换工作头

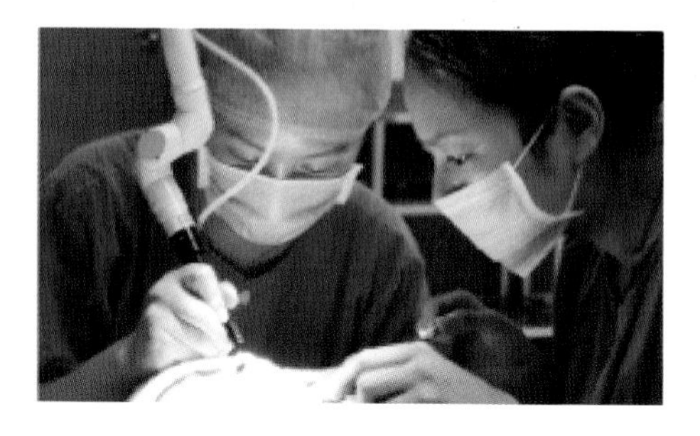

图 7-19 实施操作

（2）为防止眼睛或皮肤受到激光辐射，禁止激光照射到不锈钢表面、镜面等光滑表面上，以免激光形成反射或漫反射。

（3）仪器相关部件用75%酒精清洗消毒，待干燥后，方可使用激光。使用本仪器时，必须避免使用易燃麻醉剂或氧化性气体，如氧化亚氮和氧气。

（4）为防止刀头部位的聚焦镜片受到污染，保证清洁的手术环境，建议手术时使用吸烟器。刀头和聚焦镜片至少每3个月清洁一次。

（5）仪器使用时，35 m范围内，直视激光都会对眼睛造成伤害。仪器操作者在工作时必须佩戴安全眼镜。

（6）仪器内部会产生高压，非专业人员严禁打开仪器机箱，以免触电。

（7）一旦机器散发出非正常气味或产生不正常声响，应立刻切断电源，拔去电源线进行检查，在未排除故障前，禁止通电。

（8）机器正常使用的温度范围为5～40 ℃，湿度范围为10%～80%。

（9）为防止激光管管壁沾水结冰冻裂，在仪器要搬运前清空水箱中的水。

（10）仪器上的零部件寿命终止时，不要随意丢弃，请按照当地环保部门有关法规进行处理。

（11）仪器不使用时锁好，钥匙由专门人员保管。

（12）手术室要配备除尘或吸烟装置，避免污染。

7. 仪器养护

（1）使用时避免在周围放置医疗设备或家用电器。

（2）禁止在高温潮湿、灰尘多、油烟多、水蒸气多的环境中使用。

（3）使用中如发生故障，应立即关闭电源，必须由专业人员拆卸、维修，或送到经销商、厂家维修，不得自行修理。

（4）使用时请勿靠近墙壁，保持仪器四周有30 cm的空间，以便散热。

本项目重点提示

(1) 专业皮肤检测分析系统的作用有检测皮肤油分水分、皮肤酸碱度、皮肤黑色素及血红素、皮肤水分流失情况、皮肤弹性状况及皮肤衰老状况。

(2) 超声波美容仪的作用有减轻或消除皮肤色素沉着、消除眼袋和黑眼圈、防皱除皱，散血去淤、软化血栓，消除"红脸"、治疗炎性硬结痤疮及其愈后瘢痕等。

(3) 射频美容仪使用的禁忌证有安装心脏起搏器、有金属植入、发热、疾病晚期、出血性疾病、治疗区有严重皮肤病者以及有注射皮下填充物者和孕妇。

(4) 光子嫩肤常见功能有淡斑、嫩肤、脱毛、祛红血丝等，效果较好的是嫩肤、祛表皮斑和脱毛。

(5) 红蓝光治疗仪中红光主要功效：美白淡斑、嫩肤祛皱、修复受损皮肤、抚平细小皱纹，缩小毛孔、增生胶原蛋白。

(6) 奥桑喷雾仪的使用方法和注意事项。

能力检测

一、选择题

1. 通过皮肤检测仪观察皮肤呈现青黄色是哪种皮肤类型？(　　)

A. 油性皮肤　　B. 干性皮肤　　C. 中性皮肤　　D. 敏感性皮肤

2. 奥桑喷雾仪烧杯中的水位不得超过烧杯的哪个部位？(　　)

A. 容积的 4/5 或红色标线　　B. 容积的 2/5

C. 容积的 1/2　　D. 容积的 3/5

3. 下列哪项不是超声波导入仪的作用？(　　)

A. 减轻或消除皮肤色素沉着　　B. 防皱除皱

C. 消除眼袋和黑眼圈　　D. 补水

二、填空题

1. 超声刀美容仪的作用包括________和________等。

2. 红蓝光治疗仪红光的波长为________，蓝光的波长为________。

3. 射频美容仪的工作时间一般为________ min，每一个部位的养护时间约为________ min。

三、问答题

1. 如何使用皮肤分析与检测仪？

2. 使用超声波美容仪的注意事项有哪些？

3. 射频美容仪的作用有哪些？

4. 红蓝光治疗仪中每种光的作用有哪些？

5. 简述奥桑喷雾仪的工作原理和注意事项。

6. 简述超声刀美容仪的作用和禁忌人群。

(陈丽君　杨海腾)

参考文献

CANKAOWENXIAN

[1] 陈丽娟. 美容皮肤科学[M]. 2 版. 北京:人民卫生出版社,2014.

[2] 金玉忠,李云端. 中医学基础 [M]. 2 版. 北京:科学出版社,2008.

[3] 刘玮. 我国美容皮肤科学的现状及发展[J]. 中国医学美学美容杂志,2006,12(4):193-194.

[4] 陈景华. 美容保健技术[M]. 2 版. 北京:人民卫生出版社,2014.

[5] 刘强,程跃英,熊蕊. 美容解剖与生理[M]. 上海:上海交通大学出版社,2014.

[6] 汤明川. 美容指导・面部护理[M]. 上海:上海交通大学出版社,2009.

[7] 张秀丽,赵丽,聂莉. 美容护肤技术[M]. 北京:科学出版社,2015.

[8] 梁娟. 美容业经营管理学[M]. 2 版. 北京:人民卫生出版社,2014.

[9] 张丽宏. 美容实用技术[M]. 2 版. 北京:人民卫生出版社,2014.

[10] 杜莉. 现代美容技术[M]. 2 版. 北京:中国轻工业出版社,2011.

[11] Milady. 国际美容护肤标准教程[M]. 马东芳,译. 北京:人民邮电出版社,2016.

[12] 张卫明,袁昌齐,张茹芸,等. 芳香疗法和芳疗植物[M]. 南京:东南大学出版社,2009.

[13] 张海燕. 精油芳香疗法[M]. 北京:求真出版社,2013.

全国高等卫生职业教育创新型人才培养“十三五”规划教材（医学美容技术专业）

编委会

序

XU

中医美容是随着社会的发展和人们的审美以及健康的需求而诞生的。中医美容中使用的中草药化妆品的特点是将中草药的有效成分运用到化妆品中，充分发挥中草药的嫩肤、美白、黑发、生发、美发、美妆、祛斑、除痘、洁齿和护齿等特殊作用。编写本教材的目的是建立和完善中草药化妆品的独特理论系统，了解各类中草药化妆品的原料、制备技术、品类、选用原则，培养既具有扎实的理论基础又具有熟练的操作技能的医学美容技术中医美容方向专业人才，以满足当前美容专业人才的需要。

中医和生活美容之间的联系历史悠久。中国古代医家把对人体美的维护作为医学的任务之一，他们关注人体的修饰并合理利用中医学的方法，使修饰品和修饰手段不断完善，更符合人体健康要求，在世界美容史上占据了独特的地位。在历代各类医书中，有驻颜、悦色的作用的中草药达数百种，各种洗手面方、令面悦泽方、增白方、祛皱方、驻颜方、白牙方、染发方、香身香口方，应有尽有，甚至有发蜡、口红、胭脂配方。这些方药具有极浓的生活美容的色彩，均着眼于修饰人的容颜，使之更光彩夺目。

中医美容之中草药美容是在中医理论指导下，运用中药配制的粉、膏、液、糊等外用美容制剂，根据需要内服、外敷，并加以按摩，以滋养脏腑气血、活血通络、软坚散结、退疹祛斑，达到祛斑除皱、养颜驻容、延缓肌肤老化的美容功效。

本教材被列入全国高等卫生职业教育创新型人才培养“十三五”规划教材，反映了21世纪中草药化妆品的发展状况。本教材主要为医学美容技术及医学美容技术中医美容方向专业学生编写，中医美容的爱好者、行业从业者亦可借鉴学习。

全国卫生职业教育教学指导委员会医学美容技术专业委员会委员

全国美发美容职业教育教学指导委员会副秘书长

前言

QIANYAN

美容是一个既时尚又古老的行业。有证据表明，古埃及人很早就开始将泥土、蜡、蜂蜜和油等特殊材料制成面膜、化妆品，用以改善皮肤的健康状况。古希腊妇女利用植物的根和酵母的混合物来减少雀斑，使用含面包屑和牛奶的面膜来防治皱纹。欧洲医药之父、古希腊医生希波克拉底对美容业也有巨大的贡献，在他的推动下，皮肤护理开始进入科学的轨道。古罗马人则用水果汁、蜂蜜和橄榄油制作美容产品。指压按摩这种至今仍然广受欢迎的美容方法，被一些国家广泛运用。中国几千年以来积累的美容方法和经验更是一个伟大的宝库，亟待发掘并进一步运用。

本教材以中医美容理论知识和传统文化为基石，以中医美容从业人员在不同的工作环境和不同的工作主题下的美容实践作为主要内容，紧密结合中医基础理论来编写。旨在帮助读者掌握中医美容基础知识，了解和学习在工作过程中相关中医美容术语准确的英语表达方式和相关的情景知识，提高广大读者中医美容英语的运用能力。

本教材主要供中医美容专业使用。其中不少材料具有普遍适用性，其他专业学生及中医美容的爱好者亦可借鉴。

编　者

目录

MULU

Part One　TCM Cosmetology

Part Two　Cosmetic English Conversations

Part Three Cosmetology of TCM Conversations

Part One

TCM Cosmetology

Unit 1 Introduction to TCM Cosmetology

Section 1 TCM Cosmetic Concepts and Subject Attribute

Part Ⅰ Dialogue

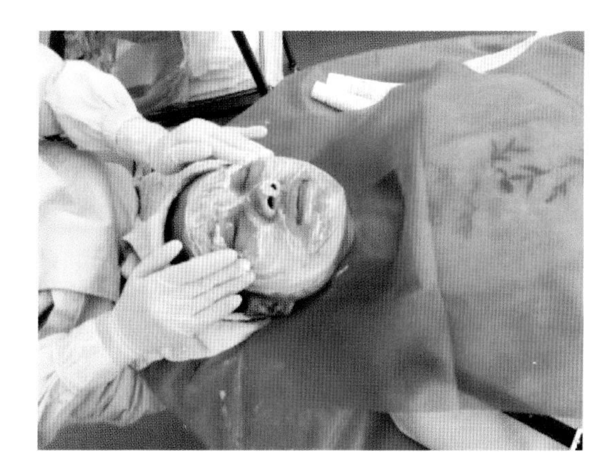

(P=Patient,D=Doctor)

D:Hello,can I help you?

P:I want to take care of my skin. Is there any way to improve it?

D:Do you have any methods for maintaining your skin?

P:I usually wear light make-up,keep a light diet and a regular routine.

D:Your habits are good. If you can combine traditional Chinese medicine (TCM) with herbal remedies,your complexion will be improved.

P:What is TCM cosmetology,doctor?

D: TCM cosmetology aims at the beauty of human body. It mainly studies the prevention,diagnosis and treatment of beauty diseases and the physical defects of beauty,and it is a kind of beauty knowledge that can prevent diseases, keep fitness, delay-aging and shape body.

P:I've heard that traditional Chinese medicine cosmetology is very popular now. I also want to use this method to maintain my skin,but I don't know if this method is safe.

D:You can assure that. The traditional Chinese medicine cosmetology is a new beauty method, and the aesthetic Chinese medicine is the product of the combination of aesthetics and Chinese medicine. Through the application of traditional Chinese medicine theory and methods, the treatment and health care of Clinical beauty is carried out.

P: Oh, that's actually a way of using traditional Chinese medicine.

D: The beauty of Chinese medicine is a new method of the basic theory of traditional Chinese medicine and the new beauty method of clinical treatment. It has the characteristics of practical, safe and small side effects.

P: That's great! I am going to use the traditional Chinese medicine cosmetology to tone my skin!

Part Ⅱ Text

TCM Cosmetic Concepts and Subject Attribute

Beauty generally means to beautify the face. A narrow sense of beauty is the beautification and modification of facial features. General beauty is the beautification of face, hair, body, and soul. Traditional Chinese medicine beauty belongs to the general beauty category. It is a beauty based on the health, and it is also according to the health of human facial features and aesthetic standards, nails, sarcoplasmic skin type, body posture, mental outlook and temperament of the results of comprehensive evaluation. And health is physical health and mental health. Physical health means that the Zang and Fu organs function normally, so that the skin is smooth, the muscles are full and the body is erect, which gives the person the aesthetic feeling of appearance. Mental health can be mental and pleasant thinking agile magnanimity, thus give a person a kind of temperamental aesthetic feeling.

Chinese medicine cosmetology is a human body building as the object, by a variety of basic and clinical discipline overlapping and into emerging discipline. It is in the basic theory of traditional Chinese medicine and human body under the guidance of aesthetic theory, using the characteristics of the methods and techniques of traditional Chinese medicine, the damage of the disease prevention, diagnosis, treatment, and damage of the physical defects of the conceal or correct theory, skill and law, has reached the prevention fitness, delay-aging, maintaining and shaping the body beauty as the main goal of specialized subject. It is an emerging discipline that is both a branch of Chinese medicine and a part of the system of aesthetic medicine. This course is a Chinese beauty professional backbone course, systematically introduces the essence of the beauty of traditional Chinese medicine basic theory, diagnostic method, the characteristics of traditional Chinese medicine beauty means and the beauty of traditional Chinese medicine clinical range of health and beauty treatment and hairdressing, etc. It is a fusion of Chinese beauty professional basic theory, method, means and clinical treatment of integrated curriculum, the teaching plan as the key to master the discipline. Through teaching, students are able to apply the theory and method of TCM to the treatment and care of beauty clinic. Chinese medicine cosmetology is a combination of aesthetics and traditional Chinese medicine, is an applied branch of medical aesthetics and a branch of traditional Chinese medicine, which is a new integrated TCM clinical discipline.

（一）对话

（P＝患者，D＝医生）

D：你好，请问有什么可以帮助你？

P：我想对自己的皮肤进行保养，有什么办法可以改善吗？

D：那你平时有什么保养皮肤的方法吗？

P：平时我都是化淡妆，饮食清淡，作息规律。

D：你的这些习惯都很好，如果能结合中医美容中草药调理皮肤的方法会很好地改善你的肤色。

P：大夫，什么是中医美容呢？

D：中医美容学是一门以人体健美为目的，研究损美性疾病及损美性生理缺陷的预防、诊断、治疗，达到防病健身、延缓衰老、维护和塑造人体神形美的一种美容知识。

P：我听说中医美容现在很流行，我也想用这种方法保养皮肤，但是我不知道这种方法是否安全。

D：这个你可以放心，中医美容是一种新兴的美容方法，中医美容学是美学和中医学相结合的产物。通过应用中医理论和方法手段进行临床美容的诊疗和保健。

P：哦，那这种方法其实就是用中医的方法来美容了。

D：是的，中医美容就是融汇中医基础理论、方法手段和临床治疗的新型美容方法，具有实用、安全、副作用小的特点。

P：太好了！那我就用中医美容的方法调理皮肤吧！

（二）课文

中医美容的概念及学科属性

美容一般是指美化容颜。狭义美容是指颜面五官的美化和修饰。广义美容是指颜面、须发、躯体、以及心灵等全身心的美化。中医美容属于广义美容范畴。它是在健康基础上的美容，是根据健康和美学标准对人的颜面五官、须发爪甲、肌质肤色、体型姿态、精神面貌、气质风度等进行综合评价的。健康状态是指身体健康与心理健康。身体健康是指脏腑功能正常，这样才能使皮肤光滑、肌肉丰满、身躯挺拔，从而给人以外形上的美感。心理健康，才能精神愉快、思维敏捷、豁达大度，从而给人一种气质上的美感。

中医美容学是一门以人体健美为目的，由多种基础、临床学科相互交叉而形成的新兴学科，是在中医学的基本理论和人体美学理论的指导下，运用中医特有的方法和技术，研究损美性疾病的预防、诊断、治疗及损美性生理缺陷的掩饰或矫正的理论、技能及规律，以防病健身、延缓衰老、维护和塑造人体神形美为主要目的的专门学科。中医美容学是一门新兴的学科，它既是中医学的分支学科，又是美容医学学科体系的组成部分。本课程是中医美容专业的主干课程，系统介绍了中医美容的基础理论、诊断方法、中医特色美容手段和中医美容临床范围的保健美容与治疗美容等篇章，它是中医美容专业中的一门融汇基础理论、方法手段和临床治疗的综合性课程，教学计划将其作为重点掌握的学科。通过教学，学生能够熟练应用中医理论和方法手段进行美容临床的诊疗和保健工作。中医美容学是美学和中医学相结合的产物，它既是医学美学的一个应用分支学科，又是中医学的一个分支学科，是一门新兴的综合性的中医临床学科。

Section 2 The Basic Content of Traditional Chinese Medicine beauty

Part Ⅰ Dialogue

(P=Patient,D=Doctor)

P:Hi,I want to know about the recent special popular Chinese medicine cosmetology.

D:Yes, you are very pay attention to maintenance, then I will introduce to you the traditional Chinese medicine. It is mainly used traditional Chinese medicine beauty, acupuncture beauty,massage beauty,sports beauty,music beauty,psychological beauty,diet beauty,etc. to adjust the body.

P:Oh,I'd like to know about acupuncture. It is amazing.

D:Acupuncture is one of the traditional Chinese medicine therapies. There is a saying that acupuncture and cupping cure half of the disease. Acupuncture and related methods can adjust the rise and fall of the body's meridians to achieve the balance of Yin and Yang. Its advantage is that the indications are wide,the efficacy is long-lasting,and the side effects are few. But patients feel slight uncomfortable. For example,there may be slight pain in some acupoints or superficial blood contact of a patient with three-edged needle.

P:Well,I still think the diet therapy is safer. Can you tell me something about it?

D:In the literature,the ancient people had the same origin of medicine. Chinese nation has been advocating"food the first necessity of man"concept. Just as the ancient saying goes, "Diet cures more than doctors". So the Chinese think that the way to keep healthy is to start with the diet. Traditional Chinese medicine has changed with the seasons. We should nourish the liver in the spring,the heart in the summer,the lung in the autumn,the kidney in the winter,and raise the spleen and stomach all the year. For example,in the fall,we should eat more white food, such as lily, snow pears, tremella, Chinese yam, wax gourd, myotonin, white radish and white fungus. According to *Materia Medica Companion*, Longan has the function of raising muscles, improving the complexion, reducing forgetfulness.

P: Thank you. Please help me plan a diet therapy.

D: All right.

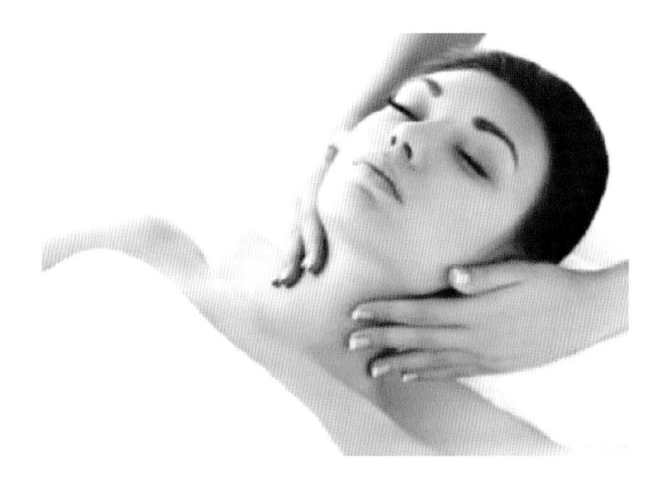

Part Ⅱ Text

The Basic Content of Traditional Chinese Medicine Beauty

Traditional Chinese medicine beauty has a long history, and many beauty methods have appeared during the thousands of years. The main of them are medicine beauty, acupuncture beauty, massage beauty, sports beauty, music beauty, psychological beauty, diet beauty, etc.

1. Medicine beauty

Traditional Chinese medicine beauty is a kind of beauty method which is used for the treatment of various kinds of beauty-damaging diseases and curing skin and delaying skin aging. It is the most abundant treatment in various methods of traditional Chinese medical cosmetology.

2. Acupuncture beauty

Acupuncture is one of the oldest traditional Chinese medicine therapies. There is a saying that acupuncture cupping cure half the disease. It is a method of using acupuncture and related methods to adjust the rise and fall of the body's meridians to achieve the balance of Yin and Yang. Its advantage is that the indications are wide. The efficacy is long-lasting. The side effects are few.

3. Massage beauty

In ancient time, massage was called pressing, according to the cheirapsis, chirismus, knead, and so on. Traditional Chinese massage has a long history, since ancient time, it has "used the medicine of the hand to refer to the needle".

Under the guidance of the basic theory of traditional Chinese medicine, the appropriate methods are used to operate on the relevant meridians and acupoints of the body aim to achieve the purpose of regulating the skin and delaying the aging.

4. Sports beauty

Sports beauty is a kind of beauty method which can promote the function of the organ and improve the metabolism of the body through various exercises to cure illness, keep fitness, and last beauty. According to the ancient saying, "food cures better than medicine and exercise cures more than food." We often say that "life dep ends on sports" "Walking after eating, long life you can expect". Thus, the movement is one of the effective measures to keep healthy, beautiful and delay aging. Chinese ancient beauty and health movement is wuqinxi, baduanjin, yijinjing, six-word qigong, tai chi, etc. The practice shows that all of these have excellent health benefits.

5. Music beauty

Music beauty is guided by the basic theory of traditional Chinese medicine and the traditional music theory, music is used as a means of beauty. According to the different physical and modern changes, choose different tones, law of section, the strength of the music, stimulate emotion, produce physical resonance, regulate visceral functions, to cure disease and to fit body and mind.

6. Psychological beauty

TCM psychology has a long history. As early as in ancient time, Shaman often used consolation and remembrance to treat diseases. The theory of the seven emotions of Chinese medicine, namely, happiness, anger, worry, thought, sadness, fear and shock, the seven emotions are overzealous and persistent, and the effect of the visceral qi and blood.

7. Diet beauty

Diet beauty is guided by the basic theory of traditional Chinese medicine, using the natural flora and fauna for food or medicine to beautify the appearance and delay senescence. Many foods are both food and medicine in our diet structure. In the literature, the ancient people had the same origin of medicine and the same root. Chinese nation has been advocating "food is the first necessity of the people" concept. According to ancient people, "Diet cures more than the doctors". So the Chinese think that the way to keep healthy is to start with the diet.

(一)对话

(P=患者,D=医生)

P:你好,最近中医美容特别流行,我想了解一下。

D:好的,您非常注重保养,那我就给您介绍一下,中医美容主要是运用中药美容、针灸美容、推拿美容、运动美容、音乐美容,心理美容、膳食美容等美容方法对人体进行调理。

P:哦,我想了解一下针灸美容,觉得比较神奇。

D:针灸美容是我国传统的古老医术之一,有句俗话"针灸拔罐,病好一半"。针灸及相关方法可以调整人体经络阴阳的盛衰,使之达到阴阳平衡,最终实现美容。其优点是适应范围广，疗效持久，副作用少。但是患者会有轻微的不适感觉。比如，用三棱针刺破患者的某些腧穴或浅表血络会有轻微疼痛感。

P：这样啊，那我还是觉得膳食美容比较安全，你能给我介绍一下吗？

D：文献记载里古人很早就有"医食同源，药食同根"之说。民以食为天，用古人的话说就是"药补不如食补""养生之道，莫先于食"。中医讲究膳食美容随四季变化而变化。春季养肝、夏季养心、秋季养肺、冬季养肾、四季养脾胃。比如，秋季要多吃白色养肺的食物，如百合、雪梨、银耳、怀山药、冬瓜、薏仁米、白萝卜、白木耳等。在《本草蒙筌》中记载龙眼具有"养肌肉，美颜色，除健忘"的功效。

P：谢谢，那请您帮我制订一个膳食美容的治疗方案吧！

D：好的。

（二）课文

中医美容的基本内容

中医美容历史悠久，源远流长。在几千年的历史进程中，出现了许多美容方法，归纳起来主要有中药美容、针灸美容、推拿美容、运动美容、音乐美容，心理美容、膳食美容等美容方法。

1. 中药美容

中药美容是通过中药的内服、外用来治疗各种损美性疾病或养护肌肤、延缓衰老的一种美容方法，是中医美容各种方法中内容最丰富的疗法。

2. 针灸美容

针灸是我国传统的古老医术之一，有句俗话"针灸拔罐，病好一半"。通过使用针灸及相关方法，调整人体经络阴阳的盛衰，使之达到阴阳平衡，最终实现美容。其优点是适应范围广，疗效持久，副作用少。

3. 推拿美容

推拿古称"按摩""按跷""跷引""案杌"等。中医推拿有着悠久的历史，自古就有"以手代药、以指代针"之说。推拿美容是在中医基础理论的指导下，运用适当的手法，在体表的相关经络和穴位上进行操作，达到调理皮肤、延缓衰老的目的。

4. 运动美容

运动美容是通过各种不同的运动锻炼方式，促进脏腑气血功能，增强机体的新陈代谢，达到疗疾、健身、美形、驻颜的一种美容方法。古人云："药补不如食补，食补不如动补。"我们也常说"生命在于运动""饭后百步走，能活九十九"。可见运动是保持健康美丽、延缓衰老的有效措施之一。我国古代美容养生运动有五禽戏、八段锦、易筋经、六字诀、太极拳等。实践证明，这些运动都起到了很好的美容养生保健作用。

5. 音乐美容

音乐美容是在中医基础理论和传统音乐理论的指导下，以音乐作为美容手段，根据不同体质及情志变化，分别选用不同音调、节律、强度的乐曲，激发情感，产生生理上的共鸣，调节脏腑功能，达到防病治病、健美身心的目的。

6. 心理美容

中医心理学历史悠久，源远流长。早在远古时代，巫医就经常采用安慰祝福、怀念追悼等方法治疗疾病。中医上有七情致病学说，七情即喜、怒、忧、思、悲、恐、惊，七种情志过度强烈和持久，会影响脏腑气血功能。

7. 膳食美容

膳食美容是在中医药基础理论的指导下，运用食物或药食两用的天然动植物来美化仪容、延缓衰老的方法。我国人民膳食组成中，许多食物既是食品，也是药品。古人早就有“医食同源，药食同根”之说。民以食为天，用古人的话说就是“药补不如食补”“养生之道，莫先于食”。

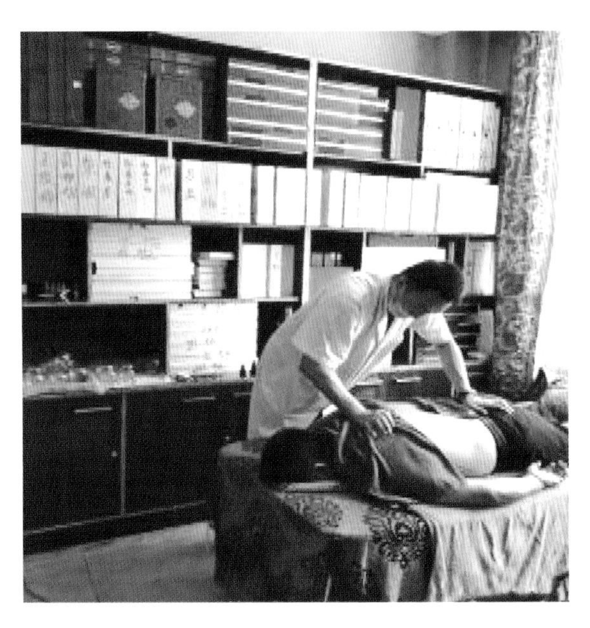

Section 3　A Brief History of the Development of Traditional Chinese Medicine beauty

Part Ⅰ　Dialogue

(P=Patient, D=Doctor)

D: Hello, can I help you?

P: I'd like to know when TCM cosmetology started.

D: I advise you to visit the traditional Chinese medicine cosmetology museum that has development period of TCMC as well as the representative works. It passed through ancient times, Xia Dynasty, Spring and Autumn period, Qin-Han Dynasties, Sui and Tang Dynasties, Song Dynasty, Ming Dynasty, Qing Dynasty and modern multiple times. For example, we know that the *Huangdi Neijing*, *Compendium of Materia Medica*, *Shen Nong's Materia Medica* involve the beauty of traditional Chinese medicines, prescription of traditional Chinese medicine, acupuncture and other methods.

P:Oh,the history is so long. What role does it have for today's beauty?

D:TCM cosmetology is more popular nowadays. It is originated from traditional Chinese medicine. Traditional Chinese medicine cosmetology is safer,healthier,long-lasting and non-toxic side effects.

P:That's great. Could you help me make a care plan?

D:All right.

Part Ⅱ Text

A Brief History of the Development of Traditional Chinese Medicine Cosmetology

In the ancient Stone Age, the caveman had made the animal bone into a pendant necklace. It shows that the primitive ancestors of China had developed a conscious awareness of the human body beauty before the five or six thousand years ago. During the Xia Dynasty, Shang Dynasty and Zhou Dynasty, traditional Chinese medicine, such as acupuncture, moxibustion and incense,was widely used,which also opened the source of TCM beauty. During the Xia Dynasty and Shang Dynasty,there appeared the text:Oracle. Oracle had bath shower wash and so on. During the Spring and Autumn period and the Warring States period,the main achievements of Chinese medicine cosmetology were the establishment of the theory of holistic view and the production of some effective beauty methods under the guidance of the whole concept. In this period of representative works,*Huangdi Neijing* has pioneered the theory of beauty of traditional Chinese medicine. *Shen Nong's Materia Medica* is a practice leader of Chinese medicine and diet beauty.

The development of traditional Chinese medicine to the Jin-Tang Dynasties,has already had the high level,gradually mature.

(1) An independent discipline has initially formed.

(2) Reform the cosmetic prescriptions and formulations.

(3) There are many kinds of cosmetic traditional Chinese medication and cosmetic food, which involves a wide range of areas.

Song and Yuan Dynasties, *General Medical Collection of Royal Benevolence & The Peaceful Holy Benevolence Formulae*, which were compiled comprehensively and systematically by official organizations,collected effective cosmetic prescriptions and added many new cosmetic treatments. The four famous doctors in Jin Dynasty and Yuan Dynasty have their own unique experiences,discussed the cause and treatment of diseases,the pure beauty experience transited to the theoretical study, the beauty of traditional Chinese medicine is more with treatment based on syndrome differentiation of features,has a certain impact on the development of beauty on later generations.

During the Ming Dynasty and Qing Dynasty,the social beauty business began to unfold, so the traditional Chinese medicine beauty science and methods, both had obvious improvement. Great medical scientist Li Shizhen's *Compendium of Materia Medica* is the greatest achievement in the history of traditional Chinese medicine in our country,provides a very valuable information for the beauty of traditional Chinese medicine research,has become unprecedented work of Chinese materia medica.

After the Third Plenary Session, the rapid development of the cause of Chinese medical cosmetology, all about aesthetics of traditional Chinese medicine and Chinese medicine cosmetology's works have been published, such as *the Aesthetics of Traditional Chinese Medicine*, *TCM Bodybuilding*, *Practical Traditional Remedies*, *the Beauty of Traditional Chinese Medicine*, *Acupuncture Cosmetology*, *Chinese Medicine Cosmetology* and so on. In 2002, the Ministry of Health issued *Medical Beauty Services Management Method*, the law explicitly pointed out that medical cosmetology department is a first-class diagnosis and therapeutic subject, cosmetic surgery, cosmetic dentistry, cosmetic dermatology and cosmetology regions for secondary medical subjects. This act established the status of traditional Chinese medicine, and will promote the further development of Chinese medicine. In the 21st century, everyone needs to be fit and healthy, and these requirements promote the development of medical beauty in our country. Traditional Chinese medicine beauty will become a sunrise industry, its development prospect is vast, traditional Chinese medicine beauty will become a brilliant wonder of the aesthetics circle.

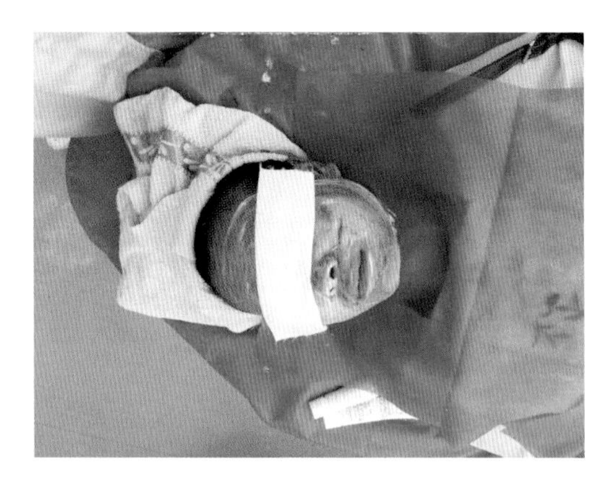

(一) 对话

(P=患者,D=医生)

D:你好,请问有什么可以帮助你?

P:我想了解一下中医美容是什么时候开始的?

D:这样建议您参观一下中医美容博物馆,博物馆里有中医美容发展的各个时期的概况及代表著作。中医美容经过了远古时期、夏商周时期、春秋-秦汉时期、隋唐时期、宋明清以及现代等多个时期。比如,我们了解的《黄帝内经》《神农本草经》《本草纲目》,这些书籍里面都涉及中医美容的中药、方剂、针灸等美容方法。

P:哦!原来中医美容历史这么悠久,那它对当今的美容有什么作用?

D:现在比较流行中医美容,中医美容源于中医学。中医美容安全、健康、持久、无毒副作用。

P:那太好了,请您帮我制订一个护理方案吧!

D:好的!

（二）课文

中医美容发展简史

在远古时期，旧石器时代，山顶洞人就已将兽骨制成骨坠、项链等装饰品，说明了我国原始先民早在五六千年前就已经产生了自觉的人体审美意识。夏商周时期，中医有了明显的分科，针刺、灸疗、热敷、熏香等中医治疗法也被普遍使用，这也开启了中医美容之源。夏商周时期已经出现了文字，即甲骨文，甲骨文有沐浴洗澡等与美容有关的文字。春秋战国时期，中医美容学的主要成就是在理论上确立了整体观，以及在整体观念指导下产生了一些卓有成效的美容方法。在这一时期有代表性的著作是《黄帝内经》——开创了中医美容的理论先河。《神农本草经》是中医药物和膳食美容的实践先导。

中医美容学发展到晋唐时期已经具有较高的水平，并逐渐走向成熟。这一时期的中医美容学具有以下几个突出的成就。

（1）初步形成了独立的学科。

（2）美容方剂和美容剂型的改革。

（3）采用的美容中药及美容食物品种繁多，涉及面广。

宋元时期，由官方组织进行全面系统的校勘编纂的《圣济总录》和《太平圣惠方》收集了更加丰富有效的美容方剂，并补充了许多美容新方，同时金元四大家亦各有独特的经验，对有碍美容的疾病的病因和证治进行了探讨，使单纯的美容经验过渡到美容理论的研究，使中医美容更加具备辨证论治的特色，对后世美容的发展产生了一定的影响。

明清时期，社会美容事业开始发展，所以中医美容学在理论和方法上都有明显的提高。伟大的医学家李时珍编写的《本草纲目》是我国中医历史上最伟大的成就，为中医美容研究提供了非常宝贵的资料，成为我国药物学的空前巨著。

十一届三中全会以后，中国的医学美容事业迅速发展，各种有关中医美学和中医美容学的著作相继出版，如《中医美学》《中医健美》《实用传统美容法》《中医美容大全》《针灸美容》《中医美容学》等。2002 年卫生部（卫计委）发布了《医疗美容服务管理办法》，该法明确指出："医疗美容科为一级诊疗科目，美容外科、美容牙科、美容皮肤科和美容中医科为二级诊疗科目"。这个管理办法确立了中医美容的学科地位，必将促进中医美容的进一步发展。进入 21 世纪，人人需要健美，这些需求促进了我国医学美容的发展。中医美容势必成为朝阳产业，其发展的前景是极其广阔的，中医美容必将成为美学界的一颗璀璨的明珠。

Unit 2 Skin Cosmetology

Section 1 The Basic Knowledge of Skin

Part Ⅰ Dialogue

(P=Patient, D=Doctor)

D: Good morning. Sit down, please. What can I do for you?

P: Good morning, with the change of the weather, I have discovered that my skin is very dry recently. I'm afraid to affect my appearance.

D: Do you know about the skin?

P: Yes, but a little. Can you tell me something about it?

D: Ok, I'll give you the basic knowledge of the skin and you will know how to take care of your skin.

P: Thank you. Please tell me right away.

D: Skin is the largest organ of the body and one of the most active organs. It covers the surface of the human body. The skin area of the adult is 1.5-2 square meters, and the total weight of the skin accounts for 16% of the body weight. The skin of eyes is the thinnest, while the skin of the palm and foot is the thickest. The skin is elastic and protective. Under normal conditions, metabolism keeps the skin alive. Skin functions: protection, feeling, excretion, absorption and so on. It can be divided into epidermal, dermal and subcutaneous tissue.

P: What caused my skin's problems?

D: Your skin is dry. Wrinkles are prone to repetitive activities on the skin. Based on the theory of TCM, your problems are caused by the lipid secretion rate decreases. The lack of moisturizing can lead to the appearance of skin removal and the relaxation of the eyes and neck. This kind of the skin should keep the moist.

P: I see. But this is the first time I do facial care, can it be complicated?

D: Don't worry. Our treatment is a combination of TCM and professional knowledge.

P: Ok, thanks.

Part Ⅱ Text

The Basic Knowledge of Skin

Actually not everyone knows their own skin. After reading this passage you will know about your skin, and then you can make better use of skin care products for maintenance. I will introduce the skin.

1. The knowledge of skin

The skin is the body's largest organ and one of the most active organs, which covers the surface of the human body. The skin area of the adult is 1.5-2 square meters, and the total weight accounts for 16% of the body weight. The skin of the eyes is the thinnest, while the skin of the palm and foot is the thickest. The skin is elastic and protective. And under normal conditions, metabolism keeps the skin alive.

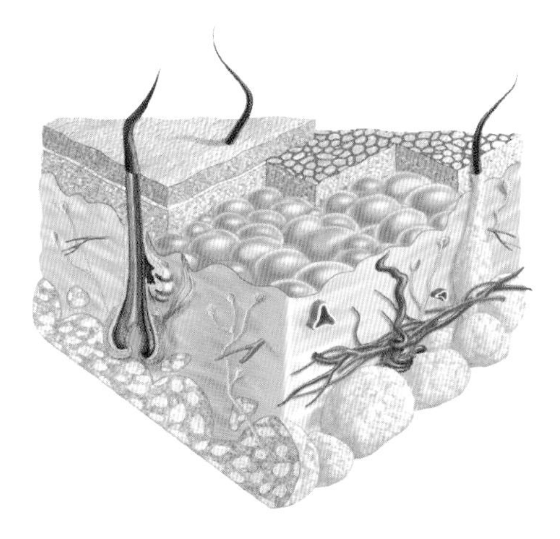

2. The basic structure of skin

(1) Collagenous fiber.

(2) Reticular fiber.

(3) Elastic fiber.

(4) Matrix.

(5) Cell.

3. The components layers of skin

Skin can be divided into three layers, epidermis, dermis, subcutaneous tissue. In addition, skin has some accessory organs (sebaceous glands, sweat glands, hair, nails).

The epidermis is the outermost layer of the skin, covering the whole body. It has a protective effect. The average thickness of the epidermis is 0.07 to 2 mm. There are no blood vessels in the epidermis, but there are many small nerve endings. They can sense the stimulation of the outside world, and produce sensations of touch, pain, cold, heat and pressure.

The dermis is located in the deep layer of the epidermis and connects downward to the subcutaneous tissue. There is no obvious boundary between the dermis and the subcutaneous tissue. The dermis consists of dense connective tissue with various connective tissue cells and a large number of collagen elastic fibers, which make the skin both elastic and resilient. Fibroblasts and mast cells are more common in connective tissue cells. The thickness of dermis is different. The dermis of palm and sole is thicker, about 3 mm or more, and the thinnest, about 0.6 mm, is in the eyelids. Generally, the thickness is between 1 and 2 mm. The dermis can be divided into papillary layer and reticular layer. The skin is thicker than the epidermis. The dermis contains a lot of elastic and collagen fibers, which make the skin elastic and tough. The dermis is rich in blood vessels and sensory nerve endings. The dermis is located between the epidermis and the subcutaneous tissue. It is mainly composed of collagen fibers, elastic fibers, reticular fibers and amorphous matrix. There are also nerves and nerve endings, blood vessels, lymphatics, muscles and skin appendages.

Subcutaneous tissue belongs to mesenchymal tissue, which consists mainly of adipocytes, fibrous septa and blood vessels. In addition, lymphatics, nerves, sweat glands and hair follicles (papillae) were also distributed in the subcutaneous tissue. The subcutaneous tissue is rich in blood vessels, which form capillaries by branches of interlobular septal arterioles, extend into fat lobules and surround each adipocyte. Capillary basement membrane contacts closely with adipocyte membrane, which is helpful for blood circulation and lipid transport. Subcutaneous tissue distributes in the dermis and myometrium, which is closely connected with the myometrium above and below the dermis, and forms the so-called adipose layer, accounting for 18% of the body weight. Its thickness varies significantly with body surface location, age, sex, endocrine, nutrition and health status.

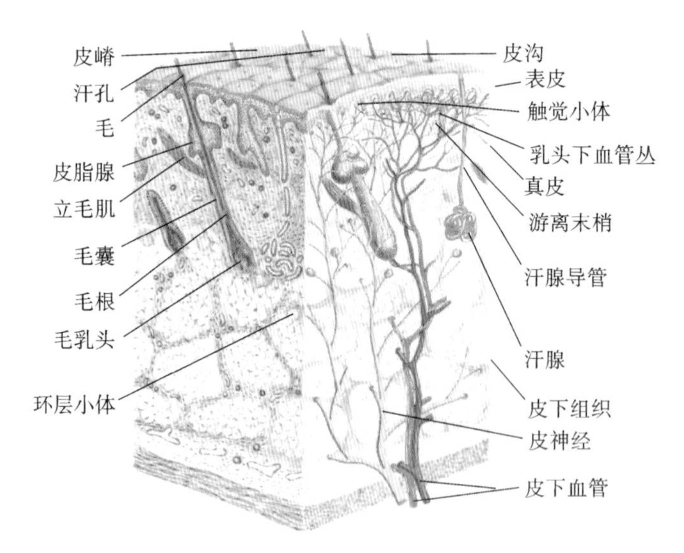

4. The functions of skin

(1) The protective function (contain corneous layer, transparent layer).

(2) The feeling function (have sensory nerve endings).

(3) The adjustment function of temperature.

(4) The action of secretion and excretion (sweat glands, sebaceous secretion of sweat and sebum).

(5) The absorptive function.

(6) Metabolism.

(7) The role of immune.

(一) 对话

(P=患者,D=医生)

D:早上好,请坐。有什么可以帮助你吗?

P:你好,最近随着天气的变化,我发现我的皮肤特别干燥,我害怕影响我的容貌。

D:你对皮肤知识了解吗?

P:是的,但是一点点,你能给我介绍一下吗?

D:好的。我给你介绍一下皮肤的基本知识,你就知道怎样保护皮肤了。

P:谢谢。请马上告诉我。

D:皮肤是人体最大的器官,也是最活跃的器官之一,覆盖于人体的表面。成人的皮肤面积为 1.5~2 m^2,总重量占人体体重的 16%。眼部的皮肤是最薄的,而手掌及脚掌的皮肤最厚。皮肤具有弹性及保护作用,在正常情况下,新陈代谢能够促使皮肤不断重现生机。

皮肤有保护、感觉、排泄、吸收等功能。一般可分为表皮、真皮和皮下组织三大部分。

P:哦,医生,我的皮肤问题是什么原因导致的?

D:你的皮肤干燥。一些重复活动的地方容易出现皱纹。你的问题是由于皮脂分泌率降低,缺少滋润会出现皮肤脱屑现象,眼部和颈部皮肤出现松弛。这种皮肤要经常进行滋润,护肤要格外细心。

P:我明白了。但这是我第一次做面部护理,会很复杂吗?

D:别担心。我们的治疗是用中医的方法结合专业的知识进行的。会对你的皮肤做好护理的。

P:好的,谢谢。

(二) 课文

皮肤的基础知识

其实并非所有人都真正了解自己的皮肤,用心看完这篇文章,你会更了解自己的皮肤,保护好自己的皮肤,然后才能更好地使用护肤品进行保养。下面我就介绍一下皮肤。

1. 认识皮肤

皮肤是人体最大的器官,也是最活跃的器官之一,覆盖于人体的表面。成人的皮肤面积为 1.5~2 m^2,总重量占人体体重的 16%。眼部的皮肤是最薄的,而手掌及脚掌的皮肤最厚。皮肤具有弹性及保护作用。在正常情况下,新陈代谢能够促使皮肤不断重现生机。

2. 皮肤的基本构造

(1) 胶原纤维。

(2) 网状纤维。

(3) 弹力纤维。

(4) 基质。

(5) 细胞。

3. 皮肤的分层

皮肤由外向内可分为三层，表皮层、真皮层和皮下组织，另外皮肤还有一些附属器(皮脂腺、汗腺、毛发、指甲)。

表皮是皮肤的最外层，覆盖全身，有保护作用。表皮的平均厚度为0.07～2 mm。表皮没有血管，但有很多细小的神经末梢，能感知外界的刺激，产生触觉、痛觉、冷觉、热觉、压力觉等感觉。

真皮位于表皮深层，向下与皮下组织相连，与皮下组织无明显界限，真皮由致密的结缔组织组成，其内分布着各种结缔组织细胞和大量的胶原纤维、弹力纤维，使皮肤既有弹性，又有韧性。结缔组织细胞中成纤维细胞和肥大细胞较多。真皮的厚度不同，手掌、足底的真皮较厚，约3 mm；眼睑等处最薄，一般厚度为1～2 mm。真皮可分为乳头层和网状层。真皮比表皮厚。真皮含有大量的弹力纤维和胶原纤维，使皮肤有一定的弹性和韧性，真皮内含有丰富的血管和感觉神经末梢。真皮位于表皮和皮下组织之间，主要由胶原纤维、弹力纤维、网状纤维和无定型基质等结缔组织构成，其中还有神经、血管、淋巴管、肌肉以及皮肤附属器。

皮下组织属于间叶组织，主要组成成分为脂肪细胞、纤维间隔和血管。此外，皮下组织内还分布有淋巴管、神经、汗腺体以及毛囊(乳头部)。皮下组织内含有丰富的血管，由小叶间隔、小动脉分支形成毛细血管，伸入脂肪小叶并围绕着每个脂肪细胞。毛细血管基底膜与脂肪细胞的细胞膜紧密接触，有助于血液循环和脂质的输送。皮下组织分布于真皮和肌膜，上方与真皮相接，下方与肌膜紧密接触，形成脂肪层，占体重的18%。其厚度因体表部位、年龄、性别、内分泌、营养和健康状态等的不同而有明显差异。

4. 皮肤的作用

(1) 保护作用(含角质层、透明层)。

(2) 感觉作用(有感觉神经末梢)。

(3) 调节体温的作用。

(4) 分泌和排泄的作用(汗腺、皮脂腺分泌的汗和皮脂)。

(5) 吸收作用。

(6) 代谢作用。

(7) 免疫作用。

Section 2 Different Types of Skin

Part Ⅰ Dialogue

(P=Patient，D=Doctor)

D：Good morning. Sit down，please. What's the matter with you?

P：With the great change of the weather，I find my skin becomes oil.

D:Do you know anything about the types of skin?

P:I'm sorry. I don't know the types of skin at all.

D:In that case,I'll introduce the skin types to you and then you will know what kind of skin you belongs to.

P:Thank you very much. Please tell me right away.

D:OK. I'll introduce you two types of skin. The first kind is dry skin: the moisture content of the cuticle less than 10%, pH>6. 5, no more sebum secretion, dry skin with lacking of grease, dermatoglyphics fine, and pore is not obvious. After washing face, you will feel that your skin is sensitive to external stimulation. Skin becomes chapped, desquamated, and wrinkles. When washing face, you can choose super fatted soap (soap with more fat).

The second skin is normal skin that is an ideal skin type. The water content of the stratum corneum is about 20%, pH is 4. 5 to 6. 5, and sebum secretion is moderate. Skin surface is smooth and delicate with out dry or greasy. It is flexible and has a strong adaptability to external stimulation. When washing face, you can choose soft soap (potash soap, high fatty acid potassium salt, which is softer than high-grade fatty acid sodium salt, so it was called soft soap).

P:Thank you very much. Is there any other types of skin?

D:Yes. Oily skin (fatty skin) is the third type of skin: the water content of the cuticle is about 20%, pH<4. 5, the sebum secretion is vigorous, the skin appearance is greasy and shiny, and the skin has enlarged pores. The surface of this kind of skin is smooth and easy to cling to dust. The color of it is dark, but it enjoy good elasticity, and has no sensitive to the external stimulation. This skin is more common in young and middle-aged and obese people, prone to acne, seborrheic dermatitis, etc. Combined skin is a skin type that combined dry, neutral and oily skin. Oily skin is in the central parts of the face (cheek, nose, nose lip and lower chin), while dry skin appears in the double cheeks, temporal and other parts. The hair traits and body's skin are generally consistent with the head and face. When cleaning, you should distinguish the oily parts and dry parts and pay attention to the oily parts.

Sensitive skin (allergic skin): skin's reaction to external stimulation is more sensitive, especially the weather is cold, heat, or windy. It is allergic to ultraviolet rays, cosmetics and easy to appear erythema, papule and pruritus in allergic body.

P:Thank you. By your introduction, I know the skin types.

Part Ⅱ Text

The Maintenance of Different Types of Skin

The maintenance of dry skin is to do more massage care to promote blood circulation. You should pay attention to use more moist, white, active moisturizers and creams.

The maintenance of normal skin care is cleaning, toning, moisturizing and massage weekly care and pays attention to daily water supply and adjusts the balance of water and oil. The choice of skin care depends on skin age and season. Choose hydrophilic skin care products in summer. Choose moisturizing skin care products in winter.

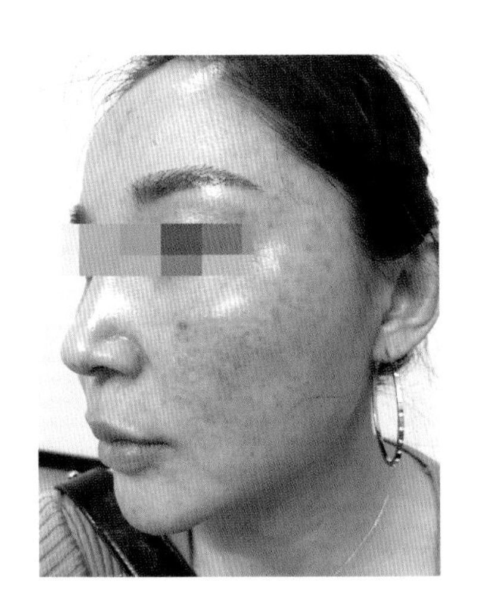

Oily skin should keep fresh and cool. You should eat less sugar, coffee, excitant food and eat vitamin B_2 or vitamin B_6 to improve skin resistance at the same time, pay attention to hydrating and deeply clean the skin to control oil secretion. Skin care cosmetics should clean and refreshing, restrain the skin fat secretion. Washing face with warm water in the daytime, choosing the cleanser that suits oily skin to keep pore and skin cleanness. Be aware of moisture.

The maintenance of combined skin should be divided into slant oily, dry, and partial normal skin. When use cosmetics, you should moisturize the dry parts, and then wipe the remaining amount to other parts. You should pay attention to replenish water, supply the nutrient composition, and adjust the balance of water and oil.

When washing sensitive skin, you should use warm and gentle facial cleanser. Using sunscreen in the morning can avoid sun damage. In the evening, using nutritious lotion to add water of the skin. Eat less food that is easily allergenic. When the skin is allergic, stop using any cosmetics immediately and observe and maintain the care of the skin. When you choose skin care products, you should test it.

(一) 对话

(P=患者,D=医生)

D:早上好,请坐。你怎么了?

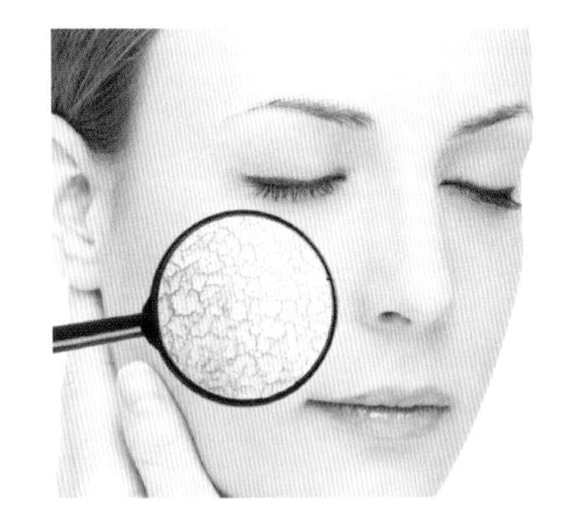

P:最近气候变化比较大,我发现我的皮肤比较油,我害怕影响我的容貌。

D:你对皮肤的类型了解吗?

P:不好意思,我对皮肤的类型一点都不了解。

D:哦。那我给你介绍一下皮肤的类型,你就知道你自己属于哪一种皮肤了。

P:谢谢。请现在告诉我。

D:好的。我先给你介绍两种类型的皮肤。第一种,干性皮肤(干燥型皮肤):角质层含水量低于10%,pH值大于6.5,皮脂分泌量少,皮肤干燥,缺少油脂,皮纹细,毛孔不明显,洗脸后有紧绷感,对外界刺激敏感,皮肤易出现皲裂、脱屑和皱纹。清洗时可选用过脂皂(加有多脂剂的肥皂)。第二种,中性皮肤(普通型皮肤):中性皮肤为理想的皮肤类型。角质层含水量为20%左右,pH值为4.5~6.5,皮脂分泌量适中,皮肤表面光滑细嫩,不干燥,不油腻,有弹性,对外界刺激适应性较强。清洗时可选用软皂。软皂是一种钾皂,含有高级脂肪酸的钾盐,这种钾盐比高级脂肪酸钠盐要软,故称为软皂。

P:非常感谢,那还有其他类型的皮肤吗?

D:是的,还有三种类型的皮肤。油性皮肤(多脂型皮肤):角质层含水量为20%左右,pH

值小于4.5,皮脂分泌旺盛,皮肤外观油腻发亮,毛孔粗大,易黏附灰尘,肤色往往较深,但弹性好,不易起皱,对外界刺激一般不敏感。多见于中青年及肥胖者,易患痤疮、脂溢性皮炎等皮肤病。混合性皮肤:干性、中性或油性混合存在的一种皮肤类型。多表现为面部中央部位(面颊、鼻部、鼻唇沟及下颌部)呈油性,而双面颊、双颞部等为中性或干性皮肤。躯干部皮肤和毛发性状一般与头面部一致。清洗时应区分油性与干性部位,重点洗油性部位。敏感性皮肤(过敏性皮肤):皮肤对外界刺激的反应性强,对冷、热、风吹、紫外线、化妆品等均较敏感,易出现红斑、丘疹和瘙痒等症状。敏感性皮肤多见于过敏体质者。

P:谢谢!通过你的介绍,我知道了皮肤的类型。

(二)课文

不同类型皮肤的保养

干性皮肤的保养是多做按摩护理,促进血液循环,注意使用具有滋润、美白、活性功效的修护霜和营养霜。注意补充肌肤的水分与营养成分,需要进行调节水油平衡的护理。护肤品应选择保持营养型的产品,清洁应选择非泡沫型、碱性度较低的产品。

中性皮肤保养的重点是注意清洁、爽肤、润肤以及按摩的周护理。注意日常补水、调节水油平衡。护肤品选择时应根据皮肤年龄、季节选择,夏天选择亲水型的护肤品,冬天选择滋润型的护肤品。

油性皮肤保养时应注意随时保持皮肤的洁净清爽,少吃糖、咖啡、刺激性食物,多吃维生素 B_2 或维生素 B_6 以增加皮肤抵抗力,注意补水及皮肤的深层清洁,控制油脂的分泌。护肤品应使用油分较少、清爽型、能够抑制皮脂分泌、收敛作用较强的产品。白天用温水洗面,选用适合油性皮肤的洗面奶,保持毛孔的畅通和皮肤清洁,注意适度的保湿。

混合性皮肤保养时应按偏油性、偏干性、偏中性皮肤的保养方法分别侧重处理,在使用护肤品时,先滋润较干的部位,再在其他部位用剩余量擦拭。注意适时补水,补充营养成分,调节皮肤的酸碱平衡。选择护肤品时,夏天参考油性皮肤的护肤品选择,冬天参考干性皮肤的护肤品选择。

应经常对敏感性皮肤进行保养。洗脸时水不可以过热或过冷,要使用温和的洗面奶洗脸。早晨,可选用防晒霜,以避免日光伤害皮肤;晚上,可用营养型化妆水增加皮肤的水分。在饮食方面,要少吃易引起过敏的食物。皮肤出现过敏时,要立即停止使用任何化妆品,对皮肤进行观察和保养护理。选择护肤品时,应先进行适应性试验,在皮肤无过敏反应的情况下方可使用。切忌使用劣质化妆品或同时使用多重化妆品,并注意不要频繁更换化妆品,不能用含香料过多及过酸、过碱的护肤品,应选择适合敏感性皮肤的化妆品。

Section 3　Problem Skin

Part Ⅰ　Dialogue

(P=Patient,D=Doctor)

D:Hello,sit down,please. Can I help you?

P:Hello. I have a lot of spots on my face,which caused me upset.

D:Can you describe the spots on your face?

P:All right. I had no skin spot, but since last year, a few small, light-colored spots appeared on my face. Whitening cosmetics has no effect. Slowly, the spots grew and the color becomes darker than before. If I didn't use cosmetics it would be obvious.

D:Do you have any other areas besides your face?

P:Only the face, the area around the eyes.

D:Besides the skin spots, do you have any other uncomfortable feeling?

P:Well, there's only a long spot on the skin that doesn't make any feeling, but it just doesn't look good.

D:Do you have a regular life? What about your temperament?

P:I usually sleep late. I am impatient and get angry easily.

D:In your description, you may have a melasma.

P:Is this a serious problem?

D:Don't worry. Melasma is a tan or dark skin discoloration. Although it can affect anyone, melasma is very common in women, especially pregnant women and those who are taking oral or patch contraceptives or hormone replacement therapy (HRT) medications. But irregular sleep and bad temper can lead to this skin problem. So you must pay attention to the regulation of life and temper.

P:Oh, I see. Thank you.

Part Ⅱ Text

1. The pigmentation problems of skin

1) Freckles

Freckles are yellowish-brown or dark-brown spots on the skin of the sun-exposed area.

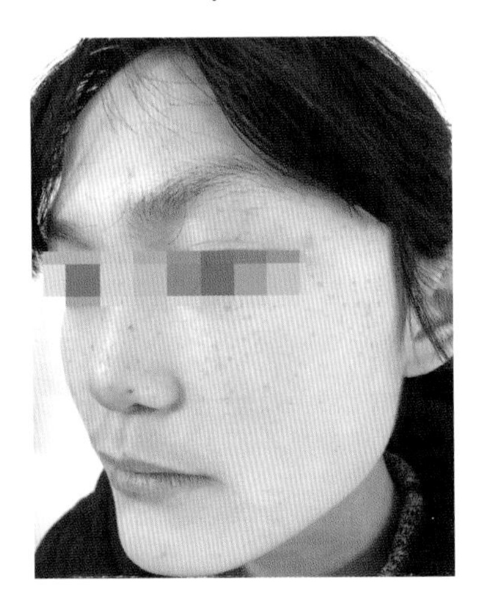

They are usually radiated around the nose, and can be seen on the side of the neck, back of the hand, chest and extremities. Most lesions are the size of needle tip to sesame and mung bean. The lesions are round, oval or irregular in shape. The lesions have clear boundary, scattered or densely distributed in groups and do not fuse with each other. Freckles mostly occur around 5 to 7 years old, a few cases of adolescent onset, women are more than men, mostly with a typical family history of inheritance.

The formation of freckles is triggered by exposure to sunlight. The exposure to UV-B radiation activates melanocytes to increase melanin production, which can cause freckles to become darker and more visible.

Freckles are predominantly found on the face, although they may appear on any skin exposed to the sun, such as arms or shoulders.

2) Chloasma

Patch of chloasma often distributes symmetrically near the eyes, forehead, nose, cheeks, lips and mouth. The color of them are more than grey, brown, dark or light brown with irregular shape and size. The disease mostly occurs symmetrically. Sometimes, it occurs unilaterally and spread to the whole face, some of which can be fused together (like a butterfly), which can be exacerbated by the day. There is no self-conscious symptoms and general malaise. Generally they appeared in young and middle aged women.

3) Senile plaques

Senile plaques are some of the flat, slightly protruding spots or patches on the skin that appear on the skin when people are old. The color is light brown, brown or black. No pain or itching. Plaques can occur in the face or in other parts of the body.

2. The problems of sensitive skin

1) Skin sensitivity

Skin sensitivity refers to the fragile skin, strong feelings, weak resistance. After being stimulated by the outside world, the skin will produce a significant response. Sensitive skin, tight pores, meticulous, dry skin, water and rough, thin skin corneum, visible microvascular and uneven flush.

2) Skin allergy

Skin allergy is an allergic inflammatory reaction after skin contact with certain external substances. The color is red or pale, mostly appear in patchy, lumpy spots, which can occur at any place of body. Patients suffer itching and tingling, and some of them will appear symptom as cold and fever. Skin allergy incidence and subsidise rate are rapid. It will not leave mark after disappearing. And it can occur in all ages.

3. Telangiectasia skin problems

Telangiectasia is small blood vessels in the skin and mucous membrane of the continuous expansion caused by congenital or acquired causes, the formation of plaque, punctual, linear damage, the color of them are red or purple red.

Telangiectasia can be divided into primary and secondary, primary is related to familial genetic, secondary may be due to the sun, improper use of cosmetics and other reasons.

4. Aging skin problems

1) White hair

White hair is depigmentation of hair, which can be divided into congenital white hair and acquired white hair. Congenital white hair is related to heredity. There are several pieces of white hair on the head after birth, which often has a family genetic history. Acquired white

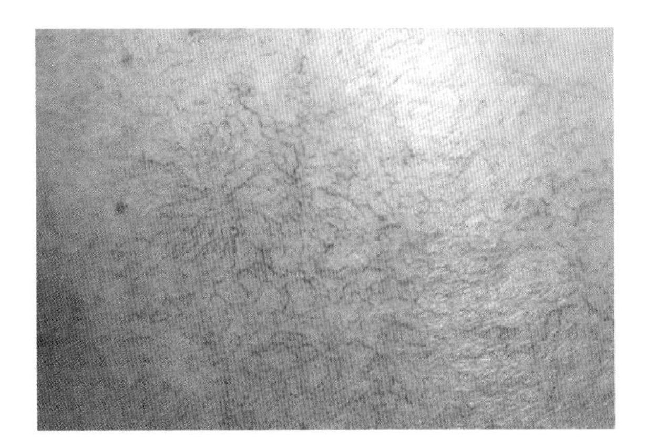

hair is related to aging or disease. It can be divided into juvenile white hair, middle-aged white hair and old white hair. Middle-aged and old white hair belong to physiological aging phenomenon, while juvenile white hair belongs to pathological category.

2) Early whitening

Early whitening is an acquired loss of hair pigmentation, which is common in young adults. In adolescents, hair gradually turns white. Generally, it begins from the top of the head and the forehead, and gradually spreads and expands. Part starts from temples, then spreads and expands to the top of the head. Some young and middle-aged white hair develops rapidly, and the whole hair turns white in a short time.

3) Wrinkles

Wrinkles are lines formed on the skin surface for various reasons. Skin wrinkles are caused by lack of moisture, reduced surface fat and reduced elasticity. Common facial wrinkles are raised eyebrow lines, crow's tail lines, corners of mouth lines, lip vertical lines, cheek twill lines and so on. The appearance of wrinkles is a sign that human function begins to decline. Generally speaking, wrinkles begin to increase in women after 28 years of age. If malnutrition or psychological burden is too heavy, wrinkles will appear earlier. Wrinkles directly affect facial appearance.

4) Eye bags

Because the eyelid skin is very thin, the subcutaneous tissue is thin and loose, so it is easy to occur edema, this edema phenomenon is called eye bags. Eye bags are generally located near the lower eyelid, which is arc continuous distribution. Young people's skin is not lax. Middle-aged people often have fine lines around the eyes. Older people can have symptoms such as eyelid ectropion, separation of eyelid and eyeball, tears and so on.

（一）对话

（P＝患者，D＝医生）

D：你好，请坐。请问您需要什么帮助吗？

P：你好。我脸上长了很多斑，这让我非常苦恼。

D：您可以具体说一下开始长斑到现在的经过吗？

P：好的，我以前皮肤没有斑，从去年过完 33 岁生日开始，脸上出现了一些小的、颜色比较浅的斑，用了很多美白化妆品，都没有效果。这些小斑逐渐增大，而且颜色也越来越深了。如果不用化妆品遮盖会非常明显。

D：请问，您除了面部长斑，其他部位长斑吗？

P：只有脸上，眼睛周围这个区域。

D：除了皮肤上长斑，您还有其他不舒服的感觉吗？

P：嗯，只有皮肤上长斑，没有什么感觉，就是看起来不好看。

D：那您平时生活规律吗？性格脾气怎么样？

P：我平时睡觉比较晚，性格比较急躁，容易生气。

D：综合您刚才的描述，您长的这种斑可能是黄褐斑。

P：啊？这个问题严重吗？

D：不用担心。黄褐斑主要是色素沉着导致的，不会影响生命和健康。黄褐斑是一种很常见的皮肤问题，尤其是孕妇及正在服用避孕药或进行激素替代疗法的患者容易出现黄褐斑。但是，您生活作息不规律，容易生气，也是导致这种皮肤问题出现的原因。所以，在生活方面和情绪控制方面您要多加注意。

P：哦，原来是这样啊，谢谢你！

（二）课文

1. 色素沉着类皮肤问题

1）雀斑

雀斑是发生在日晒部位皮肤上的黄褐色或暗褐色斑点，常以鼻部为中心呈放射状分布，也可见于颈部、手背、胸部和四肢伸侧。皮损多为针尖到芝麻、绿豆大小，形状呈圆形、卵圆形或不规则形，皮损边界清晰，散在或密集成群分布且不相互融合。雀斑多在 5～7 岁出现，少数在青春期发病，女性多于男性，多有典型的家族遗传史。

2）黄褐斑

黄褐斑是后天形成的斑片，常对称分布于眼眶附近、额部、鼻部、两颊、唇及口周等处，多呈灰褐色、褐色、深褐色、浅褐色或咖啡色，表面光滑无鳞屑。斑片散在分布，边界较清，摸之不碍手，形状不规则，大小不等。黄褐斑也可单侧发生，个别患者可波及整个面部，有的可相互融合，状如蝴蝶，日久可加重。黄褐斑无自觉症状和全身不适，在夏天斑色较重，一般好发于中青年女性。

3）老年斑

老年斑是人到老年时，在皮肤上出现的一些扁平、略突出于皮肤表面的斑点或斑块，抚之有粗糙的感觉。颜色呈浅褐色、褐色或黑色。无疼痛和瘙痒。老年斑可发生于面部，也可发生于身体其他部位。

2. 敏感性皮肤问题

1）皮肤敏感

皮肤敏感是指皮肤脆弱、感受力强、抵抗力弱，受到外界刺激后会产生明显反应。敏感性皮肤的毛孔紧闭，肤质细致，皮肤表面干燥缺水且粗糙，皮肤角质层薄，隐约可见微细血管和不均匀潮红。

2）皮肤过敏

皮肤过敏是皮肤接触某些外界物质后，在接触部位发生的过敏性炎症反应。过敏处皮肤颜色呈红色或苍白色，多为片状、块状斑或丘疹，可发生于任何部位，患者有瘙痒、刺痛感，部分患者还会出现怕冷发热的症状。皮肤过敏发病速度快，消退也快，消退后不留痕迹，容易反复发作。可发生于任何年龄段、任何季节，有些过敏体质还有遗传倾向。

3. 毛细血管扩张症

毛细血管扩张症是皮肤和黏膜上的小血管因先天或后天原因持续性扩张，形成斑状、点状、线状或星芒状损害，颜色多为红色或紫红色。

毛细血管扩张症可分为原发性和继发性两种，原发性毛细血管扩张症与家族遗传有一定关系，继发性毛细血管扩张症可由日晒、使用化妆品不当等多种原因引起。

4. 衰老性皮肤问题

1）白发

白发是毛发色素脱失症，可分为先天性白发和后天性白发。先天性白发与遗传有关，出生后头上就已经存在数根或数片白发，这种情况常有家族遗传史。后天性白发与身体衰老或疾病有关，可分为少年白发、中年白发和老年白发。中年及老年白发属于生理性衰老，而少年白发则属于病理性衰老。

2）须发早白

须发早白是后天性毛发色素脱失，常见于青少年。在青少年时头发呈渐进性变白，一般先从头顶、前额部开始，逐渐蔓延、扩展。部分从鬓角开始，再向头顶处蔓延、扩展。有的青少年白发发展迅速，整个头发在短时间内变白。

3）皱纹

皱纹是皮肤表面因各种原因形成的纹路。皮肤缺乏水分、表面脂肪减少、弹性下降就会产生皱纹。面部常见的皱纹有抬眉纹、眉间纹、鱼尾纹、口角纹、唇部竖纹、颊部斜纹等。出现皱纹是人体功能开始衰退的标志，一般女性在28岁以后开始出现皱纹并逐渐增多。如果营养不良或心理负担过重，皱纹也会提前出现。皱纹会直接影响面部的容貌。

4）眼袋

眼睑部皮肤很薄，皮下组织薄而疏松，因而很容易发生浮肿，这种浮肿现象称为眼袋。眼袋一般位于下眼睑处，呈弧形连续分布，青年人皮肤不松弛，中年人常伴眼周细纹，老年人可出现睑板外翻、睑球分离、流泪等症状。

Unit 3 Basic Skin Care Products

Section 1 Choices among the Different Skin Care Products

Ⅰ. Normal Skin

Part Ⅰ Dialogue

(B=Beautician,C=Customer)

B:Hello,welcome to our shop.

C:Hello. I am a high school student. I want to choose some skin care products,but I don't know which products work for me.

B:At first,please let me see the type of your skin. We will recommend the right products according to your skin type.

C:Okay.

B:Can you tell me your age?

C: I'm 16.

B: Oh, you are very young; you are at the beginning of youth. Your skin is very delicate smooth and elastic. Generally speaking,normal skin is drier when it is cold,and in summer skin will become a little oil. Almost no acne or clogging pores,and this type of skin is not easy to sensitive,so normal skin is the healthiest skin.

C:Thank you!

B:Normal skin should notice clean and moisturizing when choosing cosmetic. I suggest that you can use a refreshing product in spring and summer,and moisturizing product in the fall and winter.

C:Could you help me to choose a set of product?

B:I think you can choose a moisturizing deep clean cream,it can clean your skin and work well with hydration. You'd better choose a hydrating toning lotion because it is very dry in summer. You must pay attention to the hydration. You can choose a milk that have nourishing effect, which can moisturize and nourish your skin. You must use sun cream when you are outdoor. You can choose a sunscreen or sun cream.

C: Deep clean cream, toning lotion, milk and sun cream?

B: Yes. You can also use face cream in the autumn and winter. Face cream can nourish your skin. Another suggestion is that you'd better consider your eye maintenance, you can choose an eye cream to protect your eyes.

C: Okay.

B: Because normal skin is the healthiest skin. So please use skin care products to moisturize and nourish your skin.

C: Okay. Thank you!

B: In addition, I advised that you should pay attention to eat more fruit, do more exercises, it's good for your blood circulation, and then the skin will become better.

C: Thank you!

Part Ⅱ Text

Product Selection of Normal Skin

Normal skin is the most healthy, ideal skin, the pH value is between 5 and 5. 6, sebaceous glands, sweat glands secretion is moderate, not greasy, not dry and flexible, no pores, ruddy shiny, not easy to aging, delicate and elastic, not sensitive to external stimulation, no skin flaws. Cleaning the face, caring the horny, massaging the skin, masking the face, doing the basic caring, those are the most important for normal skin.

(1) Cleaning: Cleaning the skin and maintain a good skin-based, you should clean your face every morning and every night, choosing the water temperature according to different seasons. In order to make the pores in the natural diastolic state and easy to wash, generally 30-36 ℃ is appropriate. Do not use hot water to clean your face, it can make your skin rough and aging.

The program of the facial cleaning is also important. The correct way is take off right amount of deep clean cream, rub in the palm, wiped the entire face from inside to the outside in arc-shape way, the strength cannot be so big, otherwise, it can pull the skin and produce some small wrinkles. After cleaning for 3-5 minutes, wash the rest of the deep clean cream thoroughly.

(2) Horny care.

(3) Massage: Use a lot of toning lotion that contains a lot of water to massage your skin, 1-2 times a week, it can promote blood and lymph circulation, and then makes your skin maintain healthy.

(4) Mask: In order to keep the skin smooth and soft, you have to do a deep clean or use a moisturizing mask every week, it can add more water to your skin.

You have a wide choice in daytime, but you should select a product that could make the lipid membrane of the skin surface better and maintain. At daytime, you should use eye cream or eye gel, use your finger scribble the eye cream on your eyes from outside to inside in arc-shaped way.

You can choose a more refreshing cream product at night, it can nourish and moisturize your skin and make your skin more smooth and soft.

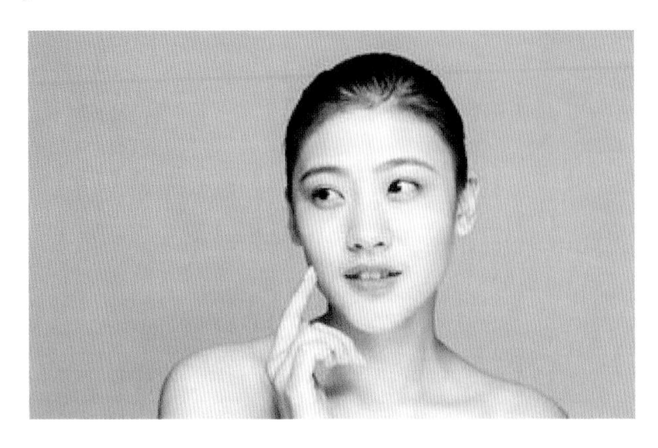

一、中性皮肤

(一) 对话

(B=美容师,C=顾客)

B:你好！欢迎光临！

C:你好！我是一名高中生,想选用一些护肤品,不知道哪些适合我?

B:请先让我看下您的皮肤类型,只有根据您的皮肤类型才能选择合适的护肤品。

C:好的。

B:可以告诉我您的年龄吗?

C:16岁。

B:16岁是花季,是青春开始绽放的年龄。您的皮肤非常细腻、光滑且有弹性,水油平衡。一般,中性皮肤在天气转冷时有点偏干,夏季偶尔会出现轻微油光,几乎没有痘痘和堵塞毛孔的情况,耐晒,不容易过敏。所以中性皮肤是最健康、理想的皮肤。

C:谢谢!

B:中性皮肤在选用护肤品时注重清洁,做好保湿工作就可以了。建议春夏用清爽型产品,秋冬用保湿型产品。

C:请帮我选择一套适合我的产品吧!

B:好的！建议您选用滋润型的洗面奶,不但有清洁皮肤的作用,还有保湿补水的作用。化妆水也以保湿补水的产品为主,因为春夏季节比较干燥,要特别注意补水。乳液也以具有滋养效果的产品为主,起保湿和营养肌肤的作用,另外,在户外活动时必须使用防晒产品,霜状或乳液状的防晒霜均可使用。

C:洗面奶、水、乳液、防晒霜就可以了吗?

B:是的。但是在秋冬季节建议再加上面霜,面霜在秋冬季节起防护、营养、滋润的作用。另外也建议您在搭配产品时也应该考虑眼部的保养,所以您可以选用一支保湿、滋润的眼部产品进行眼部皮肤的护理。

C:哦。

B:因为中性皮肤是最健康的皮肤。所以请在选用护肤品时以保湿、滋养为主。

C:好的。谢谢!

B:另外,建议您平时除了用洗面奶和补水的产品外,还要注意多吃水果,多运动,这样可以加快血液循环,对皮肤是很有好处的。

C:谢谢!

(二)课文

中性皮肤如何选择护肤产品

中性皮肤是最健康、理想的皮肤,其 pH 值为 5～5.6,皮脂腺、汗腺的分泌量适中,不油腻、不干燥、富有弹性,无毛孔,红润有光泽,不容易老化,细腻且富有弹性,对外界刺激不敏感,没有皮肤瑕疵。在护理时应以洁面、角质护理、按摩、面膜、基础护理产品为主。

(1) 洁面:中性皮肤以清洁和保持良好肤质为主,每天早晚进行洁面,清洁时的水温可根据季节的不同来选择。为了使毛孔处于自然舒张状态和便于洗净,一般以 30～36 ℃为宜,切不可长期用过热的水进行洁肤,否则容易导致毛孔变粗,皮肤粗糙、老化。

中性皮肤的洁面程序也是有讲究的。正确的使用方法是取适量的洗面奶,在手掌搓揉后,均匀地抹在整个面部,轻柔地由内向外呈圆弧状清洗,力量不要太大,以免牵拉皮肤产生细小的皱纹。在清洗 3～5 min 后,再将洗面奶完全洗净。

(2) 角质护理。

(3) 按摩:使用含水分较多的霜或液进行按摩,每周 1～2 次,这样能够促进血液循环和淋巴循环,使皮肤保持健康。

(4) 面膜:为使皮肤保持光滑柔软,每周应视皮肤的状态做 1 次深层清洁或面膜保湿,这样可及时补充皮肤的水分。

白天中性皮肤选择日霜保养的范围很大,不过还是应该选择以能帮助皮肤表面水脂质膜的添补及维护的产品为佳。晚上一定要用眼霜或眼部凝胶,使用时用指尖轻轻地将眼霜由外向内点涂于眼周,轻柔地做圆形的滑动。

晚霜可以选择较为清爽的乳液状产品,可起营养、润泽皮肤的作用,使皮肤保持光滑柔嫩。

Ⅱ. Oily Skin

Part Ⅰ Dialogue

(B=Beautician,C=Customer)

B:Hello,welcome to Haitang Chinese Medical Academy.

C:Hello,I have some acnes on my face,but I don't know what kind of reason caused this situation.

B:Your last name?

C:Mr. Li.

B:Mr. Li,please come here. I will detect your skin condition,and then we can introduce a suitable product for you.

C: OK.

B: The skin is detected by skin detector, your skin is oily skin according to detectation. The main performance of oily skin is strong oil secretion, large pores, and shiny on the nose and forehead.

C: How should I remove acnes?

B: You should pay attention to the skin cleaning and moisturizing, and then the skin's water and oil will become balance, at the same time use the skin care products that have the function of anti-inflammatory, sterilization, and shrinking pores.

C: Really?

B: Don't worry! Different skin conditions have different requirements for removing acnes. The way of the using products, and implementation of the plan will also have different effects on different types of skin. Therefore, we must make a good strategy to remove acnes.

C: Oh! Oily skin usually very oily, and have some acnes, it's really bad.

B: The biggest advantage of the oily skin is that this kind of skin is acidic and elastic, so the aging of the skin is very difficult.

C: In addition to the use of acne products, what should I pay attention to?

B: If you have acnes, do not squeeze, you'd better regulate the endocrine, mainly from the diet and exercise. To develop a good eating habit, eat more fresh fruits and vegetables, high protein foods, drink plenty of water, add more water to body, eat less irritating food, such as pepper, at the same time to participate in a variety of sports exercise, strengthen the physique, and live a scientific way of lifestyle.

C: Okay. Thank you!

B: You're welcome! I believe you will be very satisfied and happy when you see your skin condition changed.

Part Ⅱ Text

Product Selection of Oily Skin

The characteristic of oily skin is that the secretion of sebum is very strong, most people has dark skin, big pores, oily and bright skin, sometimes have the appearance of orange peel samples, the pH value is between 5.6 and 6.6.

Oily skin pore is so big because of the jam of oil and grease. The skin grease not discharge in time, and hard oil and grease clogged pores for a long time, so oily skin should do deeply clean.

Maintenance steps:

Step 1: Deep cleaning. When choosing a facial cleanser, you should choose one that have strong clean ability, which can wash skin grease, residual cosmetics and dirt. the product should be mild, nature, rich in a variety of vitamins and minerals, and moisturize the skin.

Step 2: Moisturize. Because the skin is short of water, the skin will produce a lot of grease to prevent skin dry, so after cleaning the face, you should add more water to skin. You can pat the skin with the right amount of toner to help with oil control and moisture.

When choosing convergence water or contract water, you can use contract water which contains alcohol, it can replenish water, adjust skin's pH balance, shrink pores, and inhibit oil secretion.

Step 3: Anti-inflammatory, sterilize and equilibrium moisture. You can choose hydrophilic milk which can effectively control the sebum secretion of face, shrink pores, keep the skin bright, and moist and soft dry areas of face at the same time.

When choosing moisturizing products, you can use a model moisturizing cream or moisturizing lotion, replenish moisture. In addition, you can use a acidity protect, which can balance the grease secretion and acne. A smooth and refreshing elite fluid is better than a thick elite fluid.

Step 4: Quarantine. You should choose a refreshing product, with have physical sun blocks function.

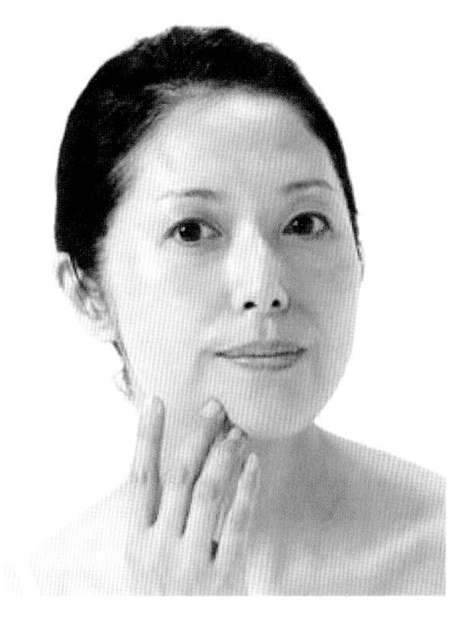

Notes on diet:

(1) Limit fatty foods and sweets, such as fatty meats, cream cakes, chocolate, and eat more vegetables and fruits.

(2) Use warm water to wash your face with a small amount of sulfur soap or borate soap every night. Remove facial grease and clean your skin.

(3) Strengthen physical exercise and improve skin immunity.

(4) Treatment with traditional Chinese medicine: when flushing, liquid, and scabs, you can be treated with heat-clearing, detoxification and diuresis. When you are itch but do not leak out, you can raise blood, moist, cure wind, clear heat.

二、油性皮肤

（一）对话

（B＝美容师，C＝顾客）

B:你好！欢迎光临海棠中医美容院！

C:你好！我最近脸上长了许多痘痘，不知道是什么原因导致的。

B:您贵姓？

C:姓李。

B:李先生。请先通过皮肤检测仪检测下您的皮肤情况，根据痘痘产生的原因才能选择合适的产品。

C:哦。

B:通过皮肤检测仪的检测，您的皮肤是油性皮肤。油性皮肤的主要表现是油脂分泌旺盛、毛孔粗大，额头、鼻翼有油光。长痘痘的原因是内分泌中雄性激素分泌过量导致毛囊皮脂腺的分泌过多。

C：那我应该怎么祛痘呢？

B：油性皮肤首先要注意皮肤的清洁，另外应加强保湿补水，使皮肤水油平衡，同时使用消炎、杀菌、收缩毛孔的护肤产品。

C：那效果好吗？

B：您放心！因为皮肤状况的差异，对于祛痘的要求是截然不同的。而一些祛痘产品的使用、祛痘计划的实施也会对不同类型的皮肤产生不同的效果。所以，祛痘策略也需要因“肤”而异、量体裁衣。

C：哦！油性皮肤爱出油，长痘！真不好！

B：油性皮肤最大的优点是偏碱性，弹性较佳，是最不易衰老的皮肤哦！

C：除了使用祛痘的产品外，我还要注意什么？

B：长了痘痘以后，千万不要挤，要以调理内分泌为主。调节内分泌主要从饮食、运动入手，要养成良好的饮食习惯，多吃新鲜果蔬、富含蛋白质的食物，多喝水，补充身体所需的水分，少吃刺激性食品，如辣椒，同时多参加各种运动锻炼，加强体质，还要有科学的生活规律。

C：好的。谢谢！

B：不客气！相信您看到皮肤的改变肯定会非常满意和开心的！

（二）课文

油性皮肤如何选择护肤产品

油性皮肤的显著特征是皮脂分泌旺盛，多数人肤色偏深，毛孔粗大、皮肤油腻光亮，甚至可以出现橘皮样油性皮肤外观，其 pH 值为 5.6～6.6。

油性皮肤毛孔粗大是由于皮肤中的油脂没有及时排出，时间久了油脂硬化阻塞毛孔而形成。因此油性皮肤要做好深层清洁、保湿补水等工作。

皮肤保养分为以下几个步骤。

第一步，深层清洁。在选择洗面奶时，要选择清洁能力强，能够洗净皮肤油脂、化妆品残余和灰尘并且性质温和清爽、富含多种维生素和矿物质、具有滋润作用的产品。

第二步，保湿补水。油性皮肤缺水，因此要分泌大量的油脂来避免皮肤干燥，所以在清洁油性皮肤后，要给皮肤补水。可以给皮肤拍打适量的爽肤水，以起到控油和保湿的作用。

选用收敛水或收缩水时，可以使用含微量酒精的收缩水，这样不仅可以补充水分，调节皮肤酸碱平衡，而且可以收缩毛孔，抑制油脂分泌。

第三步，消炎杀菌，平衡滋润。可以选用亲水性的乳液，能有效控制面部油腻部位的油脂分泌，缩小毛孔，长久保持爽洁状态，同时滋润、柔软面部干燥部位。

在选用滋润产品时，可使用清爽型的保湿乳或保湿露，及时补充水分。另外还可以用含有果酸的护肤品，这类护肤品能够平衡皮肤油脂的分泌，减少皱纹和粉刺。选择滑爽质地的精华液，比用浓稠型的精华更适合油性皮肤。

第四步，隔离防护。应选用清爽并带有物理防晒功能的产品。

油性皮肤饮食上也要注意以下几点。

（1）限制脂肪性食物和甜食，如肥肉、奶油蛋糕、巧克力等，多食蔬菜和水果。

（2）每晚用温水、硫磺香皂或硼酸皂洗脸。清除面部油腻，清洁皮肤。

（3）加强体育锻炼，提高皮肤免疫力。

（4）中医治疗。当皮肤出现潮红、渗液、结痂时，以清热、解毒、利尿为治则。仅有痒感而无渗液时，以养血、润燥、祛风、清热为治则。

Ⅲ. Dry Skin

Part Ⅰ Dialogue

(B=Beautician, C=Customer)

C: Hello!

B: Hello!

C: My face is peeling! Can you help me?

B: No problem! I'll test your skin first. Please come with me.

B: Through the skin moisture detector, your skin is low in moisture content and belongs to dry skin.

C: My skin short of water?

B: Yes. The condition of dry skin is less oil, less water, small pores, and dry. The water of skin stratum corneum is less than 10%, you can produce small wrinkle easily.

C: How do I nursing my skin?

B: One of the most important thing about dry skin nursing is to make sure your skin is hydrated. Firstly, when you choose a skin cleansing product, you can't choose a strong alkaline product. It may inhibit the secretion of sebum and sweat fluid, make the skin drier.

After thoroughly cleaning the face, you should use moisture make-up water or lotion immediately to replenish the moisture of the skin. If you have the condition, you can make a facial and nourishing facial mask once a week to promote blood circulation, accelerate cell metabolism, and increase the secretion of sebum and sweat. Use warm water to clean your skin before going to bed, then massage for 3 to 5 minutes to improve the blood circulation of your face and use the night cream appropriately. The next morning, after cleansing, use a lotion or a nourishing cream to keep your skin hydrated.

C: Give me a suit of product.

B: Ok!

Morning: water→eye cream→toner→lotion→foundation→loose powder.

Evening: deep clean cream→moisturizing facial mask→eye cream→toner→essence→moisturizing cream.

C: Okay.

B: In addition, you should pay attention to the diet recuperation, you can choose some food contains fat and vitamins, such as milk, chicken, eggs, pork liver, butter, and fresh fruits. Especially in autumn and winter, you should pay special attention to maintain, choose a cream that contain high grease to prevent the skin to be dry and desiccant, delay the aging of skin.

C: Thank you!

Part Ⅱ Text

Product Selection of Dry Skin

Dry skin is delicate, less sebum secretion, dry, white, the lack of gloss and pore is small and not obvious, capillary shallow, easy to burst, is sensitive to outside stimulation. The skin is prone to erythema. The pH value is about 5.5 to 6.0, which can be divided into two types: dry skin short of water and dry skin short of oil.

Dry skin which is short of water is due to improper care or other reasons which make the skin short of water, and make the internal water of skin and sebum out of balance. Therefore, moisturizing is very important.

Dry skin is short of oil, the sebum secretion is very little, so the skin can't lock adequate water, and is a little dry in appearance, lack luster, being sensitive to external stimulation. Therefore, when choosing to protect skin, you should choose one that could add water to your skin, and you should also consider supplementary oil.

Methods of maintenance and product choosing:

(1) When choosing skin cleansing products, you can choose a clean cream skin care product that doesn't contain alkaline substances, you can use a glycerin soap which have a small stimulation to skin. Do not use poor soap wash your face, sometimes you cannot use soap just use water to clean your face. In the morning, it is advisable to moisturizing the skin with cold cream or lotion, then adjust the skin with astringent toning lotion and massage amount of nourishing cream. In the evening, you can use sufficient amount of lotion, nutritive make-up water, nutrition frost.

(2) If your facial cleanser without nourishment composition, or your face is dry and tight after cleaning the face, you can add 2 drops of moisture in the cleanser essence (compound essential oil), or add half a bowl of hot water in the basin, drip into the 2-3 drops of rose essential oil or lavender essential oil. After fully stirring the essential oil, you should use a big towel cover your face, close your eyes, avoid essential oil fragrance and vapor stimulate your eyes, breath with your mouth or nose alternately, and lasts for 5 minutes, then wash your face with a milk, you will find an unexpected good effect.

(3) Use moisturizing lotion to replenish the moisture in your skin after cleaning your face thoroughly. The most important thing about dry skin is to make up the water. Moisturizing in time is good for skin excretion and nutrition. Finally fill the water and lock the water, don't let the water drain!

The dry skin people cannot drink caffeinated drinks, you should choose some fat foods or foods high in vitamin content, such as milk, egg, pork liver, butter, fish, mushroom, pumpkin and fresh fruit, etc.

三、干性皮肤

(一) 对话

(B=美容师,C=顾客)

C:你好!

B:你好!

C:我脸上脱皮了! 能帮我看下吗?

B:没问题! 我先检测您的皮肤情况。请跟我来。

B:通过检测,您皮肤的含水量偏低,属于干性肌肤。

C:皮肤比较缺水?

B:是的。干性皮肤表现是皮肤油分、水分较少,毛孔细小,皮肤比较干燥,皮肤角质层水分低于10%,容易产生细小皱纹。

C:那我怎么护理皮肤呢?

B:干性皮肤保养最重要的一点是保证皮肤得到充足的水分。首先在选择清洁护肤品时,不要选用碱性强的产品,以免抑制皮脂和汗液的分泌,使皮肤更加干燥。彻底清洁面部后,应立刻使用保湿型化妆水或乳液来补充皮肤的水分。有条件的话,每周可做一次薰面及营养面膜,以促进血液循环,加速细胞代谢,增加皮脂和汗液的分泌。睡前可用温水清洁皮肤,然后按摩3~5 min,以改善面部的血液循环,并适当使用晚霜。次日清晨洁面后,使用乳液或营养霜来滋润皮肤。

C:那请帮我搭配一套适合我的产品吧!

B:好的!

早晨:清水→眼霜→柔肤水→乳液→粉底→散粉。

晚上:洗面奶→补水面膜→眼霜→柔肤水→精华素→保湿霜。

C:好的!

B:另外,从饮食上调理,干性皮肤的人应选择一些脂肪、维生素含量高的食物,如牛奶、鸡蛋、猪肝、黄油及新鲜水果等。尤其在秋冬干燥的季节,要格外注意保养,选用油脂含量高的护肤霜,防止皮肤干燥脱屑,延缓皮肤的衰老。

C:谢谢!

(二) 课文

干性皮肤如何选择护肤产品

干性皮肤肤质细腻,皮脂分泌少,皮肤干燥、白皙、缺少光泽,毛孔细小且不明显,毛细血管表浅,易破裂,对外界刺激比较敏感,皮肤易生红斑。其pH值为5.5~6.0,可分为缺水性干性皮肤和缺油性干性皮肤两种。

缺水性干性皮肤是由于护理不当或其他原因造成皮肤极度缺水,皮肤内部水分与皮脂失去平衡,导致皮肤反馈性地刺激皮脂腺,皮脂腺分泌增加,造成一种“外油内干”的局面。因此,补水保湿是重点。

缺油性干性皮肤是因为皮脂腺分泌皮脂较少,皮肤因为不能及时、充分地锁住水分而导致皮肤干燥，缺乏光泽，对外界刺激比较敏感。因此，选择护肤品时除了考虑补水，还要考虑补充油分。

干性皮肤的保养方法及产品选择。

(1) 在选择清洁类护肤品时，宜用不含碱性物质的膏霜型洁肤品，可选用对皮肤刺激小的含有甘油的香皂，不要使用粗劣的肥皂洗脸，有时也可不用香皂，只用清水洗脸。早晨，宜用冷霜或乳液滋润皮肤，再用收敛型化妆水调整皮肤，涂足量营养霜。晚上，要用足量的乳液、营养型化妆水、营养霜保养皮肤。

(2) 清洁面部时，如果洗面奶没有滋润成分，或是洗面后感觉面部比较干燥或紧绷，可以在洗面奶里加入2滴保湿润肤的精油(复方精油)，或在脸盆里加入半盆热水，滴入2～3滴玫瑰精油或薰衣草精油，将精油充分搅匀后，用大毛巾将整个头部及脸盆覆盖，闭上眼睛，避免精油香味及水蒸气刺激眼睛，用口、鼻交替呼吸，维持5 min，再洗面，效果会更好。

(3) 彻底清洁面部后，应立刻使用保湿型化妆水或乳液来补充皮肤的水分。干性皮肤最重要的是补水。最后补了水还要能锁住水，不要让补的水流失！

干性皮肤的人不能喝含咖啡因的饮品，应注意选择一些脂肪、维生素含量高的食物，如牛奶、鸡蛋、猪肝、黄油、鱼类、香菇、南瓜及新鲜水果等。

Ⅳ. Mixed Skin

Part Ⅰ Dialogue

(B=Beautician, C=Customer)

B: Welcome!

C: Hello!

C: My skin must belong to mixed skin, the nose oily, dry on either side of the face, the face is yellow with a little spots and fine lines. I want to ask if you could introduce some good products for me.

B: Mixed skin between oily and dry skin, so you can use two ways and choose two sets of products to nurse to skin. T area and cheek on both sides you should nurse separately, and then you can nurse your skin more effectively.

Cleaning: you can choose detergency strong cleanser.

Facial massage: when you nurse your skin in beauty salon, oily skin part you can use degreased massage cream, then use attacking a vital point ways and knead ways to massage. Dry skin parts, you can choose nutritional massage cream or facial massage oil, it can remove your fat and increase nutrition to the dry parts, adjust to each other in improving the properties of the skin.

Mask: When choosing a facial mask, you should choose a product according to different needs and tastes, under the jaw, nose, pharynx and frontal parts you can use some special facial mask, such as fruits and vegetables facial mask.

Skin care products: for oily skin, remove excess oil, shrinks pores, and neutralize the skin pH value are most important, so you should pat in T zone to convergence pores. You should use the contraction water inhibiting oil secretion, and smear moisturizing skin care products. For the dry skin, it is important to replenish nutrition and moisture, you can choose nourishing products, such as nutrition cream smear on your cheek.

You should pay more attention to a balanced diet, eat less fat and acrimony excitant food, drink plain boiled water.

C: Yeah. Please match a set of suitable products for me.

B: No problem. In addition to the beauty salon for professional care, you can use some household products which add more water to your skin, and use some hydrating mask.

C: Thank you!

Part Ⅱ Text

Product Selection of Mixed Skin

If you are mixed skin, the most common T zone (around the forehead and brow and nose) may have a lot of the oil, and cheek place is neutral skin or dry skin.

The biggest features of mixed skin: T area oily and easily perspires, if you didn't clean your face, you will grow blain and pore will become big, but the oil of two sides of the face are very little, don't like forehead, nose. The skin on this area is more texture and delicate.

We should use different skin care products according to different parts.

Maintenance methods and products selection:

1. Clean

To clean the face every morning and every night, if you want to makeup, you should discharge makeup every night. When you choosing cleanser you should pay attention to avoid bubble type products, I advise that you'd better choose a lotion cleanser.

Washing your face gently. I often see some girls washing their face very hard, actually you don't need do it like that, wash you face gently.

2. Peel horny

If you are mixed skin, you should peel horny twice a week. T area is oil mixed skin, pore is thick, and the two sides of the face is neutral or partial dry, therefore, you can choose two masks together.

3. Toner

You should use the toner twice a day. When you choose a toner, you can't choose toner which contain alcohol, which is bad for your skin, and use only one type of toner which can moisture your skin.

4. Milky lotion

Use the milky lotion twice a day. If you find you face already very oil, don't be afraid of using milky lotion, actually this is a critical step in moisturizing and repairing. You should pay attention to moisturizing and control the oil when you choose a product. Make the skin in the appropriate water and oil balance. A kind of milky lotion is very difficult to achieve both containment and moisturizing, so I suggest you can choose two kinds of milky lotion.

5. Watery essence

The last step of skin care is using watery essence every morning. When you choose a watery essence product you should pay attention to choose one that can lighten your skin, don't give skin too much burden.

You should keep the law of work and rest time, don't stay up late, be happy, try to get outdoors, drink more water, and promote the body blood circulation renewal, timely discharge toxins and impurities. Avoid spicy, greasy, vulcanizing, sweet and excitant food, eat the food of some cool blood, such as towel gourd, wax gourd, green beans, etc.

四、混合性皮肤

(一) 对话

(B=美容师,C=顾客)

B:欢迎光临!

C:你好!

C:我的皮肤应该是混合性的,鼻子油,脸的两边偏干,整个脸都比较黄,又有点斑点,近看还有细纹。我想问一下,有什么好的护肤品可以推荐?

B:混合性皮肤因为介于油性与干性皮肤之间,所以在护理和保养品上最好是选择两套方法和两套保养品。T区与脸颊两侧分开护理,才能更有效地护理好自己的皮肤。

洁面:可选择去污力强的洗面奶。

面部按摩:在美容院做护理时,油性皮肤的部位可以用脱脂按摩膏,采用点穴、按摩手法进行按摩。而干性皮肤的部位,可选用营养按摩膏或按摩油进行面部按摩,这样既达到了去脂的目的,又可以给干燥的部位增加营养,在相互调节中改善皮肤的性质。

面膜:在选择面膜时,要根据不同的需要进行选择,在颌部、鼻部、咽部可使用干性皮肤专用面膜,比如水果蔬菜面膜。

护肤品:对于油性皮肤而言,去除多余的油脂、收缩毛孔、中和皮肤pH值是首要的,故应在T区拍打以收敛毛孔,使用抑制油脂分泌的收缩水,并涂抹保湿型护肤品。而对于干性皮肤,以补充营养及水分为首要,可选用营养滋润的产品,比如在脸颊部位涂抹营养霜。

在日常饮食生活中,要多注意饮食平衡,少食高脂肪类及辛辣刺激性食物,多喝白开水。

C:嗯,就请搭配一套适合我的产品。

B:好的。除了在美容院进行专业护理外,可以使用深层洁净的洗面奶,多用补水效果好的产品,配合做补水和清洁的面膜。

C:谢谢!

(二) 课文

混合性皮肤如何选择护肤产品

混合性皮肤最常见的是T区(额头、眉毛及鼻子周围)呈现油性肤质,而脸颊部位则为中性或干性肤质。

混合性皮肤最大的特点是T区容易出油出汗,如果清洁不干净会长痘痘,毛孔粗大。而两颊出油很少，不如额头、鼻头那么多，看上去，两颊部的皮肤纹理比 T 区细腻。

针对不同部位选择不同的护肤品。

混合性皮肤的保养方法及产品选择。

1. 清洁

清洁是每天护肤的开始，早晚各一次，如果化妆的话晚上要先卸妆。混合性皮肤在选择洗面奶时要注意避免泡沫型的，建议使用乳液型的洗面奶。

注意洗脸时动作要轻一点，经常看到很多女孩喜欢很用力地洗自己的脸，其实用不着，轻轻地把脸洗干净就可以了。

2. 去角质

混合性皮肤去角质一周最多两次。在护肤中这一步很重要。混合性皮肤的人因为 T 区较油，毛孔比较粗，而两颊可能是中性或者偏干性的皮肤，因此，可以选择两种面膜一起用。

3. 爽肤水

爽肤水一天两次。混合性皮肤在选择爽肤水时要注意不要因为追求收缩毛孔而使用含有酒精的爽肤水，并且在只使用一种爽肤水的情况下应选择以保湿为主的爽肤水。

4. 乳液

乳液的使用也是一天两次。很多人觉得自己的脸已经很油了，一直很怕涂乳液，其实这是保湿控油的关键。混合性皮肤在选择乳液时要注意其保湿和控油的作用，让皮肤恢复适当的水油平衡。一种乳液很难达到兼具控油和保湿的作用，所以建议选用两种乳液。

5. 隔离保护

隔离是早上护肤的最后一步。混合性皮肤在选择隔离产品时要注意选择质地轻盈的产品，不要给皮肤太大负担。

混合性皮肤也要保持规律的作息时间，不要熬夜，保持心情愉快，多到户外活动，多喝水，促进体内血液循环，及时排出毒素和杂物。切忌辛辣、油腻、过甜等刺激性的食物，吃一些有凉血功效的食物，如丝瓜、冬瓜、绿豆等。

Ⅴ. Sensitive Skin

Part Ⅰ Dialogue

(B=Beautician，C=Customer)

B：Good morning!

C：Morning!

C：I am sensitive skin. For years，I have been allergic to skin care products. My face is red and I don't make up for many years. Now I am getting older and my skin is getting worse and worse. Would you like help me?

B：Your skin is thin，sensitive to external stimulation，and prone to local reddish，red，and other allergic reactions. Only the right skin care products can achieve the desired results.

C：What products should I use?

B:You can choose make-up lotion containing anti-inflammatory and calm ingredients. A lot of pure natural plant extract ingredients are good anti-inflammatory and sedative, such as licorice root essence, hamamelis essence, ivy essence, aloe extract, lavender extract, allantoin, etc.

C:Ok!

B:Sensitive skin first select safe mild products which don't contain the stimulus such as flavor, preservatives, pigment, or alcohol composition. You should choose high security lotion which contains more water, combined with hydrating essence, can help more water into the cell, make cell drink lots of water.

C:Oh!

B:It's important to moisturize your skin. Apart from using of skin care products, moisturizing facial mask should be done regularly. In addition, it is necessary to use watery essence and sun block every day.

C:Ok! Thank you!

Part Ⅱ Text

Product Selection of Sensitive Skin

The characteristics of sensitive skin: thin skin, delicate white, little sebum secretion, drier, microvascular apparent. The skin has a dry function, the cuticle has no ability to retain water, and the skin surface is not complete. The skin tends to be unstable due to seasonal changes. The main symptoms are itching, burning, tingling and rashes.

The maintenance method of sensitive skin and product selection:

(1) Cleaning products. To avoid the use of alkaline ingredients and strong clean products, use the soft cleaning products with no alcohol or fragrance, and try to be gentle and clean.

(2) Gently remove the horny.

(3) Use moisturizing lotion.

(4) Moisturizer. Choose fragrance-free, the texture mild, hydrating and moisturizing strong products.

(5) Mask. Sensitive skin should be covered with hydrating mask. Hydrating mask refers to the injection of natural moisturizing factor (essence) into a facial mask which is made from various materials to make quick hydration of the face. The material of facial mask cloth is also various, and can be divided into three kinds: non-woven cloth, silk, biological fiber. The essence of silk mask cloth is the best one.

(6) Watery essence. Choose some of the less stimulating suntan products to protect your skin

from the sun.

The above methods are general anti sensitive methods, you should also pay attention to reasonable diet. Sensitive skin should pay attention to nutrition balance, eat less fat, spicy food, no smoking, no drinking. Eat more anti-allergic foods to strengthen the skin's defense capabilities. According to nutritionists, onion, garlic, broccoli and citrus contain anti-allergic compounds, those things can prevent allergies.

五、敏感性皮肤

(一) 对话

(B=美容师,C=顾客)

B:早上好!

C:早!

C:我是敏感性皮肤,好多年了,一用护肤品就过敏,脸上发红,已经素面朝天好多年,现在年龄大了皮肤越来越不好,想过来咨询一下。

B:敏感性皮肤的表现是皮肤较薄,对外界刺激敏感,容易出现局部微红、红肿等过敏现象。只有选对合适的护肤产品,才能达到预期效果。

C:那我应该选用什么产品呢?

B:应选用含有消炎、镇定成分的化妆水,许多纯天然植物提取成分都是很好的消炎、镇静剂,如甘草萃取精华、金缕梅精华、常春藤精华、芦荟精华、薰衣草精华及尿囊素等。

C:好的!

B:敏感性皮肤首先选择安全温和的产品,不含香料、色素、酒精、防腐剂等刺激成分,选择安全性高、天然保湿因子较多的爽肤水,再配合补水保湿精华,能帮助更多水分进入细胞内,让细胞快速喝到大量的水。

C:哦!

B:日常保养中加强保湿非常重要。除应使用含保湿成分的护肤品外,还应定期做保湿面膜。此外,隔离防晒霜需要每天使用。

C:好的! 谢谢!

(二) 课文

敏感性皮肤如何选择护肤产品

敏感性皮肤的表现是皮肤表皮薄,细腻白皙,皮脂分泌少,较干燥,微血管明显,角质层保持水分的能力降低,皮肤表面的皮脂膜形成不完全。因季节变化皮肤容易出现不稳定的状态。主要症状是瘙痒、烧灼感、刺痛和出小疹子等。

敏感性皮肤的保养方法及产品选择。

(1) 清洁品。要避免使用含碱性成分以及清洁力强的产品,要用不含酒精、香精的轻柔清洁产品,尽量做到温和洁面。

(2) 温和地去角质。

(3) 使用保湿水。

（4）保湿霜。选用无香味、质地温和、补水保湿能力强的产品。

（5）面膜。敏感性皮肤应该用补水面膜。补水面膜指的是将天然保湿因子（精华液）注入由各种素材制成的面膜中起到对面部快速补水的作用。面膜布的材质也多种多样，可以分为无纺布、蚕丝和生物纤维三类，其中蚕丝面膜布的精华液吸收最好。

（6）隔离、防护。选一些防晒系数小的、刺激性低的防晒产品，避免日晒对皮肤的伤害。

以上是一般抗敏防敏的方法，另外要注意合理饮食。敏感性皮肤要注意饮食营养均衡，少吃油腻、辛辣的食物，不吸烟，不喝酒。多吃一些具有抗敏功能的食物，加强皮肤的防御能力。根据营养学家研究，洋葱、大蒜、椰菜和柑橘等均含有抗敏化合物，可以预防发生过敏。

Section 2 Classification of Daily Skin Care Products

Ⅰ. Facial Cleanser

Part Ⅰ Dialogue

（B＝Beautician，C＝Customer）

B：Hello！

C：Hi！ I want to buy a facial cleanser. I do not know which one suits me.

B：Ok. I will first detect your skin，I can recommend you to buy an appropriate skin care product according to your skin type. Your skin is very white and delicate，but the skin is relatively dry. Because of water shortage，the eye will have small fine lines，so your skin belongs to dry skin，you can choose the skin care product which can moisturize and hydrate your skin.

C：Oh. Which facial cleanser is suitable for me?

B：Facial cleanser is used for cleaning and protecting your skin. Facial cleanser can be divided into two types：soft cleanser and oily skin cleanser. Soft cleanser contains mild formula suitable for neutral and dry，sensitive skin，and the formula of oily skin cleanser contains less fat. After washing your face your facial skin feels very clean，so it's suitable for oily and acne-prone skin. Neuter or dry skin can't use oily facial cleanser, because this can make skin drier and dehydrated. Therefore, dry skin should choose soft cleanser.

C: Ok.

B: This cleanser is rich in foam and is the best choice for dry skin. The product contains sweet almond essence, which can clean the dirt on the face, enhance skin vigor and make the skin moist and not dry.

C: Good! That's it!

Part Ⅱ Text

Facial Cleanser

Cleansing is the first step to beauty. Good facial cleanser can be moist, comfortable, relaxed, not tight feeling after washing a face. The quality of facial cleanser depends on its ingredients.

1. Main ingredients of facial cleanser

The main ingredients of facial cleanser: active agent, humectant, active additive, preservative, essence, etc.

Other ingredients:

(1) Hyaluronic acid.

(2) Aloe extract.

(3) Spring water.

(4) Seaweed extract.

(5) BHA.

(6) Lanolin.

2. Classification of facial cleanser

(1) Foam cleanser. Suitable for skin: oily skin.

(2) Solvent cleanser. Suitable for skin: makeup skin.

(3) Foamless cleanser. Suitable for skin: dry skin.

(4) Cosmeceutical cleanser. Suitable for skin: sensitive skin.

3. The function of facial cleanser

(1) Foam cleanser: It contains surfactant and is divided into two types: multi-foam and micro-bubble types. Active agent can resolve grease, let the skin cleaner. Soaping cleanser is one of them. But because of its obvious characteristics, it is treated differently with common active agent cleanser.

(2) Solvent cleanser: This kind of product is rely on oil and oil dissolving ability to remove oily dirt. Common solvent cleansers are some discharge makeup oil, clean frost and so on.

(3) Foamless cleanser: These products combine the characteristics of the above two types, using both the right amount of oil and some active agent.

一、洗面奶

（一）对话

（B＝美容师，C＝顾客）

B：您好！

C：你好！我想买个洗面奶，不知道哪个适合我。

B：好的。我先检测下您的皮肤，根据您的皮肤类型才能选择合适的护肤品。您的皮肤非常白皙、细腻，但皮肤比较干燥，因为缺水，眼周有小细纹，所以您的皮肤属于干性皮肤，护肤品以保湿补水的产品为主。

C：哦。那哪种洗面奶适合我呢？

B：洗面奶的主要作用是清洁、营养、保护皮肤，洗面奶可以分为两种类型：柔和洗面奶和油性皮肤洗面奶。柔和洗面奶配方温和，适合中性、干性、敏感性皮肤，而油性皮肤洗面奶配方中含油脂较少，使用后会觉得面部皮肤较干爽，所以适合油性、暗疮性皮肤。中性或干性皮肤最好不要用油性皮肤洗面奶，因为使用油性皮肤洗面奶会导致皮肤更加干燥缺水。所以，干性皮肤应选择柔和洗面奶。

C：好的。

B：这款洗面奶蕴含丰富的泡沫，是干性皮肤的最佳选择。这款洗面奶含有甜杏仁精华，能有效清洁面部的脏污，增强皮肤活力，能让皮肤水润不干燥。

C：好！就这个了！

（二）课文

洗　面　奶

洁面是美容的第一步。好的洁面产品使用后会有滋润、舒服、清爽、不紧绷的感觉。洗面奶的好坏，关键取决于它的成分。

1. 洗面奶的主要成分

洗面奶的主要成分有表面活性剂、保湿剂、活性添加剂、防腐剂、香精等。

洗面奶的其他成分有以下几种。

（1）透明质酸。

（2）芦荟萃取物。

（3）温泉水。

（4）海藻提取液。

（5）BHA。

（6）羊毛脂。

2. 洗面奶的分类

（1）泡沫型洗面奶。适合肤质：油性皮肤。

（2）溶剂型洗面奶。适合肤质：化裸妆的皮肤。

(3) 无泡型洗面奶。适合肤质:干性皮肤。

(4) 药妆洗面奶。适合肤质:敏感性皮肤。

3. 洗面奶的功能

(1) 泡沫型洗面奶:含有表面活性剂,又分为多泡沫型和微泡沫型两种。通过表面活性剂对油脂的乳化以达到清洁作用。皂剂洗面奶也是其中一类,但是由于其特征明显,所以一般会和普通表面活性剂洗面奶区别对待。

(2) 溶剂型洗面奶:这类产品是靠油与油的溶解能力来去除油性污垢的,常见的溶剂型洗面奶是卸妆油、清洁霜等。

(3) 无泡型洗面奶:这类产品结合了泡沫型洗面奶和溶剂型洗面奶的特点,既含有适量的油分又含有表面活性剂。

Ⅱ. Lotion

Part Ⅰ Dialogue

(B=Beautician,C=Customer)

B:Hello!

C:Hi! My skin is mixed skin,still have some pimples,pore is very big. I want to buy the lotion.

B:More than 85 percent of the water in the lotion is purified water, while the rest depends on the demand of different products. Common moisturizing lotion contains water-based active agent and also includes low amounts of spices, pigments, preservatives, and alcohol. More than 90 percent of the lotion on the market belongs to this category. The exfoliating lotion usually contains tartaric acid,salicylic acid and other ingredients. The toner that emphasizes convergent function often adds astringent,alcohol,etc.

C:What kind of lotion is suitable for me?

B:Different skin needs different lotion. The lotion contains salicylic acid, tartaric acid and other ingredients which are good for oily skin,acne,and horny layer thickness. But you still cannot use it for a long time. If you are dry and sensitive skin, you should avoid the lotion contains active agent,spices,preservatives,etc.

C:I see some of the ingredients in the lotion contain alcohol and spices. Do you think these will have any damage to the skin?

B:If you look at the explanation carefully, most of the lotions contain alcohol which have anti-inflammatory and calming effects. It is common in the treatment of acne. Generally speaking,lotion marks contain alcohol, but you can't smell it, when you pat the lotion on your face,you feel very cool. If you feel the smell of alcohol and you feel uncomfortable,at this time you must stop use it frequently.

C:Can a spray lotion be used to moisturize?

B:There are two types of spray lotion on the market. One belongs to the hot spring water,which is designed to be calm and soothing. For people who are allergic (such as heterotypic dermatitis) and have inflammation, volatilization of spring water can reduce the wound, reduce inflammation, and have good effect to sunburn. Another class is spray form lotion, the advantage is that you can use a little amount of lotion, avoid use the hand and this way is clean.

C: Thank you! That's all!

B: You're welcome! I hope you can choose suitable lotion!

C: Ok.

Part Ⅱ Text

Lotion

Toning water is a general term for lotion. Toner, moisturizer, lotion, although the name is different, but they all have the function of moisturizing, give skin complement moisture, soften cuticle and maintain its normal function, also has the function of antibacterial, convergence, clean, nutrition, and so on.

1. Classification of lotion

(1) Flower water: mainly extracted from different kinds of flowers, such as orchid, orange blossom and narcissus.

(2) Plant water: such as towel gourd water, cob water, small cucumber water.

(3) Active spring, mountain spring, the ice water: this water comes from high mountains. Spring water, because it contains a variety of trace minerals, so it close to skin, have composed function, anti-inflammatory effect.

(4) Yeast culture extract: rich in glycoprotein, which can activate skin and restore elasticity and compactness to skin.

(5) Deep ocean water: contains a variety of trace elements and minerals, and it contains rich magnesium, and small molecules can penetrate to the bottom of the skin, allowing the skin to moisturize from the inside to the outside.

(6) Tap water: the most readily obtained water, but the effect is pure and moisture, not too much skin care value. It is suitable for the skin without obvious aging. It is mainly used by the water shortage of the skin, because its moisturizing effect is excellent.

(7) Distilled water: this is almost no impurity, and there is no excessive mineral, but more often used to assist in the extraction of other skincare ingredients.

2. Different skin types

(1) Dry and normal skin: appropriate choice contains the lotion that contains moisture content.

(2) Oily and mixed skin: choose a lotion containing alcohol.

(3) Damaged, sensitive skin: it is advisable to use the lotion containing anti-inflammatory and calming ingredients.

二、化妆水

（一）对话

（B＝美容师，C＝顾客）

B：你好！

C：你好！我的皮肤是混合性的，还爱长痘痘，毛孔很大，今天，我想了解下化妆水。

B：化妆水中85％以上是精制过的水，其余成分则视其诉求而定。常见的保湿型化妆水中添加了水性保湿剂，还包括少量的香料、色素、界面活性剂、防腐剂及酒精。市面上大约90％的化妆水属于保湿型化妆水。去角质型化妆水中一般含有果酸、水杨酸等成分。有收敛功能的化妆水常添加收敛剂、酒精等。

C：什么样的化妆水适合我？

B：不同类型的皮肤选用不同的化妆水，含有水杨酸、果酸等治疗成分的化妆水或收敛水，对油性皮肤、有青春痘、角质层厚的人有一定效果，但不宜长期使用；干性、敏感性皮肤应避免使用含有容易引起过敏的界面活性剂、香料、防腐剂等的化妆水。

C：我看到有的化妆水成分中写的含有酒精、香料，这些成分对皮肤有伤害吗？

B：仔细看成分标识，大部分化妆水都含酒精，酒精有消炎、镇静的功效，常见于治疗青春痘的化妆水。一般来说，化妆水成分标识含酒精，却闻不到酒精味，拍上去也没有很快挥发的清凉感，代表酒精浓度较低，消费者对此不用太担心。但拍在脸上有明显的清凉感，可以闻到酒精味，出现刺激和不舒服的感觉，表示酒精浓度较高，不宜过度使用。

C：喷雾式化妆水可以用来保湿吗？

B：市面上的喷雾式化妆水可以分成两类，一类属于温泉水，以镇静、舒缓为目的。对于容易过敏（如异位性皮肤炎患者）、皮肤有炎症的人，温泉水挥发后可以收敛伤口、消炎，用于治疗晒伤。另一类只是把化妆水以喷雾形式呈现，好处是用量省，免沾手，比较清洁。

C：谢谢您！知道了这么多！

B：不客气！希望对您选择适合自己的化妆水有所帮助！

C：好的。

（二）课文

化　妆　水

化妆水是护肤用水的统称。柔肤水、保湿水、爽肤水虽然名称不同，但都以保湿为主要目的，它们给皮肤补充水分，软化角质层，保持其正常功能，还起到抑菌、收敛、清洁、提供营养等作用。

1. 化妆水的分类

（1）花水：主要根据花朵种类进行萃取，如兰花、橙花、鸢尾花、水仙等。

（2）植物水：如丝瓜水、薏仁水、小黄瓜水等。

（3）活泉水、山泉水、冰河水：此类水大多来自海拔高的山脉，甚至已结冰的泉水，因蕴含多种且大量矿物质，故其亲肤性佳，具有镇静、抗炎的疗效。

（4）酵母培养萃取液：含有丰富的糖蛋白，能活化皮肤，让皮肤恢复弹性、变得紧实。

(5) 海洋深层水:含有多种微量元素与矿物质,且其含有丰富镁质,加上微小的分子能深入皮肤底层,让皮肤从里到外水润、透亮。

(6) 自来水:最易取得的水,但效果仅为肤表补湿,没有太多的护肤价值。适合无明显老化现象的皮肤,主要供皮肤缺水困扰者使用,因为其保湿效果优异。

(7) 蒸馏水:几乎无杂质,也没有过多的矿物质,常用于协助其他护肤成分的萃取。

2. 不同肤质的化妆水选择

(1) 干性、中性肤质:宜选用含有保湿成分的化妆水。

(2) 油性、混合性肤质:宜选用含有酒精成分的化妆水。

(3) 受损、敏感性肤质:宜选用含有消炎、镇定成分的化妆水。

Ⅲ. Milky Lotion

Part Ⅰ Dialogue

(B=Beautician,C=Customer)

C:Hi! I have been a little dry lately,and I'd like to buy a bottle of milky lotion,so I'd like you to introduce a suitable milky lotion.

B:Ok! The milky lotion is a mixture of water and oil. Because water and oil are difficult to be dissolved, it is necessary to rely on emulsification technology to combine the two together. The milky lotion have a good moisturizing effect,the quality is between the make-up water and face cream,the effect is different,they cannot replace each other.

C:Which milky lotion would be suitable for my skin?

B:Different skin types have different choices. Most of the lotion is designed for dry skin,but if the lotion is labeled "refreshing",it will be less oily and more suitable for mixed or oily skin.

C:How to use it?

B:The order of smearing is very important. You can't smear it on your whole face at the same time. You can smear more lotions on the driest area first. If you haven't use eye cream, first smear some lotions around your eyes,and then two sides of the face,two sides of the mouth and jaws,and then the whole face. The dry part will get more moist.

C:How to choose a good quality milky lotion?

B:When you choose a product,you should read the product description. You can also smear some lotion on the back of your hand. If it absorbs quickly,and flow very quickly,this kind of product contains less oil. This kind of product is suitable for oil skin. If it flows very slowly,the oil content is high in this lotion.

C:Thank you!

B:You're welcome!

Part Ⅱ Text

Milky Lotion

The water in the milky lotion is very high,it can instantly moisturize skin,and replenish moisture for dry skin. Milky lotion can prevent the water flow and protect the skin from losing water. Milky lotion also contains grease, so it has certain moisture effect, but not as good as face cream. In generally, the function of milky lotion is locking in water and moisturizing the skin, which is a daily use of skin care products.

Milky lotion generally composed of oil, water, emulsifier and other basic substances, for easy storage, it usually has some fungicides, antiseptic, etc. Some also use thickening agent to make it look thick, make the appearance more delicate. If it smells aroma, that is because it has essence, essential oil and other materials.

1. Different skin types

(1) Normal skin without oil: Suggestion: choose the foundation skin cream with moisturizing effect to make skin reach ideal moist state.

(2) Oily and mixed skin in T area: Suggestion: select the moisturizing lotion with light texture, and control the oil content to the appropriate amount when adding moisture to the skin.

2. Classification of milky lotion

According to the specific efficacy, it is divided into moisturizing lotion, oil control moisturizing lotion and whitening lotion.

三、乳液

(一) 对话

(B=美容师,C=顾客)

C:你好！我最近脸有点干,想买一瓶乳液,所以想让你介绍一种适合我的乳液。

B:好的！乳液是水分与油分的混合物。因为水分和油分难以融合,所以需要依靠乳化技术将二者合而为一。乳液的滋润效果很好,质地介于化妆水和面霜之间,因其功效各不相同,所以无法互相替代。

C:请问哪种乳液适合我的皮肤呢?

B:不同类型的皮肤应选择不同的乳液。乳液大多是为干性皮肤设计的,但是如果乳液上注明是“清爽型”的话,就说明它的含油量少一些,更适合混合性或偏油性的皮肤使用。

C:乳液怎样正确使用呢?

B:乳液涂抹的顺序有先有后。乳液并不是要同时全脸涂抹,而是哪里最干就先涂抹哪里,如果还没有使用眼霜,就先涂眼周,之后是两颊、嘴角两侧、上下颚等,这些部位涂完之后再涂全脸,这样干燥的部位就得到了双重的滋润。

C:怎么选购质量好的乳液呢?

B:选择时,除了阅读产品说明书,还可将乳液涂抹在手背上,若其吸收迅速,液体的流动性好，说明油脂含量较低，非常适合出油的皮肤使用。如果在手背上流动性差，吸收没那么快，则说明油脂含量较高。

C：谢谢！

B：不客气！

（二）课文

乳　液

乳液的含水量很高，可以瞬间滋润皮肤，为干燥的皮肤补充水分。乳液可以在皮肤表面形成轻薄透气的保护膜，以防止水分流失，从而起到极佳的保湿效果。乳液中还含有油脂，所以具有一定的滋润度，但其滋润作用不如面霜。总的来说，乳液的作用是保湿、锁水和滋润，是每日必用的护肤品。

乳液一般由油、水、乳化剂等基本物质组成，为便于存放还会加入一些杀菌剂、防腐剂等添加剂。有的也会使用增稠剂稍微增稠，使外观看起来更加细腻，要是有香味，则是加入香精、精油等添加剂的缘故。

1. 不同肤质的选择

(1) 不干不油的中性皮肤：建议选择具有保湿补水功效的基础护肤乳，从而让皮肤达到理想润泽状态。

(2) T 区油腻型混合皮肤：建议选择质地轻薄的清爽补水型乳液，在补充皮肤水分的同时将油分控制在适当的分泌量。

2. 乳液的分类

按具体的功效划分为保湿乳液、控油保湿乳液和美白乳液。

Unit 4 Chinese Herbal Cosmetics

Section 1 Overview of Chinese Herbal Cosmetics

Part Ⅰ Dialogue

A:Hello!

B:Hello!

A:It looks like your skin is getting better now. Have you used any special cosmetics?

B:I've used some Chinese herbal cosmetics which I bought recently. They are pretty good.

A:Chinese herbal cosmetics? New products?

B:No,they have a long history.

A:Is there any advantage of Chinese herbal cosmetics?

B:Plenty of safe and effective,non-toxic side effects,extensive and lasting effect.

A:Wow,that's good! What are the types of these cosmetics?

B:There are many different types,such as freckle removal,whitening,moisturizing,acne removal,etc. Everyone can choose the proper cosmetics according to their skin.

A:Which type do you use?

B:Because I always have acne on my face,I choose to get acne removal cosmetics.

A:Well,I'll have a try. I'd like to buy a set of whitening cosmetics.

B:Hum,that's great.

A:Thank you very much!

B:You are welcome!

Part Ⅱ Text

The Concept of Chinese Herbal Cosmetics

Chinese herbal cosmetics are based on the basic theory of traditional Chinese medicine. which are mainly made of Chinese herbs or added some effective ingredients of TCM in chemical synthetic substances. These cosmetics make users more attractive by cleaning body,beautifying looks and improving the appearance.

Under the guide of TCM theory,Chinese herbal cosmetics have distinctive features,such as obvious function, strong pertinence. Traditional Chinese medicine is mainly based on prevention. This feature is also prominently reflected in traditional Chinese medicine cosmetics. It has the definite cosmetic effect. Being green and natural, safe and reliable, TCM has human clinical experience for thousands of years, accumulated a lot of unique and significant effects of single and compound Chinese medicine products. Compared with synthetic chemicals, Chinese medicine is safer and more reliable because it originates from nature, pure and mild, with little toxic and side effect. With complete category and various forms, Chinese herbal cosmetics cover almost all kinds of modern cosmetics, even more varied, moreover these cosmetics can adapt to different body parts, different efficacy needs and different usage preferences.

(一)对话

A:你好!

B:你好!

A:你看起来皮肤变好了,你用了什么特殊化妆品吗?

B:我用了中药化妆品,新买的,特别好用。

A:中药化妆品? 新产品吗?

B:不是,已经有很久的历史了。

A:那中药化妆品有什么优点呢?

B:优点可多了,安全有效,无毒副作用,作用广泛,效果持久。

A:呀,这么好啊! 那大概有哪些种类呢?

B:种类也有很多,有祛斑的、美白的、润肤的、祛痤疮的,每个人可以根据自己的肤质选择合适的化妆品。

A:那你用的是什么化妆品啊?

B:因为我脸上一直有痤疮,所以我选择的是祛痤疮的化妆品。

A:嗯嗯,真好。那我也去试一下,我要去买一套美白的化妆品。

B:嗯,可以。

A:嗯,谢谢你!

B:不用谢!

(二)课文

中药化妆品的含义

中药化妆品是指以中医药基础理论为指导,由中药制成或是在化学合成物质中添加中药或中药有效成分而成,具备清洁身体、美化外表、改善外貌、提升魅力作用的物质。

以中医药基础理论为指导配制生产,这应该是中药化妆品不同于其他化妆品的一个显著特点,故其功能明显,针对性强。中医以预防为主,这一特性也显著地体现在中药化妆品上,有确切的美容养颜效果。中药有着悠久的历史传承,绿色天然,安全可靠,中药拥有几千年的人体临床应用经验,留下了许多作用独特、效果显著的单品和复方,并且中药来源于大自然,纯正温和,毒副作用小,相对于化学合成品更加安全可靠,品类齐全,剂型多样,中药化妆品几乎涵盖了现代化妆品的各种类别,目前中药化妆品的剂型更加丰富多样,不同的身体部位、不同的功效需求和不同的使用偏好都有其相应的产品可供选择。

Section 2 Application of TCM in cosmetics

Part Ⅰ Dialogue

A:Oh,my skin is so rough lately. What should I do?

B:Let me see. It seems to be a bit dark,rough,and some melasma.

A:What should I do? It's not beautiful at all. Hey,why is your skin so nice and smooth and ruddy?

B:Because I often do facial masks! Here,let me show you my masks.

A:Wow,so many kinds! What is the difference?

B:You see,the main ingredients of these masks are extracted from traditional Chinese medicine. The main component of this mask is extracted from ginseng,and it can moisturize and anti-aging. This mask is mainly extracted from the pearl,which can fade the spots and whiten the skin.

Part Ⅱ Text

The Significance of Chinese Herbs in Cosmetics

Because of querying the stimulation of chemical raw materials and the safety of animal extracts,plant extracts have become the first choice in cosmetics industry. Chinese herbal extracts have developed rapidly among plant extracts. Natural plant cosmetics are the most primary trends in cosmetics research and development at home and abroad in the future. And Chinese herbal medicine is the best material for functional cosmetics. Nowadays,not only the Asians have a higher acceptance of traditional Chinese herbs, but also European and American countries are changing their health concepts, beginning to pay more attention to the application of traditional Chinese medicine in cosmetics. In the sales market, many cosmetics have Chinese herbal ingredients,because the cosmetics added with Chinese herbal medicine have shown obvious advantages. For example, angelica has a strong role in promoting blood circulation,which can inhibit tyrosinase activity effectively,and prevent the tyrosine oxidation to form pigment, therefore, it has a better effect on melasma and other pigmented dermatosis. The alkaloid in motherwort can kill and restrain all kinds of bacteria and fungi. The addition of such traditional Chinese medicine in cosmetics,which has not only been effective, but also has no dependence, is loved by consumers. Many countries have carried out research on traditional Chinese medicine cosmetics. Various natural actives contained in Chinese herbal medicines have been widely used in modern cosmetics and played various roles.

Examples of Chinese Herbs Used in Cosmetics

1. Ginseng

It is the root of herbaceous perennial ginseng in Araliaceae. Ginseng can reinforce vigour, strengthen the spleen and benefit the lung, help produce saliva and slake thirst, calm the nerves and prolong life by taking ginseng for a long time.

Ginseng, functioning as beauty and hairdressing, is widely used in cosmetics. The main cause of skin aging is bad circulation, decreased metabolism and weakened skin elasticity. Because ginseng contains a variety of ginsenoside, amino acids, vitamins and minerals, when it is added into the skin care products, it can help to promote the blood circulation of the capillaries, increase skin nutrition, prevent arteriosclerosis, regulate the skin moisture balance and so on. Ginseng can postpone skin aging, prevent skin dehydration and increase skin elasticity, so as to keep the skin delicate luster, prevent and reduce skin wrinkles. Ginseng active substances also inhibit the reduction property of the melanin, so that the skin could be white and smooth. Ginseng added in shampoo can dilate the capillaries of head, increase hair nutrition, improve hair toughness, reduce hair loss and hair broken, and protect the wounded hair.

Taking ginseng orally not only strengthens the body, but also helps to fight against aging for skin care and beauty.

2. Pearl

Pearl can be used as a fine decoration, but also as a medicinal material. It has the functions of calming the nerves, calming the liver to stop endogenous wind, removing liver fire to improve eyesight, and detoxifying to improve skin. Meanwhile pearl can nourish the skin, delay the wrinkles appearance whether it is taken by internal or external as a fancy cosmetic. It has been documented as early as *the Compendium of Material Medica*: pearl powder coated, a good moisturizing face looked. At present, many cosmetics sold in the market include pearl powder.

3. Angelica

Angelica is a traditional blood activating drug, which can regulate the menstrual

function to relieve pain and moisten dryness for relaxing bowels. Angelica can promote blood filling and flowing, so that it makes the face ruddy and glossy, the skin delicate and elastic. Angelica contains a variety of microelements, which plays important roles in nutritional skin and rough prevention. Angelica can also be used in the treatment of acne, chloasma and so on. Angelica has been widely used in skin care products, and can be taken orally through cooking, medicated wine and decoction.

4. Coix seed

Coix seed, also known as barley, is a kind of medicinal and edible plant. The function of coix seed is tonifying spleen and antidiarrheal, which can be used for the treatment of edema and dysuria. Coix seed has very high nutritional value. Compared with rice, it contains a larger number of fat, protein and total amino acid, including many kinds of trace elements. Besides anticancer, hypoglycemic, analgesia, antipyretic, and enhancing immunity, coix seed has beauty effects, such as remove stains, verruca plana and soften skin. There are special coix tablets for the removal of senile plaques, coix beauty tea, instant coix extract. Now we have produced some moisturizing cream with coix seed extract.

5. Ganoderma lucidum

Ganoderma lucidum is a kind of fungi plant, which is recognized by traditional Chinese medicine experts as a kind of precious medicinal materials to nourish, strengthen and prolong life. *Ganoderma lucidum* can not only relieve cough, expel toxins, but also help for anti-aging and agerasia. Li Shizhen in *the Compendium of Materia Medica* believes that good quality

Ganoderma lucidum can keep youth and be good for living longer. Modern studies have shown that *Ganoderma lucidum* contains a variety of trace elements, of which magnesium and zinc have anti-aging effect. Polysaccharide components in *Ganoderma lucidum* can protect cells and delay cells aging. *Ganoderma lucidum*, as blood scavenger, can eliminate melanin and brown pigment in the blood and control the formation of freckles and senile plaques.

6. Polygonum multiflorum

Polygonum multiflorum was found to have effects of protecting nerve cells, resisting skin lipid peroxidation, lowering blood fat and anti-aging. Modern medical research has proved that Polygonum multiflorum has the function of dilating blood vessels and alleviating spasm. It can make skin cells, brain cells and hair get enough blood nourishment. Therefore, the long-term use of Polygonum multiflorum not only refreshes people's spirit, but also promotes ruddy and lustrous complexion, black and shiny hair.

7. Bee products

Bees collect natural pollen to make pure natural materials such as honey, royal jelly and propolis, etc. These bee products can improve the human body immunity, enhance the body metabolism, enhance the cell vitality, beautify the face, and delay aging. Especially, they can promote skin metabolism and wound healing, even ease the discomfort of menstruation.

8. *Cordyceps sinensis*

Cordyceps sinensis, also known as Chinese caterpillar fungus, or *Cordyceps*, is a parasite fungus that lives in the bat moth larva.

Modern research of pharmacology indicates that *Cordyceps* contains 7% cordycepic acid, 28.9% carbohydrate, 8.4% fat, 25% protein and a large number of vitamins. 82.2% of fat is unsaturated fatty acid, which is fungal parasitic products. Cordycepic acid is mainly used to treat cerebral edema and keep out ARF. In general, *Cordyceps* is a tonic that helps the body build strength, improve the organic function, strengthen immunity and bring longevity. *Cordyceps* has a good cosmetic effect on human skin.

9. *Lithospermum*

The main function of *Lithospermum* is cooling blood, removing stasis and detoxification, which can accelerate the metabolism of acne and scar. So *Lithospermum* can remove acne and acne marks and can diminish inflammation. *Materia Medica for the Origin* mentioned: *Lithospermum* is helpfull to people with acne hidden, wanted to come out, red and dry, and acne out and closed, purple and black, even with black furuncle. If the pox has already spread red and lively, and the two stools are tuned, then the purple velvet is used instead, in the blood heat is not clear, in order to promote blood and contain the meaning of rising. If red, slippery, and white trap, avoid it. When it is grey and slippery, it is suitable to visit the shikonin of the insect department.

10. *Poria cocos*

Common names include poria, tuckahoe, and fuling. *Poria cocos*, a parasitic fungi in pine roots, is used as a medicinal mushroom in Chinese medicine. *Poria cocos* is sweet, light and plain in taste. It has the functions of promoting water and dampness, benefiting the spleen and stomach, and tranquilizing the mind. The ancients called *Poria cocos* "Four Seasons God Drug", because it has a wide range of efficacy, regardless of the four seasons, it can be compatible with a variety of drugs, regardless of cold, temperature, wind, dampness and other diseases, can play its unique role. According to the records in "*Bencan Pinhui Jingyao*", *Poria cocos* can remove the long sores and spots on the face, and has a good effect on removing the black spots caused by women after childbirth.

In addition to the above traditional Chinese medicine, the commonly used beauty Chinese medicine is *Astragalus mongholicus*. *Acanthopanax senticosus*, it has anti-aging, beauty, fitness, known as ginseng, delaying and lightening skin aging, reducing pigmentation, nourishing the beauty of the *Rehmannia glutinosa*. Yan's Ophiopogon japonicus, long serving light body is not old, not hungry, and moisturizing; and Huangjing, peach kernels, almonds, cypress kernels, Chuanxiong, aloe, chrysanthemum, winter melon kernels, *Angelica dahurica*, Fangfeng, Xinyi, *Schisandra chinensis*, *Xanthium sibiricum*, peach blossom, asarum, etc.

（一）对话

A:呀,我最近皮肤粗糙了好多,怎么办?

B:我看看,好像是有点黑了,比较粗糙,还有些黄褐斑。

A:那我该怎么办,一点都不漂亮了,咦,你皮肤怎么这么好,又光滑又红润。

B:因为我经常做面膜啊！来,我给你看看我的面膜。

A:哇,好多种,这些有什么不同呢?

B:你看这些面膜的主要成分是从中药里提炼而来的。这个面膜的主要成分是从人参里提炼出来的,可以保湿、抗衰老;这个面膜主要成分是从珍珠里提炼出来的,可以淡化色斑,美白。

（二）课文

中药应用于化妆品的意义

随着人们对化学原料的刺激性、动物提取物安全性的质疑声越来越大,植物提取物已经成为当今美容行业制造化妆品的首选成分。在植物提取物中,中草药提取物的应用发展迅速。天然植物类化妆品是今后国内外化妆品研发的最主要趋势,而中草药则是功能性化妆品的最好原料。目前,不仅亚洲人对传统中草药的接受程度较高,而且欧美等国家的健康观念也在改变,开始重视传统中药在化妆品中的应用。目前在销售市场上,含有中药成分的化妆品很多,原因是添加中药成分的化妆品已经显示出明显的功效优势。比如:当归具有较强的活血作用,能有效地抑制酪氨酸酶的活性,防止酪氨酸酶氧化形成色素,故对黄褐斑等色素沉着性皮肤病有较好的疗效;益母草中的生物碱能很好地杀灭和抑制各种细菌、真菌。诸如此类添加中药成分的化妆品,既有良好的功效又没有依赖性,因此受到广大消费者的喜爱。很多国家开展了对中药化妆品的研究,中草药中所含的各种天然活性物,已经被广泛地应用于现代化妆品。

化妆品中常用的中药

1. 人参

人参为五加科多年生草本植物的根茎。人参能够大补元气,补脾益肺,生津止渴,安神益智,久服能够轻身延年。

人参具有美容、美发的作用,在化妆品中被广泛应用。人体皮肤衰老的主要原因是血液循环不良、新陈代谢减慢、皮肤弹性减弱。由于人参中含有多种人参皂苷、氨基酸、维生素及矿物质,将人参中的有效成分添加在护肤品中具有促进毛细血管的血液循环、增加皮肤的营养供应、防止动脉硬化、调节皮肤水分平衡的作用。所以它能延缓皮肤衰老,防止皮肤干燥脱水,增加皮肤的弹性,从而起到维持皮肤光泽柔嫩,预防和减少皮肤皱纹的作用。人参活性物质还具有抑制黑色素的还原性能,使皮肤洁白光滑。

人参应用在洗发剂中能使头部的毛细血管扩张,可增加头发的营养,提高头发的韧性,减少脱发、断发,对损伤的头发具有保护作用。

人参内服,既可强身,也能起到抗衰老及护肤美容的作用。

2. 珍珠

珍珠既可作为精美的装饰品,还可作为药材,具有安神定惊、镇肝熄风、清肝明目、解毒生肌等功效。同时还可作为美容佳品，无论是内服还是外用均对皮肤有滋养保健、延缓皱纹产生的作用。早在《本草纲目》中就有记载：珍珠……涂面，令人润泽好颜色。目前市售的许多化妆品中都加有珍珠粉。

3. 当归

当归是传统的补血活血药，能够调经止痛、润燥通便。当归可以使气血充盈，血流通畅，故使面色红润、光泽、皮肤细嫩富有弹性。当归中含有多种微量元素，对营养皮肤、防止皮肤粗糙起着重要作用。当归还可以用于粉刺、黄褐斑、雀斑等的治疗。当归外用已被广泛用于护肤霜，也可煎煮、泡酒或烹制药膳内服。

4. 薏苡仁

薏苡仁是药食两用植物，又称薏米。薏苡仁能够健脾渗湿，除痹止泻，用来治疗水肿、小便不利。薏苡仁有很高的营养价值，它含有的脂肪、蛋白质、总氨基酸量等超过了大米，并含多种微量元素。薏苡仁不仅具有抗癌、降糖、镇痛、解热、增强免疫等作用，而且有美容作用，能够祛色斑、除扁平疣、柔嫩肌肤。目前有专门祛老年斑的薏苡片、薏苡仁美容茶、速溶薏仁精等。目前亦有商家生产添加薏苡仁提取物的润肤霜。

5. 灵芝

灵芝是一种菌类植物，是历代中医药学家公认的滋补强壮、延年益寿的珍贵药材。灵芝不仅具有润肺止咳、护肝解毒等功效，而且具有抗老防衰、驻颜的作用。李时珍在《本草纲目》中认为灵芝“好颜色，久服轻身不老延年”。现代研究表明，灵芝含有多种微量元素，其中镁、锌有延缓衰老的作用；灵芝所含的多糖成分能消除体内自由基，保护细胞并延缓细胞衰老；灵芝还是血液的清道夫，能清除血液中的黑色素、褐色素，控制雀斑、老年斑的形成。

6. 何首乌

何首乌具有保护神经细胞、抗氧化、降血脂、抗衰老、美容驻颜等功能。现代医学研究证明，何首乌有扩张血管和缓解痉挛的作用，能使皮肤细胞、脑细胞和头发获得足够的血液滋养。故长期用何首乌不仅使人精神焕发，还可促使面色红润有光泽，头发乌黑发亮。

7. 蜂蜜制品

蜜蜂采集纯天然花粉，酿成蜂蜜、蜂王浆、蜂胶等纯天然物质，长期服用能够提高人体免疫力，增强机体新陈代谢，增强细胞活力，美容养颜，延缓衰老，并对促进免疫系统的抵抗力，加速皮肤更新，缓解妇女经期不适，促进伤口愈合具有显著功效。

8. 冬虫夏草

冬虫夏草简称虫草，又名冬虫草或虫草。现代药理研究表明，冬虫夏草含有约 7％的虫草酸，28.9％的碳水化合物，约 8.4％的脂肪，约 25％的蛋白质，其中脂肪中 82.2％为不饱和脂肪酸，是冬虫夏草真菌的寄生产物。虫草酸主要用于治疗脑水肿和阻止急性肾衰竭。一般来说，冬虫夏草为滋补强壮的名贵中药，能够改善器官功能，增强免疫力，延长寿命。冬虫夏草对人体皮肤有很好的美容作用。

9. 紫草

紫草主要功能为凉血、活血化瘀、解毒透疹，因此能加速痘印和瘢痕的新陈代谢。紫草具有显著的祛痘、祛痘印和消炎的作用。《本草求原》中载：紫草，痘疹隐隐，欲出未出，色赤干枯，及已出而便闭、色紫黑者宜之，痘夹黑疔亦宜。若痘已齐布红活，二便通调，则改用紫草茸，……于血热未清，用以活血而寓升发之义也。若红活，二便滑，及白陷者，忌之。至灰滞而便滑，则又宜虫部之紫草茸，宜参观之。

10. 茯苓

茯苓，俗称云苓、松苓、茯灵。其原生物为多孔菌科真菌茯苓的干燥菌核，多寄生于马尾松或赤松的根部。主产于云南、四川等地。茯苓味甘、淡，性平，入药具有利水渗湿、益脾和胃、宁心安神之功用。古人称茯苓为“四时神药”，因为它功效非常广泛，不分四季，将它与各种药物配伍，不管寒、温、风、湿诸疾，都能发挥其独特功效。据《本草品汇精要》中记载茯苓可以很好地祛除人脸上长的疮、斑，以及对祛除妇女产后所生的黑斑有很好的疗效。

除上述中药外，常用美容的中药还有补气驻颜的黄芪；具有抗老、美容、健身作用，号称“赛人参”的刺五加；延缓皮肤老化、减少色素沉着、养血美颜的地黄；久服轻身不老、润肤悦颜的麦冬等；以及黄精、桃仁、杏仁、柏子仁、川芎、芦荟、菊花、冬瓜仁、白芷、防风、辛夷、五味子、苍耳、桃花、细辛等。

Unit 5　Living Beauty

Living beauty is a non-invasive medical treatment, which uses non-medical means such as cosmetics, health care products and non-medical instruments.

The beauty of life is derived from the service industry such as fashion, image design, makeup, modification techniques, barber and massage. It includes the following contents: image design, sales of cosmetics and makeup techniques, hair design, skin care and massage. Here we focus on skin care.

生活美容是运用化妆品、保健品和非医疗器械等非医疗手段，对人体进行的诸如皮肤护理、按摩等带有保养或保健性质的非侵入性的美容护理。

生活美容是从时装、形象设计、化妆、修饰技巧、理发和按摩等服务行业发展而来的，因此生活美容包括以下内容：形象设计、化妆品销售及化妆技巧、发型设计、皮肤护理、保健按摩。这里我们重点讲述皮肤护理。

Section 1　Normal Skin Care

Part Ⅰ　Dialogue

(B=Beautician, C=Customer)

B: Hello! What can I do for you?

C: As I grow older, my skin does not look as good as before. Therefore, I want to do skin care.

B: Let me help you do skin analysis and judge your skin type.

(Skin analyzer is working...)

B: Your skin is good, normal skin, which is the most perfect type.

C: What is the most perfect skin?

B: Your skin is healthy and perfect type, which is more common among preadolescent girls. This skin has a moderate amount of sebum secretion. The skin is not oily, not dry, ruddy, delicate, and elastic. The skin has proper thickness. It has small pores and is not sensitive to outside stimulation.

C: If there is no problem with normal skin, why is my skin getting old in recent years?

B: Normal skin is relatively healthy compared to other skin types, but its tolerance of the outside world changes with age. In summer it can be slightly oily while in winter it will be a little dry. After the age of thirty it will turn into dry skin.

C: How should I care my skin?

B: The key is the cleaning and moisturizing. The most important factor in skin is age, followed by sunlight, climate, and air pollution. So with aging even normal skin will get worsen, but moisturizing can delay skin aging. The focus of maintenance should be on cleaning, refreshing, moisturizing and weekly massage.

C: What kind of skin care products can I choose?

B: You can choose according to skin age and seasons. In summer you can choose according to hydrophilicity while in winter by moisture.

Part Ⅱ Text

Care Methods of Normal Skin

Skin maintenance is divided into domestic daily care and care in the beauty salon.

Routine skin care products are as follows:

(1) Cleanser: clean the skin surface dust.

(2) Cleansing cream: thoroughly clean the skin surface makeup, open pores, deeply clean the residue and metabolites in pores.

(3) Lotion: regulate skin pH balance, replenish moisture, make the skin soft and comfortable.

(4) Oil emulsion: regulate oil and water balance, supplement nutrition, lock the moisture, keep skin moist and shiny.

(5) Cream: ①Forming a cortical membrane on the skin surface to effectively protect the skin. ②Replenish nutrition to prevent rough skin. ③Moisturized, make skin softer.

(6) Essence: ① Highly replenish nutrition. ② Promote the metabolism of the skin. ③Regulate skin sebum secretion.

(7) Massage cream: promote blood circulation of skin, replenish nutrition, make the skin more dynamic, healthier and more flexible, and the function of the skin more smooth.

1. Domestic daily care

Domestic daily care can be done in the morning and evening.

In the morning, the nursing process is: cleanser→lotion→massage cream→lotion→oil emulsion→sunscreen→wet powder→dry powder→paint eyebrow→eye shadow, mascara→lipstick→lip gloss→blush.

Night nursing process is: cleansing cream→moisturizing cleanser→lotion→mask→lotion→oil cream→essence.

Normal skin should be applied more lotion whether in the morning or in the evening.

2. Care in the beauty salon

Procedures:

(1) Preparation work. The beautician does a good job in personal hygiene, equipment and the circuit, and puts all the skin care products, tools, equipment on the small cart,

inspects the instrument, and disinfects all tools with 75% alcohol. Put customer's coats and other items in the store, carry or store valuable items. Put a clean towel on the bed first, and two towels for covering the head and chest.

(2) Removing makeup. Take the right amount of cleanser, apply it evenly on the face and neck with cotton pads or fingertips, gently massage in circles. Clean the nose in a spiral way from the outside to the inside. Remove the foundation on the neck with tissue paper or cotton pads.

(3) Cleansing. First, wet face with warm water, take proper amount of cleanser and apply it evenly on the face and massage. Keep it on face for one and an half minutes and wash it with clean water.

(4) Steaming and cold spray. Moisture the skin and soften the skin cutin by using cold spraying machine, which is good for the absorption of the essence.

(5) Removing cutin. Remove horny skin cutin according to the thickness and skin color. If the horny layer is too thin, the skin will become sensitive and has poor resistance.

(6) Use beauty equipment or facial massage.

(7) Mask. Choose the mask with good moisturizing effect.

(8) Basic maintenance. Nourish the skin with lotion and cream.

(9) Finish work and organize supplies.

(一)对话

(B＝美容师,C＝顾客)

B:您好,请问有什么可以帮您?

C:我觉得我近来年龄大了,皮肤没有以前那么好了,所以想做下护理。

B:让我来帮您做下皮肤检测,看看您的肤质。

(皮肤检测仪检测中……)

B:您的肤质还不错,属于中性皮肤,也是最完美的肤质。

C:什么是最完美的肤质?

B:最完美的肤质就是说您的这种中性皮肤是健康理想的皮肤。这种皮肤多见于青春发育期前的少女,皮脂分泌量适中。皮肤既不干燥也不油腻,红润细腻。富有弹性,厚度适中,毛孔较小,对外界刺激不敏感。

C:中性皮肤既然没有问题,为什么我这几年的皮肤会变老?

B:中性皮肤较其他类型皮肤健康,但是中性皮肤对外界环境的耐受力会随着年龄的增长发生变化。夏天皮肤稍呈油性,冬天皮肤常会偏干。三十岁后中性皮肤可变为干性皮肤。

C:我的皮肤应该怎么护理?

B:中性皮肤护理的要点是做好清洁,注意保湿。影响皮肤最首要的因素是年龄,其次是阳光、气候、大气污染。所以随着年龄的增长即使是中性皮肤也会出现问题,而保湿可以延缓皮肤衰老。保养重点在于注意日常清洁,以保持皮肤清爽、滋润,并注意以周为单位定期进行按摩护理。

C:我应该选择什么样的护肤品呢?

B:根据皮肤年龄、季节选择不同的护肤品,夏天选择亲水型护肤品,冬天选择滋润型护肤品，选择范围比较广泛。

（二）课文

中性皮肤的护理方法

皮肤的养护分为家居日常护理和美容院护理。

常规的护肤品有以下几种。

（1）洗面奶：清洁皮肤表面的灰尘。

（2）清洁霜：彻底清洁皮肤表面的彩妆，打开毛孔，深层清洁毛孔内的残留物、代谢物，使皮肤更加洁净。

（3）润肤水、柔肤水：调节皮肤 pH 值，补充水分，使皮肤柔和、有舒适感。

（4）乳液：调节水油平衡，补充营养，锁住水分，使皮肤更加滋润，有光泽。

（5）营养霜：①在皮肤表面形成皮脂膜，有效保护皮肤。②补充营养，防止皮肤粗糙。③保湿，让皮肤更柔软。

（6）精华液：①高效补充营养。②促进皮肤新陈代谢。③调节皮脂分泌，改善皮肤。

（7）按摩霜：促进皮肤血液循环，补充营养，使皮肤更有活力，更健康，更有弹性，皮肤功能更加顺畅。

1. 家居日常护理

家居日常护理分为早上护理和晚上护理。

早上护理的流程如下：洗面奶→润肤水→按摩膏→润肤水→乳液→隔离霜→湿粉→干粉→眉笔→眼影、睫毛膏→口红→唇彩→腮红（其中后面七个步骤是在需要化妆的情况下进行的护理）。

晚上护理的流程如下：清洁霜→洗面奶→润肤水→面膜→润肤水→乳液→精华液。

中性皮肤的护理，不论在早上还是在晚上，都要加强补水。

2. 美容院护理

美容院一般护理程序。

（1）准备工作。美容师做好个人卫生，将全部所需的护肤品、美容工具、仪器整齐地放在随手可取的小推车上，检查仪器电路是否通畅，并将用品、用具使用 75% 的酒精消毒；将顾客的外衣等物品放在存放处，贵重物品嘱顾客随身携带或暂存。先放一条清洁毛巾在床头，再分别用两条毛巾将顾客的头部和胸部包盖好。

（2）卸妆。取适量的清洁霜，用化妆棉或指尖均匀地涂于面部、颈部，以打圈的方式轻柔按摩。鼻部以螺旋式由外向内轻揉，颈部由下向上清洁，最后用面巾纸或化妆棉拭净，直至面巾纸或化妆棉上无粉底颜色。

（3）清洁。面部先用温水打湿，再取适量的洗面奶在手心打圈后（如果是泡沫型洗面奶要打出泡沫），均匀地涂抹于面部并按揉。洗面奶在面部停留时间约为 1.5 min，最后用清水冲洗干净。

（4）蒸面，面部冷喷。使用冷喷机进行面部冷喷，软化皮肤角质，有利于皮肤对护肤精华的吸收。

（5）脱角质。根据皮肤角质层的薄厚、面部肤色情况适当地选择脱角质产品。注意：角质层太薄的皮肤抵抗力差，对外界环境敏感。

(6) 使用美容仪器护理或者进行面部按摩。

(7) 敷面膜。选取保湿效果好的面膜。

(8) 基本保养。滋润营养,利用爽肤水、润肤霜保养、滋润皮肤。

(9) 结束工作,整理用品、用具。

Section 2 Problem Skin Care

The problem skin is divided into five categories: dry skin, oily skin, sensitive skin, telangiectasia skin and aging skin.

我们的皮肤除了相对正常的中性皮肤外,问题性皮肤大致分为五类:干性皮肤、油性皮肤、敏感性皮肤、毛细血管扩张类皮肤、衰老性皮肤。

Ⅰ. Dry Skin Care

Part Ⅰ Dialogue

(B=Beautician, C=Customer)

C: Can you tell me what type of skin do I have, dry skin or oily skin?

B: Let me see. Yours is dry skin.

C: What does dry skin looks like?

B: Dry skin is delicate and thin without obvious pores. It has less and even sebum secretion without a greasy look. It looks clean, delicate and beautiful. The skin is not prone to acnes and has strong adhesion. The makeup can last longer in the face.

C: Sounds good.

B: But dry skin cannot withstand external stimulation. When stimulated there will be red tide, and it's easy to age due to burning. Wrinkles can easily appear around the eyes and mouth. It is prone to freckles and dry, fragile and sensitive.

C: How to care dry skin?

B: Moisture is the key to the care of dry skin. Clean the skin with warm water before bed and massage for three to five minutes to improve the blood circulation of the face. Then apply appropriate amount of night cream. In the next morning, apply emulsion or cream after cleaning the face to insure skin moist.

Part Ⅱ Text

Care Methods of Dry Skin

Dry skin looks delicate and thin without obvious pores. It has less and even sebum secretion without a greasy look. The skin is dry and looks clean, delicate and beautiful. The skin cuticle moisture is less than 10%, with less sebum secretion.

Wrinkles can easily appear around the eyes and mouth. It is prone to freckles and dry, fragile and sensitive. Dry skin can be divided into two categories, lack of water and lack of oil. The skin will worsen under the circumstances of wind, sun, dry air and air-conditioned environment. Wrinkles will appear if there is not enough care. Therefore, proper care is essential to recover its normal function to preventing aging at an early age.

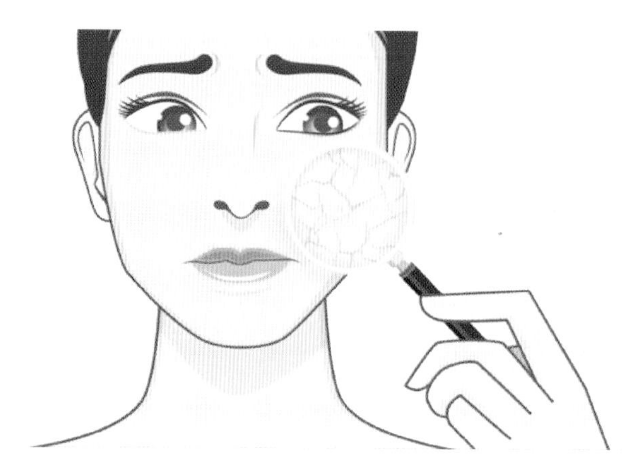

1. Dehydrated dry skin

Some people even don't know they have dry skin, because there is no obvious problem to the sebaceous glands. Their skin lacks water because of improper care or other reasons. Thus, the balance between the skin moisture and sebum is broken, which in turn stimulates the secretion of sebaceous glands, forming an atmosphere of "oily outside and dry inside". So the key to the problem is to hydrate your skin. Many people control oil blindly when they see their face glossy. They think that they have oily skin, which is actually wrong. In fact, water-deficient dry skin is the most taboo to use strong oil control products and oil absorbent paper!

How to care for dry skin at ordinary times? Because strong oil control products and oil absorbent paper can only temporarily degrease, there is no oil protection on the face, sebaceous glands began to work crazily again, and soon, oil light reappeared. Water supplement is the kingdom of dry skin care for water deficiency. As long as the skin is not dry, the shiny look will naturally disappear.

2. Dry skin lacking oil

People with this type of skin often know it because of their sebum gland secretion. Their skin looks dry without timely hydrating and locking water. The skin looks dull and sensitive. The key is to replenish water and oil at the same time, because the sebaceous glands of this type of skin cannot produce enough oil to lock the water. If so, the water will disappear quickly, creating the phenomenon of "the more water, the drier skin".

In conclusion, here are dry skin care tips.

(1) The most important thing is to ensure adequate skin moisture. The first choice when cleaning the skin should not be the products containing alkaline substances. Washing

the face with soaps containing glycerin and do not use crude soaps, or wash face with clean water in case the secretion of sebum is inhibited and the skin is made drier. Emulsion or lotion had better be applied in the morning to moisturize the skin. Then apply enough lotion and nutrition cream. In the evening, apply enough emulsion, lotion and nutrition cream.

(2) When cleaning face, you can add 2 drops of moisturizing essence oil if your cleanser doesn't contain moisturizing ingredients or add half a pot of hot water in the wash basin, 2-3 drops of rose essence oil or lavender essence oil. Then stir it. Cover a big towel over your face, close your eyes, make sure the essence oil or water vapor will not irritate your eyes. Breath with mouth and nose in turn for 5 minutes. Then clean your face again. Finally you can get a surprising effect.

(3) The key to dry skin care is to ensure the skin to get sufficient water. Clean the skin before bed, then massage for 3-5 minutes to improve the blood circulation and apply proper amount of night cream. The next morning, apply lotion or nutrition cream to maintain the skin moisture. Hydrate skin thoroughly by replenishing water to cells. Finally, make sure the water is locked in case of water loss.

Many people only pay attention to the use of skin care products, while ignoring the daily habits of life. Remember: people with dry skin don't drink caffeine beverages, eat more food rich in vitamins, such as milk, bananas, carrots and so on. Remember to drink at least eight cups of water. Drink less water before bed and put a basin of water at the bedside.

(4) Recommendation of nursing products.

①Use a skin cleanser suitable for dry skin both day and night: facial cleanser, toner, etc.

②1-2 times special nursing every week: facial cleanser, skin killer, massage cream, mask, toner, night cream and so on.

③Clean with cleanser without soap but with more moisturizing ingredients.

④The use of cosmetics with more moisturizing ingredients and good moisturizing effect can make the blood circulation of facial skin good. Massage can be used to accelerate skin metabolism.

⑤First aid tip for dry skin is apply the spa mask.

一、干性皮肤的护理

(一) 对话

(B=美容师，C=顾客)

C：你能告诉我，我的皮肤属于什么类型吗？干性还是油性皮肤？

B：让我看一看。你是干性皮肤。

C：干性皮肤是什么样子的？

B：干性皮肤肤质细腻，较薄，毛孔不明显，皮脂分泌少、均匀，没有油腻感。皮肤比较干燥，看起来比较干净、细腻且美观。这种皮肤不易生痤疮，且附着力强，化妆后不易晕妆。

C：干性皮肤听起来还不错。

B：但干性皮肤经不起外界的刺激。受刺激后皮肤会出现潮红，甚至灼痛；容易老化起皱纹，特别是在眼周、嘴角处；易长雀斑，缺水，脆弱和敏感。

C：干性皮肤怎么护理？

B：干性皮肤护理最重要的一点是保证皮肤得到充足的水分。睡前可用温水清洁皮肤，然后按摩 3～5 min，以改善面部的血液循环，并适当地使用晚霜。次日清晨洁面后，使用乳液或营养霜来保持皮肤的滋润度。

（二）课文

干性皮肤的护理方法

干性皮肤的肤质细腻、较薄，毛孔不明显，皮脂分泌少、均匀，没有油腻感。皮肤比较干燥，看起来比较干净、细腻且美观。干性皮肤的角质层水分低于 10%，皮脂分泌量少。

干性皮肤最明显的特征是皮脂分泌量少，皮肤干燥、白皙、缺少光泽，毛孔细小且不明显，容易产生细小的皱纹，毛细血管表浅，易破裂，对外界刺激比较敏感，皮肤易生红斑。干性皮肤可分为缺水性干性皮肤和缺油性干性皮肤两大类。在寒风烈日、空气干燥的环境和持续在空调环境工作情况下，干性皮肤缺水的情况会更加严重。如果干性皮肤长期不加以护理会产生皱纹，所以干性皮肤必须通过适当的皮肤护理促使其恢复正常的生理功能，以防止未老先衰。

1. 缺水性干性皮肤

这类皮肤的顾客有些根本不知道自己属于干性皮肤，因为他们的皮脂腺没有问题，只是由于护理不当或其他原因造成皮肤极度缺水。皮肤内部水分含量与皮脂分泌量失去平衡，导致皮肤反馈性地刺激皮脂腺分泌增加，造成一种“外油内干”的局面。所以说，补水才是王道（很多人看到自己满脸油光就盲目控油，认为自己是油性皮肤，其实是不对的）。其实，缺水性干性皮肤最忌讳用强效控油产品和吸油纸！

干性皮肤平时要如何护理呢？因为强效控油产品和吸油纸只能暂时去油，脸上没有了油脂的保护，皮脂腺又开始疯狂工作，不一会儿，油光重现。补水、补水、再补水，才是缺水性干性皮肤护理的王道。只有皮肤不缺水，油光才能自然而然地消失。

2. 缺油性干性皮肤

这类皮肤的人通常都知道自己是干性皮肤，因为他们皮脂腺分泌皮脂较少，皮肤因为不能及时、充分地锁住水分而导致皮肤干燥，皮肤缺乏光泽，对外界刺激比较敏感。缺油性干性皮肤需要注意在选择护肤品时不能单纯考虑补水，还要考虑补充油脂。因为这类皮肤的皮脂腺先天不足，不能分泌足够的油脂，只单纯补水，皮肤没有锁水能力，补得快，蒸发得也快，就会造成“越补越干”的恶性循环。

总而言之，干性皮肤的护理要点有以下几点。

（1）干性皮肤护理最重要的一点是保证皮肤得到充足的水分。首先在选择清洁类护肤品时，宜用不含碱性物质的膏霜型洗面奶，可选用对皮肤刺激小的含有甘油的香皂，不要使用粗劣的肥皂洗脸，有时也可不用洗面奶，只用清水洗脸，以免抑制皮脂和汗液的分泌，导致皮肤更加干燥。早晨，宜用乳液滋润皮肤，再用收敛性化妆水调整皮肤，涂足量营养霜。晚上，要用足量的乳液、营养性化妆水、营养霜。

（2）清洁面部时，如果洗面奶没有滋润成分，或是洗面后感觉面部比较干燥或紧绷，可以在洗面奶里加入2滴保湿润肤的精油(复方精油),或在脸盆里加入半盆热水,滴入2~3滴玫瑰精油或薰衣草精油,将精油充分搅匀后,用毛巾将整个面部覆盖,闭上眼睛,避免精油香味及水蒸气刺激眼睛,用口、鼻交替呼吸,维持5 min,再洗面,会得到意想不到的效果。

(3) 干性皮肤保养最重要的一点是保证皮肤得到充足的水分,睡前可用温水清洁皮肤,然后按摩3~5 min,以改善面部的血液循环,并使用适量的晚霜。次日清晨洁面后,使用乳液或营养霜来保持皮肤的滋润。干性皮肤重要的就是补水,但是补水要彻底,要做到给细胞补水。这样才能对皮肤的排泄和营养起到一定作用。最后补了水,还要锁住水,不要让补的水流失!

许多人只注重护肤品的使用,忽略了日常生活中的习惯,切记:干性皮肤者谢绝含咖啡因的饮料,宜多吃富含维生素A的食物,如牛奶、香蕉、胡萝卜等。每天喝至少8杯水,睡觉前可以喝少量水,另外最好在床旁放一盆水。

(4) 护理产品推荐。

①早晚使用1次适合干性皮肤的洁肤护肤品:洗面奶、爽肤水等。

②每周做1~2次特殊护理:洗面奶、去死皮素、按摩霜、面膜、爽肤水、晚霜等。

③用不含皂类及含有较多滋润成分的洁面品清洁。

④使用含滋润成分较多及保湿效果好的化妆品可令面部皮肤的血液循环良好,按摩可以加速皮肤的新陈代谢。

⑤干性皮肤急救方法是敷水疗膜。

Ⅱ. Oily Skin Care

Part Ⅰ Dialogue

(B=Beautician,C=Customer)

C:Hello,my skin is always greasy,which often ruins the makeup. How should I do the skin care?

B:Don't rush to care the skin after getting up in the morning. Use a blotting paper to press the nose. If there is a lot of oil in the absorbent paper,it shows your skin is oily.

C:Yes,it appears lots of oil in my face,especially around the nose when getting up in the morning. Even if the face is cleaned,oil will secrete quickly again.

B:According to your skin conditions,I think you have oily skin.

C:Then what are the characteristics of oily skin?

B:Oily skin features dark skin color,large coarse pores and vigorous sebum secretion. In the most serious cases,orange peel skin can even appear. Oily skin is weak acid with a pH value ranging 5.6 to 6.6. This kind of skin is very easy to be infected. Such poor skin is common in adolescence. Oily skin can also be divided into two types:general oily and super oily. Oily skin also has advantages:not sensitive,and can keep "young" for a long time. With the increase of age,the secretion of oil will be gradually reduced,and some even turn into normal skin.

Part Ⅱ　Text

Care Methods of Oily Skin

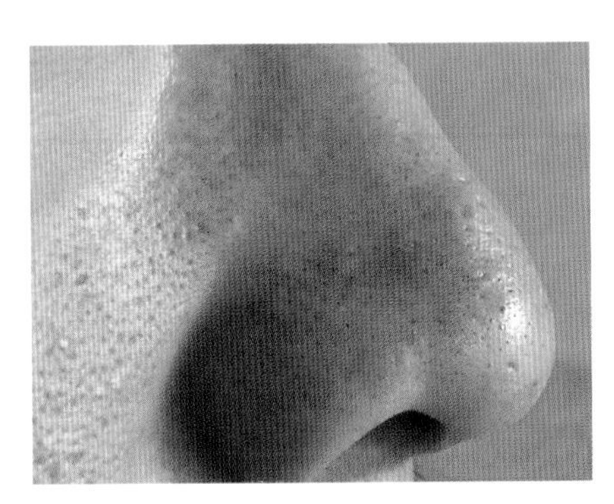

Oily skin features a shiny look, and the major feature of it is large pores and always oily in spite of your good care. In fact, the care of oily skin is very simple. You just need to do some basic procedures and pay attention to some small details and then your skin will not be so oily.

1. Cleansing

Cleansing is of great importance to the oily skin as the first step of the three basic skin care steps. The following steps will be of good effect if the first step is well done. The characteristic of oily skin is the vigorous oil secretion, especially the T zone. So choosing the cleansing products with strong cleansing effect is very important. Keep the water at 40 ℃. Use strong soluble cleanser to remove the sebum and adjust skin pH level. After washing the face apply astringent lotion. Some too astringent cleansing products will destroy the skin's sebum film, so the choice of products had better be foam and gel cleanser.

Do not wash face too many times per day. Three times a day will be proper. Too much cleaning will cause dehydration of skin. When cleaning, apply cleanser on the palm and rub until foam appears, then carefully clean the T zone. As for the zone with acne, gently rub the face in circles. Then clean the face with water repeatedly.

Apply the foam cleanser on the palm of the hand and rub until foam appears. Then apply it evenly in the face and massage in circles. When choosing cleanser, two factors should be considered.

(1) Whether it can remove the oil thoroughly.

(2) You don't feel strained in the face.

2. Moisturizing

Remember the principle of controlling oil and moisturizing at the same time. Actually the more oily the skin is, the more lack of water it is. So do remember to choose the oil-control emulsion or cream or moisturizing lotion. You had better not apply skin care products when go to sleep to insure the normal excretion of skin. After cleaning the skin at night, massage the skin properly to improve the skin blood circulation and adjust the physiological functions of the skin. To clean the skin pores thoroughly, do facial massage and facial mask once a week.

3. Mask

Mask care is very important, but you should know that oily skin mask mainly focus on moisturizing or cleansing. One time a week.

Oily skin is in need of more care to the pores, oily grease can cause accumulation of waste into the pores, which will form acnes and blackheads after some time. Masks with moisturizing and cleansing include seaweed mask and aloe mask.

4. Choose suitable hydrating products according to different seasons

According to the seasons, we need to apply hydrating masks in dry seasons to keep the balance between water and oil. For oily skin, you had better use as less makeup as possible, especially oily makeup, which may block the pores. Moisturizing skin care products are better choices.

5. Diet

Why is oily skin so oil? The first reason is gene. Second, it has something to do with our diet. Such food as chocolate, coffee, cream and other high calorie substances and seafood, spicy food, alcohol and other irritating foods will stimulate the secretion of a large amount of oil. Eat less food containing fat or high carbohydrate. Don't drink alcohol or eat too much spicy food. You should eat more fruits and vegetables to improve the rough and oily skin.

二、油性皮肤的护理

(一) 对话

(B=美容师,C=顾客)

C:您好,我的皮肤总是油腻腻的,化妆之后容易晕妆,我应该怎么护理?

B:早晨起床后先不要急于护理皮肤。先用一张吸油面纸在鼻子上轻按一下,如果吸油纸上显示有很多油,那就显示你的皮肤是油性的。

C:是的,每天早上起来面部油腻,尤其是鼻子周围。洗完脸后皮肤用不了多久就分泌出油性物质。

B:根据您的皮肤状况,判断您属于油性皮肤。

C:油性皮肤有什么特点?

B:油性皮肤的肤色一般偏深、毛孔较粗较大、易堵塞、皮肤油腻、皮脂分泌量大。严重者面部可出现橘皮现象。皮肤一般呈弱酸性,pH 值为 5.6~6.6。这类皮肤很容易被感染,抵抗细菌的能力较差。此类皮肤常见于青春期发育的年轻人。油性皮肤还可细分为一般油性和超油性。油性皮肤的优点是对外界环境不敏感并且能够长期保持“年轻”。随着年龄的增长,油脂的分泌也会逐渐减少,有些人甚至会转变为中性皮肤。

(二) 课文

油性皮肤的护理方法

油性皮肤,脸上总是泛油光,其主要特点是毛孔粗大,感觉怎么护理都油。其实,油性皮肤的护理非常简单,只要做好基础步骤,并注意一些小细节,就不用担心油光满面了。

1. 清洁

作为基础护肤三部曲中的第一步,清洁对于油性皮肤来说相当重要。如果将第一步做好，那么后续相应的工作就能更轻松地解决。油性皮肤的特点是易出油，尤其是T字部位，所以在做清洁时一定要选取清洁力强的产品，如此才能将油光扫除。洗脸的水温应在40 ℃左右，使用水溶性、清洁力强的洗面奶，主要用于清除油脂和调整皮肤酸碱值。洗完脸后可用一些孕妇专用收敛水。清洁力过强的产品会破坏皮肤的皮脂膜，所以在选择产品时要选择泡沫类、凝胶类清洁产品。

每天洗脸次数不宜过多，一天三次即可。洗脸次数过多不仅达不到控油的目的而且还会导致面部皮肤脱水。洗脸时，将洗面奶放在掌心上搓揉起泡，再仔细清洁T字部位，对于长痤疮的地方，则用泡沫轻轻地划圈，然后用清水反复冲洗。

在清洁的过程中，掌握好步骤与方法也相当重要，将泡沫型洗面奶放在手掌中搓揉出泡沫，然后均匀涂于面部，进行画圈状的按摩，最后用清水洗净。在选择洗面奶时要注意考虑以下两点因素。

(1) 洗面奶是否能够将皮肤的油光扫除。

(2) 洗脸后，面部不紧绷。

2. 润肤

油光泛滥的油性皮肤在选用润肤产品时一定要谨记控油保湿的原则。皮肤的水油平衡原理告诉我们越油的皮肤实际上越缺少水分，所以在选择产品时一定要选择控油型的乳液、面霜或保湿型的乳液。入睡前最好不用护肤品，以保持皮肤的正常排泄。晚上洁面后，也可进行适当的按摩，以改善皮肤的血液循环，调整皮肤的生理功能。可每周做一次按摩和面膜，以达到彻底清洁皮肤的目的。

3. 面膜

面膜护理非常重要，但是需要注意，油性皮肤使用的面膜以纯保湿或者具有清洁功效的面膜为主，一周一次即可。油性皮肤更需要对毛孔进行护理，油光泛滥会使油脂及废弃物堆积到毛孔，长久就会形成痤疮及黑头。具有保湿或者清洁功效的面膜有海藻面膜、芦荟面膜等。

4. 根据不同季节选择合适的补水产品

根据季节，油性皮肤的人在干燥的季节一样需要进行补水的护理，以保持水油平衡。油性皮肤在使用化妆品方面，宜少不宜多，特别是油性化妆品会导致皮肤更加油腻，出现毛孔堵塞，最好选用保湿润肤的护肤品。

5. 饮食

为什么油性皮肤会如此油呢？第一个原因是基因。其次，与我们日常的生活饮食习惯有着不可分的关系。巧克力、咖啡、奶油等高热量物质以及海鲜、辛辣食物、烟酒等刺激性食物都会刺激皮肤分泌大量油脂。饮食上要注意少食高脂肪、高碳水化合物的食物，忌烟及过食辛辣食物，多食水果、蔬菜，以改善皮肤的油腻粗糙感。

Ⅲ. Sensitive Skin Care

Part Ⅰ Dialogue

(B=Beautician，C=Customer)

C：Hello，my skin is always red，easy to dry and it is sensitive in spring. How should I care my skin?

B: According to your skin conditions, you have sensitive skin.

C: Then how about care?

B: As for daily care, you can attach importance to hydration because sensitive skin is due to lack of water. Then its ability of resisting external stimulus declines. At the same time, cleaning products should be mild. The suggested skin care steps are cleanser, lotion, cream, moisturizing essence, sunscreen.

C: How about care in the beauty salon?

B: I suggest you going to the salon once a week and doing replenishment and desensitization at intervals.

C: How long can I see results?

B: You can see corresponding effect after each care. At the beginning, you will find that the skin is not so dry, and there appears better gloss. After one or two months, you can find your face is not as red as before. However, the skin care should be done both at home and in the beauty salon. So I suggest you take care in the morning and evening and pay attention to the selection of skin care products. When in our salon, we promise to provide you with the most professional and best service. I bet your skin will improve with our joint efforts.

C: OK, I got it! Thank you.

Part Ⅱ Text

Care Methods of Sensitive Skin

Sensitive skin is a common skin problem. A survey shows that 2 or 3 out of 100 belong to the sensitive skin. Skin sensitivity refers to the skin which is easily irritated by external environmental changes. We should pay attention to the following aspects.

1. Maintain the cuticle

The thin cuticle is the cause of sensitive skin. Therefore the first principle is to safeguard the cuticle from harm. Do appropriate cleaning. Don't use soap. You had better use emulsion or non-soap lotion, which can help to adjust the pH value of our skin. Never use exfoliating products and facial scrub.

2. Strengthen the protection

The epidermis of sensitive skin is very thin and lacks UV defense ability, which is prone to age. Therefore, pay attention to the use of sunscreen. Besides, don't apply the sunscreen directly in the face. You should apply it after the basic skin care products.

3. Full moisture

The thin stratum corneum cannot lock enough water. People with sensitive skin is more easily affected by the lack of water in the skin whether in the air-conditioned rooms in summer or the dry climate in winter. Thus, moisturizing is of great importance, which can be realized by using moisturizing lotion with moisturizing ingredients and applying moisturizing masks regularly. During the change of seasons, do remember to change your skin care products.

4. Reduce half of the nutrition

For sensitive skin, high concentration and good effect means high risk and high sensitivity. So people with this type of skin should dilute the products before using them, especially the essence. In addition, don't use curative products and choose non-curative products which will not increase the burden of the skin.

5. Reduce irritation

Don't expose the skin to the wind or sun too much. Don't eat excitant food. Stop the use of skin care products and cleaning products and clean the skin with clear water every day. After one week, use low sensitivity products to reduce the damage. The skin may recover with its own self-healing ability.

6. The matters needing attention in the life of sensitive skin

(1) Pay attention to diet and eat less allergic foods (such as seafood, silkworm chrysalis, etc.).

(2) Careful selection of cosmetics. We should choose suitable products according to our own situation and avoid abuse.

(3) Avoid abrupt changes in the climate affecting the skin.

(4) Enhance skin resistance.

(5) Ensure adequate sleep and nutrient intake.

7. Find suitable skin care products

(1) Choose appropriate cleaning products. Powerful deodorant or antimicrobial soap can make skin very dry. Use non-spicy soap and avoid irritating skin care products, such as abrasive facial mask and granular scrub. These irritating skin care products will remove the outermost protective layer of skin. For thin and sensitive skin such as eyelids, alcohol-containing shrinkage and toner should be avoided. Oily makeup remover is better than oil-free makeup remover when removing eyes makeup.

(2) Choose appropriate emollient products. Moisturizing products can moisturize the skin surface and form a protective film between skin and cosmetic products. Alpha Hydroxy acid can be used to exfoliate necrotic, dull skin cells and grow new skin cells. But before use, test on the arm, pay attention to the pH value of the products, the concentration of fruit acid or other ingredients to determine the degree of skin irritation.

(3) Choose sunscreen with physical ingredients. The sunscreen containing physical shielding components, such as zinc oxide and titanium dioxide, should be selected. Avoid using chemical sunscreens, such as those containing p-aminobenzoic acid and cinnamic acid, which can irritate sensitive skin.

(4) Use hypoallergenic products. It should be noted that hypoallergenic products are not completely safe, but are less likely to cause allergic reactions to sensitive skin. Sometimes oily foundation is less irritation to sensitive skin than water-based foundation, because oily foundation contains a small amount of preservatives. But if acne is easy to grow, you should choose hypersensitive water-based foundation.

三、敏感性皮肤的护理

（一）对话

（B＝美容师，C＝顾客）

C：您好，我的皮肤总是发红，容易干燥，春天还容易过敏，应该怎么护理？

B：根据您的皮肤状况，判断您属于敏感性皮肤。

C：那应该怎么护理呢？

B：日常护理应该注意补水，因为敏感性皮肤都是干燥缺水的，缺水会导致皮肤抵抗外界刺激的能力下降，所以才会敏感。同时，清洁产品应该选用温和无刺激的。皮肤的防护也很重要，建议您日常护理的流程应该是洗面奶、润肤水、乳液、精华、防晒。

C：美容院护理呢？

B：我建议您每周来美容院进行一次护理，补水护理和脱敏护理可以间隔进行。护理的流程是洗面奶、按摩霜、面膜、润肤水、乳液、面霜等。

C：多久能看到效果？

B：每周护理一次都会有相应的效果。刚开始你会发现皮肤没那么干了，光泽度会变好，一至两个月后会发现皮肤没那么红，但是敏感性皮肤还是需要美容院护理和日常护理相辅相成，如果只做一种护理的话效果不会那么好，所以我建议您每天早晚都进行护理，注意护肤品的选择和使用方法。每周来美容院我们会给您提供最专业的护理和最细致的服务。相信我们共同努力，您的皮肤一定会越来越好的。

C：好的，我知道了，谢谢你。

（二）课文

敏感性皮肤的护理方法

敏感性皮肤是一种常见的皮肤问题。调查显示，100个人当中有2～3个人的皮肤是属于敏感性皮肤。敏感性皮肤是指容易受刺激而引起某种程度不适的皮肤。当外界环境发生变化时，皮肤无法调适，会出现不舒服的感觉以及过敏现象。平时护理时应注意以下几个方面。

1. 保护角质层

角质层薄是造成皮肤敏感的主要原因，因此保养的首要原则就是保护角质层，使其不受伤害。注意不可过度清洁，不要选用皂性洗剂，最好使用乳剂或非皂性的洗剂，这有助于调节皮肤的酸碱值。至于磨砂膏、去死皮膏等产品更应该敬而远之。

2. 加强防护

敏感性皮肤的表皮层较薄，缺乏防御紫外线的能力，容易老化，因此，应该注意防晒霜的使用。防晒霜最好不要直接涂在皮肤上，应该在进行基础保养后，再涂防晒霜。

3. 充分保湿

敏感性皮肤的角质层浅薄，无法锁住足够的水分。皮肤敏感的人在夏天的空调房中或是在冬天干燥的气候中，能够更敏锐地感觉到皮肤缺水、干燥。因此，日常护理中加强保湿非常重要。除使用含保湿成分的护肤品外，还应定期敷保湿面膜。在季节更替时，需要更换不适用的护肤品。

4. 滋养减半

现代的护肤品强调的是高效性，因此要求其活性成分能够透过皮肤，作用到皮肤深层才能产生高效果。对于敏感性皮肤，高浓度、高效果就是高风险、高敏感。因此这类皮肤的人在使用护肤品（尤其是精华液之类的高浓度护肤品）时，应先将其稀释一半后再使用。另外，敏感性皮肤不适合疗效性太强的产品，使用不给皮肤增加负担的非疗效性产品，才是使皮肤恢复健康的良方。

5. 减少刺激

皮肤一旦出现干燥、脱屑或发红状况，说明皮肤的健康状况已亮起红灯。要让皮肤尽快复原，最好的方法就是减少刺激，不过度受风吹、日晒，不吃刺激性食物，停止当前一切护肤品、清洁品的使用，让肌肤只接触清水等。每天只用温水清洁皮肤，持续一周，然后再使用低敏系列的产品，在降低伤害后，皮肤运用本身的自愈能力，可以自行恢复健康。

6. 敏感性皮肤在生活中需要注意以下事项

(1) 注意饮食，少吃易过敏的食物（如海鲜、蚕蛹等）。

(2) 化妆品的选用要慎重。应根据自己的情况，选用适合自己的产品，避免滥用。

(3) 避免气候的突然变化对皮肤的影响。

(4) 增强皮肤的抵抗力。

(5) 保证充足的睡眠以及营养的摄入。

7. 找到合适的护肤品

(1) 选用合适的清洁类护肤品。强力除味或抗菌肥皂能使皮肤变得非常干。要使用无香料的肥皂，避免刺激强烈的护肤品，如研磨性面膜、有颗粒的磨砂膏等，这些有刺激的护肤品会去掉皮肤最外面的保护层。对眼睑等处薄且敏感的皮肤，应避免使用含有酒精的收敛水、爽肤水；眼部卸妆时，使用油性的卸妆乳比无油的好。

(2) 选择合适的润肤产品。润肤产品能够湿润皮肤表层，可在皮肤与彩妆产品之间形成一层保护膜。可以使用果酸，果酸可以使坏死的、无光泽的皮肤细胞脱落，长出新的皮肤细胞。但在使用前要在手臂上进行测试，注意产品的 pH 值，果酸的浓度或其他成分决定皮肤受刺激的程度。

(3) 选择物理成分的防晒霜。应选择含有物理遮蔽性成分的防晒霜，如含有氧化锌、二氧化钛等成分的防晒霜。避免使用化学成分的防晒霜，如含有对氨基苯甲酸、肉桂酸等成分的防晒霜，这些化学成分均会刺激敏感性皮肤。

(4) 使用低过敏性产品。要注意低过敏性产品也不是完全安全的，只是比较不容易引起敏感性皮肤的过敏反应而已。有时，油性粉底比水性粉底较不容易刺激敏感性皮肤，因为油性粉底含有少量防腐剂。但如果易长痤疮，还是应选择低过敏性的水性粉底。

Ⅳ. Skin Care of Telangiectasis

Part Ⅰ Dialogue

(B=Beautician, C=Customer)

B: Hello, what can I do for you?

C: There is red blood silk in my face. What should I do?

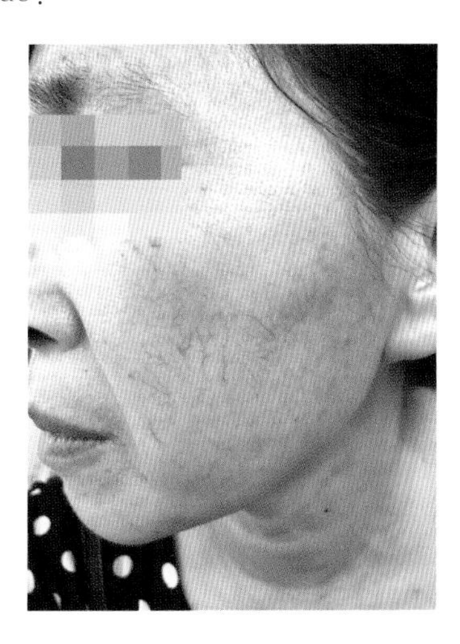

B: In daily life, we often see people have facial redness and clearly dilated capillaries, some of which are in the color of red or purple. This is commonly known as the telangiectasis. According to TCM theory, pathogenesis can be divided into two types: blood heat and blood stasis, the color of the former is red and the color of the latter is purple. You are diagnosed with capillary telangiectasis.

C: What are the causes of it?

B: There are a number of causes for it, such as plateau climate, gene, frostbite and so on.

C: Then how to cure it?

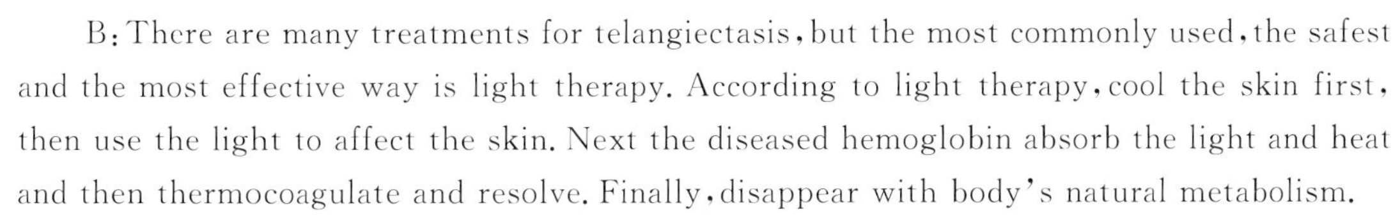

B: There are many treatments for telangiectasis, but the most commonly used, the safest and the most effective way is light therapy. According to light therapy, cool the skin first, then use the light to affect the skin. Next the diseased hemoglobin absorb the light and heat and then thermocoagulate and resolve. Finally, disappear with body's natural metabolism.

C: Will it be very painful?

B: During treatment, your skin has the obvious sense of burning and stinging, after the treatment, the burning pain will last for 30 minutes or so. The color of the red blood silk worsens. But don't worry. It's normal. We will do some repair care according to your situation.

C: How soon can I see the effect?

B: There will be effect after only once. If you want to have better effect, you need to cure for a course of treatment which includes five times at an interval of 28 days.

C: OK, I see. Thank you.

B: You are welcome.

Part Ⅱ Text

Care Methods of Telangiectasis

Many women who care much about beauty often feel troubled for their blushing faces. However, many people don't know that red blood silk not only affects the beauty but also affects their health. It seriously affects the absorption of nutrients, resulting in the symptoms of insufficient skin nutrient supply and causing rough, dry and premature skin aging. In TCM, it is called "red face", which attributes to heart sutra stagnation and poor blood circulation.

In order to solve the problem of telangiectasis, cosmeceuticals are suggested to promote blood circulation, clear blood dirt, restore the elasticity of blood vessels so as to achieve the purpose of curing telangiectasis. In addition, as for daily care, we should pay attention to moisturize and enhance the barrier function of the stratum corneum because such kinds of patients have very thin cuticle and dry skin. In terms of treatment instrument, it has a giant leap, that is the laser treatment machine with a wavelength of 585 nm. The machine makes use of the selective absorption of certain wavelength laser of oxyhemoglobin. Besides, strong pulse electronic rejuvenation instrument can greatly improve telangiectasis, which is proved to be a safe and convenient treatment means. After treatment, the skin keeps intact, which really brings hope to all the patients.

Daily diet. Potassium in the body can alleviate the harmful effects of sodium and promote the excretion of sodium, reduce the blood pressure and has certain effects on the expansion of capillaries. The foods rich in potassium are acacia honey, beans, tomatoes, milk, kelp, fresh mushrooms, green leafy vegetables and oranges, apples, bananas, pears, pineapples, kiwi fruits, walnuts, hawthorns and watermelons.

四、毛细血管扩张类皮肤的护理

(一) 对话

(B=美容师,C=顾客)

B:您好,请问有什么可以帮您?

C:我脸上一直有红血丝,请问该怎么调理?

B:日常生活中我们经常看到一部分人面部皮肤泛红,并且肉眼就能看见一条条扩张的毛细血管,部分呈红色或紫红色,表现为斑状、点状、线状或星芒状的损害,这就是毛细血管扩张,俗称红血丝。中医把毛细血管扩张的病因分为血热和血瘀两种,血热型毛细血管扩张颜色偏红,血瘀型毛细血管扩张颜色偏紫,通过诊断,您应该属于血热型毛细血管扩张。

C:为什么会这样呢?

B:毛细血管扩张的病因很多,与高原气候、遗传、冻伤等都有关系。

C:那怎么治疗呢?

B:毛细血管扩张的治疗方法很多,但目前最常用和最安全有效的方式是光疗法。光疗法是在表皮冷却的前提下,光作用于皮肤,病变的血红蛋白吸收光热,瞬间热凝固且分解,随人体自然代谢。

C:光疗法会很疼吗?

B:治疗中,您的皮肤会出现明显的热刺感,治疗后,火辣辣的感觉会持续 30 min 左右,红血丝颜色加重呈紫红色。不过您放心,都是在正常人可以接受的范围,我们也会根据您治疗后的情况给您选择冰敷等后期修复护理。

C:那做多久会有效果?

B:一般做一次光疗法就能看到红血丝变少,颜色变淡,但想要一个好的效果需要一个疗程左右,一个疗程是五次,28 天做一次就可以。

C:好的,我知道了,谢谢。

B:不客气。

（二）课文

毛细血管扩张类皮肤的护理方法

许多爱美的女士常常因自己潮红的面庞而感到十分困扰。可许多人也许不知道，红血丝不仅仅会影响你的美丽，而且会影响你的健康，会严重影响皮肤汲取营养，导致皮肤养分供养不足，出现粗糙、干燥和过早衰老的症状。中医理论中称其为"红赤面"，其原因归结为心经郁热、血行不畅。

要解决毛细血管扩张建议使用具有活血化瘀、清除毛细血管血垢、恢复血管弹性、去腐生肌、修复角质层、美白等功效的药妆，从而达到根治毛细血管扩张的目的。另外，日常护理应注意补水和增强皮肤角质层的屏障功能，因为毛细血管扩张类患者的角质层薄，皮肤干燥，屏障功能较差，所以日常护肤品的选择方面应着重具有补水和健全角质层屏障功能的护肤品。

仪器治疗方面，毛细血管扩张治疗上出现了一个飞跃，那就是用波长为 585 nm 的激光治疗机治疗。激光治疗血管性皮肤损害的原理主要是利用氧合血红蛋白对一定波长激光的选择性吸收，从而导致血管组织的高度选择性破坏。另外强脉冲光子嫩肤仪也可以明显改善毛细血管扩张，实践证实仪器治疗是一种安全、简便的治疗手段。治疗后皮肤完好无损，给广大患者带来了福音。

日常饮食方面。钾在体内能缓解多余钠的有害作用，促进钠的排出，对毛细血管扩张有一定的预防作用。含钾的食物有槐花蜂蜜、豆类、番茄、乳品、海带、鲜蘑菇及各种绿叶蔬菜、橘子、苹果、香蕉、梨、菠萝、猕猴桃、核桃、山楂、西瓜等。

Ⅴ. Aging Skin Care

Part Ⅰ Dialogue

(B＝Beautician，C＝Customer)

B：Hello!

C：Hello!

B：What can I do for you?

C：There are many wrinkles on my face and the skin is less compact. Would you like to give me some advice on how to improve it.

B：Would you mind telling me your age?

C：I am 38 years old.

B：You look really younger than your age! Your skin condition is very good at your age. Aging is a normal physiological phenomenon, which features facial wrinkles. But don't worry. With the development of science and technology, the problem of wrinkles can also be solved. Let me see if your wrinkles are true or false.

C：How to distinguish between the true and false wrinkles?

B：Quite simple, stretch the wrinkles with your thumb and index finger in the direction of the vertical direction of its growth. If the wrinkles disappear, it means these are false ones. If the wrinkles are still there, it means these are true ones.

C: Then my wrinkles are true wrinkles, because they didn't disappear.

B: Yeah.

C:. What should I do?

B: There are many solutions, such as acupuncture, filling injection or using anti-aging instruments. Acupuncture therapy is the use of 1 to 1.5 "cun" needles to stimulate the dermis of the skin in the wrinkled areas to accelerate the production of collagen. Filling injection is to inject hyaluronic acid and collagen into the wrinkled areas. Besides, there are different kinds of instruments to remove wrinkles such as optical, ultrasonic and radio instruments.

C: Ok, I know. Thank you.

Part Ⅱ Text

Care Methods of Aging Skin

There are many ways to treat aging skin, usually including non-surgical and surgical methods.

1. Non-surgical methods

Non-surgical methods include drug therapy, chemical stripping method, microwave therapy, cosmetics, enzyme and so on, which can be used for the light and moderate levels of facial wrinkles.

(1) Drug therapy is to improve the nutritional condition of the skin by regulating the biological activity of the skin cells.

(2) The major effects of chemical peel is to peel aging epidermal layer, promote the growth of basal cells, repair the aging collagen fiber and improve the elasticity and tension of the skin. The major stripping agents are retinoic acid and chloroacetic acid. Retinoic acid is cell differentiation inducer, which promotes basal cell hyperplasia, thickens dermis papillary, irritates the growth of blood capillaries, regulates the function of immune system, reduces fine wrinkles, and improves face gloss. The effects of retinoic acid therapy are the increase of fibroblasts and collagen type Ⅰ. The potential complications of chemical peel are pigmentation and superficial scar, which requires experience and good formula.

(3) The principle of microwave facial wrinkles removing is microwaves of different wave lengths act on different layers of the skin. It can help to recover the elasticity of the skin and stimulate the proliferation of collagen fiber. Besides, it can promote gland activities, microcirculation and the metabolism.

(4) The principle of anti-aging for cosmetics: First, it is to clear excess free radicals according to the anti-aging theory. Such active ingredients which are rich in free radicals are vitamin C, vitamin E, and coenzyme Q. Second, protect against UV according to the light aging theory. Ultraviolet radiation can cause skin erythema and delayed melanin pigmentation, destroy the moisturizing ability of skin, make the skin become rough and increase wrinkles. At the same time, cosmetics can promote the metabolism of the skin cells

and supplement collagen and elastin, moisturize and repair the barrier function of the skin.

(5) The principle of enzyme: the enzyme can eliminate the waste in the body and balance the intestinal bacteria. Longevity enzyme can promote metabolism, that is, superseding the old cells with new ones, which requires the joint function of all kinds of enzymes.

2. The surgical methods

The surgical methods include subcutaneous injection, dermabrasion and facial wrinkle removing.

(1) The subcutaneous injection includes autologous fat filling, subcutaneous collagen injection and implants, the purpose of which is to reduce wrinkles by filling. However, the effect will not last too long because the fat and collagen will be absorbed in half a year.

(2) Dermabrasion is commonly used for photoaged skin. After the surgery, the skin histology is changed and collagen will grow. According to immunohistology, dermabrasion can improve the light aging skin.

五、衰老皮肤护理

（一）对话

（B＝美容师，C＝顾客）

B：您好，欢迎光临。

C：你好。

B：有什么可以帮您？

C：我脸上出现了很多皱纹，皮肤也不如以前紧致了，想问问有没有什么改善的方法。

B：方便告诉我您的年龄吗？

C：我今年 38 岁。

B：您看起来真年轻！按照您的年龄，这种皮肤状态已经很好了。衰老是一种正常的生理现象，最明显的表现就是面部长皱纹，但也不用担心，随着科技的发展，皱纹也是可以解决的皮肤问题，让我看看您是真性皱纹还是假性皱纹。

C：怎么分辨真性皱纹和假性皱纹呢？

B：非常简单，用我们的拇指和食指沿着皱纹生长方向的垂直方向将皱纹拉伸，如果纹路消失，证明是假性皱纹，如果纹路还在，就属于真性皱纹。

C：那我的应该属于真性皱纹，纹路没有消失。

B：是的。

C：应该怎么办呢？

B：皱纹目前解决的方法很多，可以用中医针刺疗法、注射填充或者用其他抗衰老的仪器。中医针刺疗法就是用 1～1.5 寸的毫针刺激皱纹局部皮肤的真皮层，加速皮肤新陈代谢，使其产生胶原蛋白的方法来除皱。注射填充就是给皱纹局部注射玻尿酸、胶原蛋白等来填充除皱。除皱的仪器有很多，有光学除皱仪、超声波除皱仪、射频等。

C：好的，知道了，谢谢。

（二）课文

衰老皮肤的护理方法

治疗皮肤老化的方法很多，通常有非手术方法和手术方法。

1. 非手术方法

非手术方法包括药物疗法、化学剥脱法、微波疗法、化妆品、酵素等，可用于轻、中度的面部皱纹。

（1）药物疗法主要是对皮肤细胞进行生物活性调控以改善皮肤营养状况。

（2）化学剥脱法的主要作用是除去老化的表皮角质层细胞，促进基底细胞增生，修复老化的胶原纤维，提高皮肤的张力和弹性。常用的剥脱剂有视黄酸和三氯乙酸。视黄酸是细胞分化诱导剂，调整角质形成细胞的增殖、分化，抑制角化过程，溶解角质，起剥脱作用，并促进基底细胞增生，真皮乳头层增厚，基底膜锚状纤维增加，毛细血管增生；刺激朗格汉斯细胞增生，调节皮肤免疫功能，减少细小皱纹，改善皮肤表面的光泽度。应用维甲酸治疗经检测发现真皮乳头层Ⅰ型胶原形成增多。三氯乙酸适用于中等深度的化学剥脱。运用三氯乙酸治疗光老化皮肤皱纹，结果剥脱后皮肤内成纤维细胞增多，Ⅰ型胶原总量增加。皮肤化学剥脱的潜在并发症是色素改变和浅表瘢痕，使用时需有经验和良好配方。

（3）微波面部除皱的原理是不同波长的微波作用于皮肤和皮下各组织，可促进恢复皮肤弹性活力，刺激胶原纤维增生修复。此外，还可通过电离渗透作用，促进皮肤吸收水分、营养，促进腺体活动、微循环和新陈代谢。

（4）化妆品对抗衰老的原理：一是根据自由基衰老学说清除过量的自由基，拥有清除自由基功能的活性原料以维生素C、维生素E、辅酶Q为代表。二是根据光老化学说预防紫外线。日光中的紫外线可引起皮肤红斑和延迟性黑色素沉着，破坏皮肤的保湿能力，使皮肤变得粗糙，皱纹增多。同时，化妆品还有促进皮肤细胞新陈代谢，补充胶原蛋白和弹性蛋白，保湿和修复皮肤屏障功能等作用。

（5）酵素（酶）对抗皮肤老化的原理是酵素可消除体内废物，平衡肠道细菌；长生酵素能增强机体新陈代谢的能力，新陈代谢过程需要各种酵素共同参与完成。

2. 手术方法

手术方法包括皮下填充法、擦皮术和面部除皱等。

（1）皮下填充法包括自体脂肪注射、皮下胶原注射和种植体植入，目的是利用皮内填充物减少面部皱纹，但效果不持久，脂肪、胶原于半年后往往被吸收。

（2）擦皮术常用于光老化皮肤，皮肤磨削至真皮乳头层。擦皮术后皮肤组织发生改变，胶原增生，真皮乳头层成纤维细胞增多。免疫组织学观察，可见Ⅰ型胶原产生增多，表明皮肤磨削可改善光老化皮肤。

Unit 6　Special Cosmetology

Section 1　Manicure

Ⅰ. The Basic Idea of Manicure

Part Ⅰ　Dialogue

(M=Manicurist,C=Customer)

C:Can you tell me what manicure is?

M:Manicure is a kind of work that beautifies a finger (toe), also known as nail art design.

C:What is the feature of manicure?

M:Manicure is the process of disinfecting, cleaning, maintaining and beautifying the fingers(toes),according to the different hands,the different nail types,the costume color and the requirements. It has the characteristics of diversity of expression.

Part Ⅱ　Text

The Basic Concepts and Characteristics of Manicure

Manicure,also known as nail beauty,is to stand out in the design of the overall image of the new project and elegant dress with manicured hands. Manicure is the symbol of body and the position. Manicure shows good accomplishment and elegant temperament. With the development of the society and the improvement of material level,manicure is not only the exclusive enjoyment and demands of women, but the men's demands for manicure are also increasing. Manicure is a set of theory and practice in the integration of subjects. According to different shapes of the hands,nail quality and different requirements of customers,using professional manicure tools and equipment materials. There is mainly in the nail surface of beautification design. According to the scientific and technical procedures, the hands and fingernails are cleaned,beautified and designed. Manicure is a very fashionable technology right now,So we

must follow the trend and pay attention to the combination of theoretical, practice and aesthetics.

一、美甲的基本概念

(一) 对话

(M=美甲师,C=顾客)

C:你能说说什么是美甲吗?

M:美甲是一种对指(趾)甲进行修饰美化的工作,又称甲艺设计。

C:美甲的特点是什么?

M:美甲是根据不同的手形、指(趾)甲的形状、服装色彩和要求,对指(趾)甲进行消毒、清洁、保养、修饰美化的过程。美甲具有表现形式多样化的特点。

(二) 课文

美甲的基本概念及特点

美甲也称为指甲美容,是从整体形象设计中脱颖而出的新项目。典雅的服饰,再加上一双精心修饰的手,是身体和地位的象征,体现了良好的修养和优雅的气质。随着社会的进步及物质水平的提高,美甲已不仅是女性独有的享受及需求,男性对美甲的要求也在持续增加。美甲是一门集理论和实践于一体的科目,主要是在指甲表面进行美化设计,根据人的手形,指甲的形状、质量及要求,运用专业的美甲工具、设备、材料,按照科学技术操作程序,对手部及指甲表面进行清洁护理、保养修整及美化设计的工作。美甲是目前很时尚的一种技术,必须紧跟流行趋势,所以要注意理论、实践和美学相结合来进行教学。

Ⅱ. The structure of the nails

Part Ⅰ Dialogue

(M=Manicurist,C=Customer)

C:Do nails grow every day?

M:The daily growth of fingernails is 0.1 mm. Toenail daily growth rate is only about 1/3-1/2 of the nails. So the growth cycle of toenail is 3-6 months.

C:What affects the growth of nails?

M:The growth rate of the fingernails is affected by age, health, season and mechanical stimulation, and the frequent friction will speed up the growth of the nails. People usually have more opportunities to use their hands than their feet.

C:Does the growth of fingernails related to nutrition?

M:The growth of nails requires the continuous supply of amino acids, including sulfur amino acids, to form keratin. Normal growth of nails requires vitamin B, calcium and phosphate ions. Vitamin A and D deficiency can cause crispy nail.

Part Ⅱ Text

Basic Concept of Nails

1. The standard for healthy nails

Healthy nails should be smooth, bright, full, round, pink. The surface has no spots and bumps and ribbed. The moisture content is about 18%, presenting smooth arc, firm and elastic.

2. Nail composition

The nail is composed mainly of three parts: the nail tip, the nail body, and the nail root.

(1) The nail tip. Nail tips are also called nail presides, which are the part of the nail that separate from the nail bed. There is no support at the bottom and lack of moisture and oil, so it is easy to crack.

(2) The nail body. The nail body is also known as a "cap", which is made up of three or four layers of hard scaly keratin, without blood vessels or nerves.

(3) The nail root. The root of the nail is very thin and soft in front of the methyl group, at the base of the nail, which acts like the root of the crop.

(4) The nail groove. The nail groove is a depression between the outer frames of the nail, if it is too dry, it is prone to spines.

(5) Nail meniscus. Nail meniscus is semicircle, attached to the joint of nail and nail bed. The color is white and turbidity. It reflects the color of immature nail cells.

(6) Nail foundation. Nail foundation refers to the root of the nail, which contains blood vessels called capillary and lymphatic, which acts like a soil, and is known as the source of the growth of the nail.

(7) Nail skin. Nail skin is the skin of the base of the nail to protect nail root. Normal skin is soft.

(8) Nail core. The nail core is the thin layer of skin on the front edge of the nail. It is sensitive and causing the nail to atrophy when the core is damaged.

(9) Free margin of nail. Free margin of nail is the connection between the nail body and the nail, and it is called the smile line.

二、指甲的结构

(一) 对话

(M=美甲师,C=顾客)

C:指甲每天都会生长吗?

M:指甲每天生长 0.1 mm,趾甲每天生长速度仅为指甲的 1/3~1/2,故指(趾)甲的生长周期为 3~6 个月。

C:指甲的生长和什么有关系?

M:指甲的生长速度受年龄、健康、季节和机械刺激的影响,经常摩擦就会使指甲的生长速度加快。人一般用手活动的机会比脚多。

C:指甲的生长和营养有关系吗?

M:甲的生长需要包括含硫氨基酸在内的氨基酸的不断供给,形成角蛋白。正常甲的生长需要维生素B、钙和磷酸离子。维生素A和维生素D缺乏可引起脆甲。

(二)课文

指甲的基本概念

1. 健康指甲的标准

健康的指甲表面光滑、亮泽、饱满、圆润、呈粉红色,表面无斑点、凹凸及棱纹,含水量约18%,呈平滑的弧形,坚实且有弹性。

2. 指甲的构成

指甲主要由三个部分构成,即甲尖、甲体和甲根。

(1) 甲尖:也叫指甲前缘,是指甲面从甲床分离的部分,由于指甲下方没有支撑,缺乏水分及油分,所以容易裂开。

(2) 甲体:也称“甲盖”,由3~4层坚硬的鳞状角质蛋白重叠而成,不含血管和神经。

(3) 甲根:位于指甲根部,在甲基的前面,极为薄软,其作用类似农作物的根茎。

(4) 甲侧沟:位于指甲两侧外框之间的凹陷处,如果该处皮肤太干燥,易长肉刺。

(5) 甲半月:附在甲根与甲床连接处,呈白色,浑浊状,半圆形。甲半月能够反映未成熟的甲体细胞的颜色。

(6) 甲基:位于指甲根部,含有毛细血管及淋巴管,其作用类似土壤,被称为指甲的生长源泉。

(7) 指皮:指甲根部的皮肤,正常的指皮松软且柔润,其作用是保护甲根。

(8) 指芯:指甲前缘下的薄层皮肤,敏感,指芯受损时会导致指甲萎缩。

(9) 游离缘:甲体与甲尖之间的连线,也称“微笑线”。

Ⅲ. The Color and Health of Nails

Part Ⅰ Dialogue

(M=Manicurist, C=Customer)

C: What color is a healthy nail?

M: Healthy nails are smooth, flat, translucent and even pale pink.

C: What's the reason for white nails?

M: If the nails are often white, they indicate that there is not enough blood in the body and a sign of anemia.

C: What is the cause of red nails?

M: A bad blood circulation can also lead to red nails.

C: Why is the nail purple?

M: This is a characteristic of heart disease and blood disease, which reflects the lack of oxygen in the blood or the abnormality of certain components.

Part Ⅱ Text

Nail Color and Health

Nails are a barometer of human health, because the twelve meridians are on the fingers. Normal nails are clear, smooth, translucent, fair and pale red. The nail has a set of radians and has a crescent of gray. Once the body appears the disease, namely through the nerve, the blood vessel, the meridian and so on to the nails. The color and shape of the nails will change.

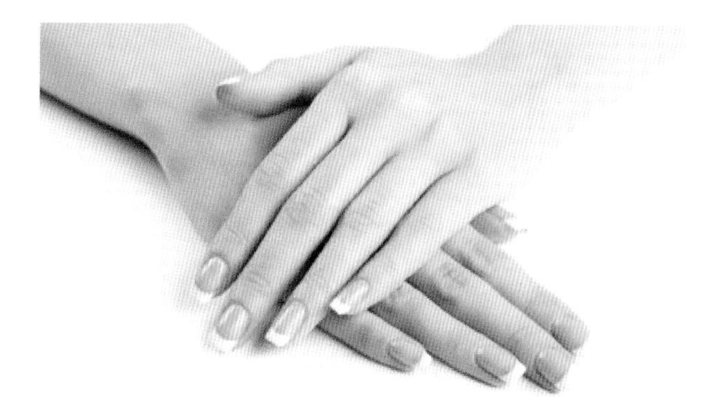

1. The nails are white

(1) If the nails are often white, they indicate that there is not enough blood in the body and a sign of anemia.

(2) If nail is pewter without light, that means the body has ulcer bleeding or chronic blood loss, such as hookworm disease.

(3) If the most of the nails are white, the normal pink is reduced to a small strip near the nail tips, this may be a sign of cirrhosis.

(4) If the distal part of the white fingernail is reddish brown, It may be in azotemia due to chronic renal insufficiency.

(5) If there is a cross-cut white line on the nail, it may be toxic to metals such as arsenic and lead, or in Hodgkin's disease, pellagra, and so on.

(6) If there are two white lines across the nail, they often indicate a decrease in white protein in the blood, often seen in the hypoalbuminemia of chronic kidney disease.

(7) If the surface of the deck appears dotted or white, often because of nutritional disorders. It may be a sign of chronic liver disease, cirrhosis and kidney disease.

(8) The nails are usually grey and white, and may be a sign of heart failure in advanced pulmonary disease. In addition, the nail all white may be congenital, some is the occupation causes. White patches and vertical stripes appear on the old nail, and it appears periodically, which is a common variation of the old nail, not pathological, not to worry.

2. The nails are yellow

(1) The nail is yellow, which generally indicates that the liver has a problem, mostly jaundiced hepatitis, and also in chronic hemorrhagic disease.

(2) Hypothyroidism, nephrotic syndrome, and beta-carotene, onychomycosis can cause yellow nails.

(3) If not only yellow nail thickening with lateral bending, but also slow growth, less than 0.2 mm per week, combined with the primary lymphedema and the pleural effusion, this is the yellow nail syndrome.

(4) If you see a nail that is shaped like a brass, it's caused by an autoimmune alopecia that we still know very little about. The disease can cause partial or total hair loss.

(5) If you find yellow around your fingertips, beware of malignant melanoma. In addition, because of the long-term use of tetracycline drugs, degenerative changes of the old people, and long-term smoking, nails can be yellow, these are not pathological nails.

3. The nails are red

(1) Near the base of the nail is scarlet, while the middle of the nail is a pale white with the symptoms that most people suffer from cough and hemoptysis. Conversely, close to the tip of the nail is pink or red, and nearly half of the protective film is white, which may be a sign of chronic renal failure.

(2) The nails are all scarlet. It is a symbol of early tuberculosis and intestinal tuberculosis. Pressing the nail, if the blood color recovers quickly, the disease is light. If the blood color restores slowly, the disease is longer.

(3) The nails are deep red and the color of the pressure is constant, suggesting that there is a serious inflammation of the internal organs.

4. The nails are purple

(1) This is a characteristic of heart disease and blood disease, which reflects the lack of oxygen in the blood or abnormalities in certain components.

(2) If the purple is alternating with pale color, it can be seen in raynaud's disease.

5. The nails are cyan

(1) Patients with acute abdominal pain have a sudden onset of cyan in their extremities.

(2) The fetus dies from a pregnant woman, and the nail will continue to be cyan.

(3) In addition, some people have observed that the nails appear to be cyan stasis, which

can prompt poisoning or early cancer.

(4) Nails are violaceous color, often seen in lobar pneumonia, severe emphysema and other lung diseases.

6. The nails are green

(1) Some or all of the nail decks are green, which is related to the long exposure to soap and detergent.

(2) Sometimes it can also be caused by the spread of Pseudomonas aeruginosa or the green bacteria.

7. The nails are blue

(1) Patients with diphtheria, lobar pneumonia, acute intestinal infection, and obstruction of esophagus are ultramarine.

(2) In the case of liver degeneration, copper has a metabolic disorder and sometimes blue nails.

(3) To eat freshness vegetables can cause the enterogenous cyanosis, which can make the normal low blood red protein oxidation or methemoglobin lose oxygen ability to cause tissue hypoxia, resulting in skin cyanosis and blue nails, but it is also important to note that certain medicines such as the quinoline of sulfurate nitrate can also cause blue nails.

(4) The blue half-moon shaped nail root, appear this kind of nails, might mean that patients had impaired blood circulation, heart disease or raynaud's syndrome, and sometimes associated with rheumatoid arthritis or systemic lupus erythematosus.

8. The nails are grey

(1) It can be seen in systemic disease, myxedema, rheumatoid arthritis, or hemiplegic.

(2) If malnourished, nails will become thicken or atrophy, with pigmentation or greyness.

(3) If the lower part of the thumb is grey and wavy, it is often seen in glaucoma.

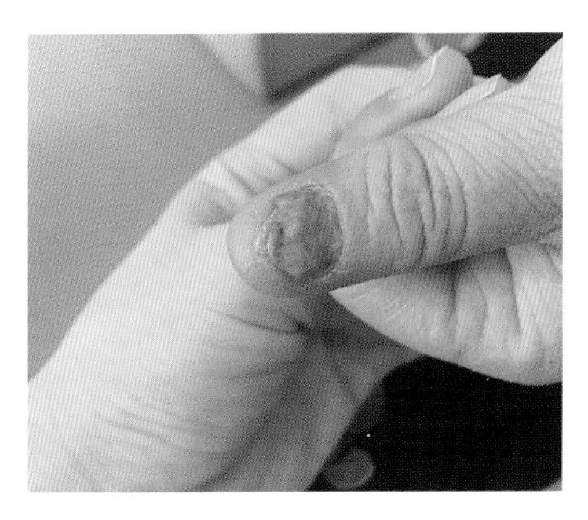

9. The nails are black

(1) Most of the injuries caused by trauma may cause bleeding under the nails and beginning to be fuchsia.

(2) Nail bed melanin increases and heavy metal silver composure, can produce black brown fingernail.

(3) Nails can be black or blue when the onychia is infected with Pseudomonas aeruginosa.

(4) If there is chronic renal failure, it is common to have a pronounced black on the distal end of the nail.

(5) If lack of vitamin B_{12}, adrenal insufficiency, gastrointestinal polyps syndrome, or long-term exposure to coal tar, nails can also be turned to grayish black.

(6) When nails are a black or brown, or freckles shaped, it most commonly occurs in thumbs and big toes, especially a nail and its surrounding tissues are also brown or black, which could mean people with a kind of malignant tumor, named melanoma.

(7) When the nail root grows several root black lines, usually only in the middle of the nails, indicate that the body is or has been canceration, should immediately go to the hospital to further examination, diagnosis, early treatment.

三、指甲的颜色与健康

(一) 对话

(M=美甲师,C=顾客)

C:健康指甲是什么颜色?

M:健康的指甲光滑、平整、半透明,呈均匀的淡粉色。

C:指甲发白是什么原因?

M:如果指甲外表经常是白色,表示身体里的血液不太充足,有贫血征兆。

C:指甲发红是什么原因?

M:阴虚火旺,血液循环差可能导致指甲发红。

C:指甲呈紫色是什么原因?

M:这是心脏病、血液病的一个特征,反映血液缺氧或血液内某些成分异常。

(二) 课文

指甲的颜色与健康

指甲是人体健康的晴雨表,这是因为人的十二经脉会于指端。正常人的指甲光洁、平滑、半透明,呈均匀的淡红色,甲端有定型的弧度,甲根有一月状灰白色的甲弧。身体一旦出现病状,即会通过神经、血管、经络等反映到指甲上,出现指甲的形状、颜色的异常变化。

1. 指甲呈白色

(1) 如果指甲颜色为白色,表示身体里的血液不太充足,有贫血征兆。

(2) 指甲白蜡色无光泽,正是消化道溃疡出血,或有钩虫病等慢性失血症的表现。

(3) 指甲大部分为白色,正常的粉红色减少到只有靠近指尖的那一小条,可能是肝硬化的征兆。

(4) 毛玻璃样白色指甲的远端部分呈红褐色,可见于因慢性肾功能不全而出现的氮质血症。

(5) 指甲上出现横贯的白色线条，可见于砷、铅等金属中毒，或见于霍奇金淋巴瘤、糙皮病等。

(6) 指甲上有两条横贯的白色线，提示血中的白蛋白减少，多见于慢性肾病的低蛋白血症。

(7) 甲板表面出现点状或丝状白斑，常为营养障碍，多为慢性肝病、肾脏疾病的征兆。

(8) 指甲平时为灰白色，可能是肺结核晚期、肺源性心脏病心力衰竭的征兆。此外，灰白甲有的可能是先天性的，有的是职业性的。在老年人指甲上出现白色斑块和纵向线条，呈周期性出现，这是老年人指甲常见的变化，不是病态，不必顾虑。

2. 指甲呈黄色

(1) 指甲变黄，一般表示其肝脏出了问题，常见于黄疸型肝炎，也见于慢性出血性疾病。

(2) 甲状腺功能减退、肾病综合征、胡萝卜素血症，可引起黄甲，甲癣也能致黄甲。

(3) 如果指甲发黄变厚、侧面弯曲度大，而且生长缓慢，每周生长小于 0.2 mm，再合并胸腔积液、淋巴水肿，则表示出现黄甲综合征。

(4) 如果看到形同锤打过的黄铜模样的指甲，这是由一种目前我们还知之甚少的自身免疫性脱发导致的，这种病可使人部分或全部脱发。

(5) 如果发现指尖周围出现黄色，则要警惕恶性黑色素瘤。长期服用四环素类药物者、长期吸烟者，指甲可呈黄色，老年人由于指甲退行性变可使指甲呈浅黄色，这些都不属于病态指甲。

3. 指甲呈红色

(1) 靠近甲根处的指甲为绯红色，而甲体中部、前端为淡白色者，大多患有咳嗽、咯血症。反之，接近指甲尖处呈粉红色或红色，而接近护膜处呈白色，这可能是慢性肾功能不全的征兆。

(2) 整个指甲为绯红色，为早期肺结核、肠结核的征兆。如压迫指甲，血色恢复快的病情轻，血色恢复慢的病程较久。

(3) 指甲呈深红色，压之颜色不变，提示可能某内脏器官有严重的炎症存在。

4. 指甲呈紫色

(1) 指甲呈紫色是心脏病、血液病的特征，反映血液缺氧或血液的某些成分异常。

(2) 若紫色与苍白色交替出现，可见于肢端动脉痉挛症。

5. 指甲呈青色

(1) 急腹症患者四肢厥冷，指甲会突然发青。

(2) 胎儿死于腹中的孕妇，指甲会持续性发青。

(3) 指甲出现青色瘀斑，可提示中毒或早期癌病。

(4) 指甲呈青紫色，多见于大叶性肺炎、重度肺气肿等肺部疾病。

6. 指甲呈绿色

(1) 甲板部分或全部变绿，多与长期接触肥皂、洗涤剂有关。

(2) 感染铜绿假单胞菌或绿色曲菌也可引起。

7. 指甲呈蓝色

(1) 白喉、大叶性肺炎、急性肠道传染病、食管异物阻塞的患者，其指甲呈青蓝色。

(2) 肝豆状核变性导致铜代谢紊乱,有时也可出现蓝甲。

(3) 食用不新鲜的蔬菜可引起肠源性发绀,导致低铁血红蛋白氧化或高铁血红蛋白失去输氧能力,造成组织缺氧,从而发生皮肤发绀及蓝甲,但应注意某些药物如硫磺、亚硝酸盐、伯氨喹啉等也可导致蓝甲。

(4) 指甲根部呈蓝色半月状,出现这种指甲,可能意味着患者有血液循环受损、心脏病或雷诺综合征,蓝甲也与类风湿性关节炎或系统性红斑狼疮有关。

8. 指甲呈灰色

(1) 可见于全身性疾病、黏液性水肿、类风湿性关节炎或偏瘫患者。

(2) 营养不良患者,指甲会变厚或萎缩,且有色素沉着或呈灰色。

(3) 拇指下端呈灰色波浪状常见于青光眼患者。

9. 指甲呈黑色

(1) 多数由外伤引起。受伤指甲下出血,开始为紫红色,久之变为紫黑色。

(2) 甲床黑色素增加,重金属银沉着,可产生黑褐色指甲。

(3) 甲下或周缘有铜绿假单胞菌感染的甲沟炎存在时,指甲可呈黑色或蓝色。

(4) 慢性肾功能不全患者,常见指甲远端有明显的发黑。

(5) 维生素 B_{12} 缺乏、肾上腺皮质功能减退、胃肠息肉综合征,或长期接触煤焦油等,指甲也可变成灰黑色。

(6) 当指甲呈一片黑色或褐色,或者呈雀斑状,最常发生于拇指和踇趾,尤其是出现指甲及其周围组织也呈褐色或黑色时,这就意味着可能患有一种恶性肿瘤——黑色素瘤。

(7) 当指甲根部生长出数根黑色线条,通常只长在指甲的中部,提示体内正在或已经发生了癌变,应即刻去医院进行进一步检查,明确诊断,及早治疗 。

Ⅳ. Nail Shape and Disease

Part Ⅰ Dialogue

(M=Manicurist,C=Customer)

C:What shape is a healthy nail?

M:The surface of a healthy nail is full of circular arc.

C:What is the reason that nail appears horizontal grain?

M:It may be gastrointestinal disease,but it's also related to measles and mumps.

C:What's the reason for the floccus cloud?

M:Indigestion,constipation,a sign of roundworm.

C:What is the reason that nail appears vertical stripe?

M:It's the lack of vitamins or poor sleep.

Part Ⅱ Text

Nail Shape and Disease

(1) Nails protruding and extending to the point of flesh, indicating that chronic respiratory systems may develop diseases,such as emphysema and tuberculosis.

(2) That the nail stripes are disordered and dark brown is a response to dehydration and early kidney deficiency.

(3) Nails appear horizontal lines, may be gastrointestinal diseases, also related to measles mumps.

(4) Nails are floccus. There is often a sign of roundworm.

(5) Nail striping is a sign of lack of vitamins.

(6) Nail parts fall out, indicating a metabolic disorder.

(7) The nail is flat and spoonful with light and fragile, that is caused by iron deficiency anemia.

(8) The sudden thickening of the nails and peeling of the desquamation around nails indicate that there may be tumors in the respiratory tract or digestive tract.

(9) The middle of the nail and the whole nail becomes flat or spoonful, which often associated with the lack of iron, syphilis, thyroid disorders and rheumatism.

(10) The nail has an irregular concave spot, which is common in many patients with psoriasis.

(11) The concave of a line of nails, often caused by cluster baldness.

四、指甲形状与疾病

(一) 对话

(M=美甲师,C=顾客)

C:健康指甲是什么形状的?

M:健康指甲的表面呈饱满的圆弧形。

C:指甲出现横纹是什么原因?

M:提示可能存在胃肠疾病,也和麻疹、腮腺炎有关。

C:指甲出现絮状白云是什么原因?

M:消化不好,便秘,常是有蛔虫的迹象。

C:指甲出现竖的条纹是什么原因?

M:多是缺乏维生素、睡眠质量不好导致的。

(二) 课文

指甲的形状与疾病

(1) 指甲凸起并向指肉中卷伸,为肺气肿、结核病等慢性呼吸系统疾病的征兆。

(2) 指甲条纹紊乱,并呈深褐色,是脱水和肾虚初期的症状。

(3) 指甲出现横纹,表示可能存在胃肠疾病,也和麻疹、腮腺炎有关。

(4) 指甲出现絮状白云,常是有蛔虫的迹象。

(5) 指甲出现竖的条纹,多是缺乏维生素的表现。

(6) 指甲部分脱落,表明新陈代谢紊乱。

(7) 指甲平坦,凹陷成匙状,无光且脆弱,多是缺铁性贫血所致。

(8) 指甲突然变厚和指甲周围脱皮，表明呼吸道或消化道可能存在肿瘤。

(9) 指甲中间下陷，整片指甲变平或呈匙状，常和缺铁性贫血、梅毒、甲状腺功能障碍、风湿病等有关。

(10) 牛皮癣患者的指甲常出现不规则凹点。

(11) 指甲出现成行的凹状麻点，常与患簇状秃发症有关。

Ⅴ. Manicure tools

Part Ⅰ Dialogue

(M=Manicurist, C=Customer)

C: Do you have any professional tools for nails?

M: According to the types of manicures, the tools are differentiated.

Part Ⅱ Text

Manicure Tools

Nail tools are not disposable. They must be sterilized every time when they are used, so the size of the tool is appropriate in a disinfectant container.

(1) Crystal plier: Crystal pliers are only used when making crystal nails, and can not be replaced with finger plier, otherwise the crystal nails can be easily broken.

(2) Nail clipper: Nail clippers are mainly used to trim all types of nails, including crystal nails and natural nails.

(3) Finger plier: Finger pliers are used to cut off unwanted nail skin.

(4) Plastic or bristle brush: Plastic or bristle brushes are used for hand care cleaning nails and crystal nails.

(5) Nail file: Nail files are used to trim natural nails.

(6) Sand stick: Sand sticks are used to push the leather.

(7) Scissor: Scissors are used for cutting fiber products such as nylon, silk and

fiberglass nail.

(8) Small tweezer: Small tweezers are used to clip nails and diamonds or to clip their fingernails for pruning.

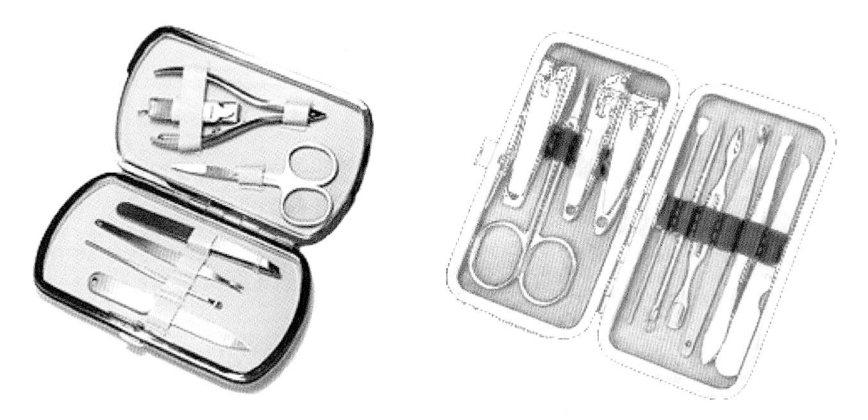

1. Nail scissor

(1) Classification: Nail scissors have big size or small size. Nail scissors have a flat head and a bevel face, depending on the shape of the front end.

(2) Usage: After washing your hands, you can cut out the desired length with a flat nail scissor. If the nail groove's on both sides of the nail are long in the direction of the nail, using bevel nail scissor to cut the nails on both sides.

(3) Caution: Whether you cut your nails with a flat nail scissor or a bevel nail scissor, you can't cut them too deep. If the nail is often cut deeper, the bed will become shorter and shorter, which will affect the beauty of the nail, especially women. Do not cut off the corners of the nail while fixing the square nails.

2. Nail file

(1) Classification: Nail file with steel file and color file, which is commonly referred to as flower file.

(2) Usage: Use a steel file or a flower file to cut the length of the nail and grind it into the desired shape.

There are usually six types of nails: square, squircle, oval, pointed, round, trumpet shape. Manicurists can advise customers to choose their own shape according to the customer's hand shape.

(3) Caution: When using the steel file or flower file grinding nails, it is important to note that on either side of the nail grinding to fine. The shape of the front end must be rounded.

3. Manicure bowl

(1) Classification: There are imported and homemade bowls. The professional bowl should be just the shape of a hand which fits nicely into the shape of the bowl.

(2) Usage: We pour the foam or warm water into the bowl and soak the left hand first, five minutes later change the right hand, which will clean your nails and soften the skin.

(3) Caution: Cold water and too hot water should not be put in the manicure bowl.

4. The skin softener

It is a kind of milky liquid, which can accelerate the softening.

(1) Usage: Use a towel to dry the soaked hands and apply them evenly on the surface of the nails.

(2) Caution: Do not apply the softener to a cover to prevent the nail from being softened.

5. Push skin stick

(1) Classification: Push skin sticks are divided into wooden push rod, steel push rod and push leather bar.

(2) Usage: The professional salon use more for steel bar, flat head with ellipse aging refers to the side of the finger skin hand direction, in order to make a cover appears slender. Use the other end of the scraper to scrape off the cutin from the nails.

(3) Caution: Apply force to the skin in moderation, not too hard, so as not to damage the methyl, otherwise it will affect the growth of the nails.

6. Leather plier

(1) Classification: The leather pliers are made of stainless steel with scissors (curved fingers) and pliers.

(2) Usage: Using a leather tongs to cut off the dead skin and spines that have just been removed, making the fingers look neat and neat.

(3) Caution: When using the pliers, you should be careful not to pull it, and cut it directly, so as not to damage the skin, and not too deep.

7. Nail art massage oil

(1) Classification: Nail art massage oil is also called nourishment or nail oil, can be divided into massage oils including apricot, vitamin A, vitamin E and other nutrients.

(2) Usage: Take a small amount of massage oil and apply it around the skin with a little massage. It moisturizes the skin and prevents the spines from growing around the nails, making the skin soft and protecting the healthy and bright nails.

(3) Caution: The use of nutrient oil daily should not be too much, otherwise it will be too greasy.

8. Polishing file

(1) Classification: There are three-polish bar and four-polish block.

(2) Usage: Generally, in accordance with black, white, grey order, the use of polishing, black surface may throw to cutin on the surface of the nails, white face can throw the nails surface thinner, grey surface can be cast surface bright. Nails can appear crystal shine after the procedure.

(3) Caution: If the customer's nails are thinner, do not use the coarser side of the polished file, otherwise the nails will be thinner and thinner. Don't rub it back and forth, because the heat generated by friction is uncomfortable.

9. Bottom oil

(1) Classification: Base oil has calcium base oil and protein bottom oil, moisturizing bottom oil, etc.

(2) Usage: After the nails are polished, the oil is added on the bottom. According to the customer's nails to choose base oil, such as soft nail can be used to add calcium oil. Applying bottom oil before nail polish can prevent nails become yellow, nutrition effect.

(3) Caution: The oil must be coated before applying nail polish in professional nails.

10. Nail polish

(1) Classification: common oil and quick-drying oil.

(2) Usage: ①Dark nail oil painting: Dark nail oil is not too much, otherwise it will be thick and uneven. Applying 2 to 3 times, each time thin, the effect will be better. ②Light nail oil painting: Light colors such as pink polish series improper use is easy to show traces of daub is uneven, special attention should be paid when the first layer in besmear nail oil brush dips in quantity and the gradient of the polish, and as soon as possible in the first layer when working with the second floor, it is very important. ③Pearlescent nail oil painting: Pearl shell oil is easy to do, take a bit more on the brush dips in nail, apply as soon as possible, otherwise it will appear uneven, so brush should be used vertically, to avoid leaving traces and put on both sides, after besmear in the middle.

(3) Caution: When coating oil, we do not apply to the skin of the finger and the nail groove. It will be messy. It will affect the breath of the nail.

11. Bright oil

(1) Classification: Bright oil is divided into ordinary bright oil and UV bright oil.

(2) Usage: Bright oil on dry nail polish, which protects nail polish and elongation. When making crystal nail, use anti-yellow, anti-UV bright oil.

(3) Caution: The oil should not be too thick.

12. Manicure equipment

Besides necessary workbench, chair, storage ark, illuminant, and so on, manicure professional equipment still have the following goods.

(1) Cushions: A towel wrapped in a sponge for the arms of a manicure.

(2) Drill: Used to clean the front of the nails.

(3) Humidifier: Care for dry split hands.

(4) Glass bowl: Soak crystal nails.

(5) Hand bowl: Soak your hands.

(6) Foot bowl: For foot care.

(7) Scuff: Remove the calluses from your feet.

(8) Finger model: Practice crystal nail polish.

13. Nail art staple

(1) Bottom oil: Transparent or milky white, before applying nail polish, can enhance the adhesion of nail polish.

(2) Nail polish: Dark pigment, chosen according to need.

(3) Soften frost: Usually contains glycerin, which removes the aging cuticle and keeps the skin moist and used for massage.

(4) Bleach: Hydrogen peroxide or citric acid, used to remove stains from crystal nails.

(5) Nail polish thinner: For diluting thicker nail polish, don't use nail polish remover.

(6) Nail polish essence: The ability to make natural nails stronger and replace base oil in foot care.

(7) Bright oil: Used to protect nail polish and keep it shiny. The thicker the oil is, the longer the drying time, the higher the gloss.

In addition, because professional nail needs strict aseptic operation, therefore, such as disposable paper towel, cotton balls, scraper, 75% alcohol, hemostatic wipes, antibacterial agent, baking soda, fungicide, and corrosion inhibitor are indispensable.

14. Nail polish supplies

(1) Polished sponge: It is used in conjunction with dried pink nail or cream nail polish to polish the nails. Always go in one direction while polishing, avoid the back-and-forth, and the sponge must be replaced.

(2) Sander: Similar to a rectangular sponge, the surface is covered with sandpaper and used with oil to polish the crystal nails (only for the polishing of crystal nails).

(3) Size 100 sander: The granule is coarser and is used for a large area of the crystal nail service, as well as the shape of the crystal nail.

(4) Size 180 sander: The finer particles are applied to the top of the nail skin and the top of the crystal nail, making it smoother.

(5) Foot sand-board: Remove calluses from the feet.

(6) Sand stick: Used to remove the bumps on natural nails and stains on the skin.

(7) Nail file: For natural nails only, one end is thicker and the other is thinner.

五、常用的美甲工具

(一)对话

(M=美甲师,C=顾客)

C:美甲有专业的工具吗?

M:根据美甲的种类来区分工具。

(二)课文

美 甲 工 具

美甲工具不是一次性用品,每次使用前必须进行消毒,因此,工具的大小以能放进消毒液容器中为宜。

(1) 水晶钳:仅在做水晶指甲时使用,不能用指皮钳代替,否则易造成水晶指甲破裂。

(2) 指甲剪:主要用于修剪所有类型的指甲,包括水晶指甲和天然指甲。

(3) 指皮钳:用于剪去多余的指甲皮。

(4) 塑料或鬃毛刷子:用于手部护理时清洁指甲及水晶指甲。

(5) 指甲锉:用于修整天然指甲的前缘。

(6) 砂棒:用于推指皮。

(7) 剪刀:用于裁剪纤维制品,如尼龙、丝绸和玻璃纤维甲。

(8) 小镊子:用于夹持指甲片、钻石,或夹住指甲皮以便修剪。

1. 指甲剪

(1) 分类:可以按指甲剪的大小分类,也可按指甲剪前端的形状来分,有平头和斜面两种。

(2)使用方法:在洗净双手后,先用平头指甲剪剪出所需的长度,如果指甲两侧的甲沟太深,且指甲往甲沟方向生长,应用斜面指甲剪剪掉两边的指甲。

(3) 注意事项:在剪指甲时不管是用平头指甲剪,还是斜面指甲剪,都不可剪得太深,如果经常把指甲剪得较深,那么甲床会变得越来越短,这样会影响指甲的美观,尤其是女性。在修方形指甲时指甲前端的两个角不要剪去。

2. 指甲锉

(1) 分类:指甲锉分钢锉和彩色锉条两种,彩色锉条也就是通常所说的花锉。

(2) 使用方法:将剪好长短的指甲用钢锉或花锉按先两侧后前端的顺序修磨,修磨成所需的形状。

通常指甲有六种形状:方形、方圆形、椭圆形、尖形、圆形、喇叭形。美甲师可以根据顾客的手形建议顾客选择适合自己的形状。

(3) 注意事项:在使用钢锉或花锉修磨指甲时,一定要注意指甲两侧的修磨要精细,前端的形状一定要圆润。

3. 泡手碗

(1) 分类:泡手碗有进口和国产两种,专业泡手碗应该刚好是一只手的形状,将手放在上面正好与碗的形状相吻合。

（2）使用方法：将泡手液或温水倒入泡手碗中，先浸泡左手，5 min 后再换右手，这样既可以清洁指甲，又可以松软指皮。

（3）注意事项：泡手碗内不可倒入凉水和太热的水。

4. 指皮软化剂

指皮软化剂是一种乳白色的液体，可加快指皮软化速度。

（1）使用方法：将浸泡过的手用毛巾擦干，将指皮软化剂均匀地涂在指甲表面。

（2）注意事项：不要将指皮软化剂涂在甲盖上，防止甲盖被软化。

5. 推皮棒

（1）分类：推皮棒分为木推棒、钢推棒和推皮砂棒。

（2）使用方法：专业美容店用的多为钢推棒，用椭圆扁头的一面将手指上老化的指皮往手心方向推动，以使甲盖显得修长。再用另一头的刮刀刮净残留在指甲上的角质。

（3）注意事项：推指皮时用力应适度，不可用力过猛，以免损伤甲基，影响指甲的生长。

6. 指皮钳

（1）分类：指皮钳一般是用不锈钢材料制成，有剪刀形（弯的指皮剪刀），也有钳子形。

（2）使用方法：用指皮钳剪去刚推完的死皮、肉刺，使手指显得美观整齐。

（3）注意事项：使用指皮钳时应注意不可拉扯指皮，应直接剪断，以免损伤指皮，且不可剪得太深。

7. 美甲按摩油

（1）分类：美甲按摩油也叫营养油或甲缘油，可按成分分为含杏仁成分、含维生素 A 及含维生素 E 等营养物质的美甲按摩油。

（2）使用方法：取少量美甲按摩油涂在修剪过的双手指皮周围，用手指稍加按摩。它能滋润指皮，防止指甲周围长肉刺，使皮肤柔软，并保护指甲健康、亮泽。

（3）注意事项：营养油可每天使用，用量不宜太多，否则会显得太过油腻。

8. 抛光锉

（1）分类：有三面抛光条和四面抛光块两种。

（2）使用方法：一般按照黑、白、灰的使用顺序依次抛光，黑色面可抛去指甲表面的角质，白色面可把指甲表面抛得更细，灰色面可把表面抛亮，经过这三道程序后指甲会显得晶莹亮泽。

（3）注意事项：如果顾客的甲盖较薄，不可用四面抛光块中最粗糙的那一面抛，否则指甲会越抛越薄。抛光时切勿来回摩擦，因为摩擦产生的热度会令人不适。

9. 底油

（1）分类：底油有加钙底油、蛋白质底油和保湿底油等。

（2）使用方法：指甲抛光后上底油时，应根据顾客的指甲质地来选择底油，如顾客指甲较软即可用加钙底油，如需上甲油，可在涂甲油之前上底油，可防止指甲变黄，起到营养作用。

（3）注意事项：专业美甲中涂甲油之前必须涂底油。

10. 甲油

（1）分类：甲油分普通型和快干型。

（2）使用方法：①深色甲油的涂法：深色甲油第一遍涂的量不宜太多，否则会厚重、不均匀。涂 2～3 遍，每一遍薄一些，效果会较好。②浅色甲油的涂法：粉色等浅色甲油系列使用不当很容易露出涂抹不均匀的痕迹，在涂第一层时需特别注意甲油的蘸取量和刷甲油的倾斜度，并在第一层未干时尽快涂第二层，这一点非常重要。③珠光甲油的涂法：珠光甲油容易干，在刷上蘸取稍多一些甲油，尽快涂好。否则会显得不均匀，因此刷子应直立使用，为避免留下痕迹，应先涂两边，后涂中间。

（3）注意事项：涂甲油时不可涂到指皮、甲沟上，否则容易显得脏乱，也会影响指甲的呼吸。

11. 亮油

（1）种类：亮油分普通亮油和 UV 亮油。

（2）使用方法：亮油涂在干后的甲油上面，能保持甲油的亮泽和延缓脱落时间。做水晶甲后，要使用防黄、防 UV 的亮油。

（3）注意事项：亮油不可涂得太厚。

12. 美甲专业设备

美甲专业设备除了必需的工作台、椅、储物柜、光源等用品外，还包括以下物品。

（1）垫枕：以毛巾包裹海绵，用于托垫顾客的胳膊。

（2）钻子：用于清洁指甲的前缘。

（3）加湿器：用于干裂手的护理。

（4）玻璃碗：用于浸泡水晶指甲。

（5）泡手碗：用于浸泡手。

（6）脚盆：用于脚的护理。

（7）磨脚板：用于去除脚上的老茧。

（8）手指模型：用于练习做水晶指甲。

13. 美甲必备用品

（1）底油：透明或乳白色，涂甲油前使用，可增强指甲油的附着力。

（2）甲油：含深色素，根据需要选用。

（3）软化霜：通常含有甘油，能够去除老化的角质层，保持皮肤的润泽，用于按摩和对干裂手的护理。

（4）漂白剂：含有过氧化氢或柠檬酸，用于去除水晶指甲上的污渍。

（5）甲油稀释液：用于稀释较黏稠的甲油，切勿用洗甲水代替。

（6）指甲精华素：能使天然指甲更坚固，可在脚护理中代替底油。

（7）亮油：用于保护甲油，保持其光泽。亮油越黏稠，干燥时间越长，光泽度越高。

此外，由于专业美甲需要严格完善的无菌操作，因此一次性纸巾、棉球、刮刀、75%的酒精、止血巾、抗菌剂、小苏打、杀菌剂、防锈剂等都是必不可少的。

14. 指甲打磨用品

（1）抛光海绵：与干粉指甲或膏状甲油配合使用，用于指甲的打磨抛光。抛光时要始终沿一个方向进行，切忌来回打磨，海绵用后必须更换。

（2）打磨砂块：类似长方体海绵，表面贴有砂纸，与甲油配合使用打磨水晶指甲（仅用于水晶指甲的打磨）。

(3) 100 号打磨砂条:颗粒较粗,用于大面积的水晶指甲打磨服务,也用于修整水晶指甲的形状。

(4) 180 号打磨砂条:颗粒较细,用于指甲皮周围和水晶指甲顶端的打磨,使其更加光滑平整。

(5) 脚砂板:用于去除脚上的老茧。

(6) 砂棒:用于去除天然指甲上的凸起和皮肤上的污点。

(7) 指甲砂锉:仅用于天然指甲,一端颗粒较粗,另一端较细。

Ⅵ. Real Nail Care

Part Ⅰ Dialogue

(M=Manicurist,C=Customer)

C:What is real nail care?

M:Real nail care is to make your nails bright,clean and smooth.

C:All right. I know. Thanks!

Part Ⅱ Text

Hand Care and Nail Care

1. Hand care

It is important to choose the right hand care products,but the right approach is more important. Regular,comprehensive and persistence care of the hands can make the skin moist and greasy,the fingers flexible and soft,the advice is best to do once every week.

As the first step,soften horniness:Drip into the right amount of olive oil in warm water,immerse hands finish,keep for 15 minutes,can improve dry rough skin,soften cutin,midway if water cool please add hot water appropriately.

Second,deeply clean:Take a small amount of peeling cream or scrub containing protein mixture of hand care lotion,then gently massage the palm and wrist,especially around the nail prone to hard skin agnail of parts and tucks and knuckles,can remove dead skin and promote cell metabolism, and can be taken advantage of the trim nails, nails are not easy to break. If you don't have a scrub, you can use a cleansing cream with salt. Rinse off with warm water after you have finished.

The third step, relieve and repair: With the soothing hand repair cream smear on the hand, pay attention to select the products that contain vitamins and protein, can help to promote cell metabolism and quickly improve skin elasticity, make skin soft and moist.

The fourth step, the hand massage: Evenly apply massage frost, start from the palm of the finger to massage to the base of the finger, the movement should be gentle. Then massaging the palm of the hand and using the knuckles to press the points in the palm of the hand. Use your forefinger and middle finger to hold your finger and rotate it from the roots to the tips. Also pay attention to the mouth and joints, for about 10 minutes. Meanwhile, apply the oily moisturizing lotion to the knuckle and rough position on the back of the hand, which can soften the rough skin and joint position, and also nourish the protection.

Fifth, deep care: Get started after the film, with tin foil, hot towels or cotton gloves packages about 10 minutes, help to consolidate the subcutaneous tissue and deeply moisturize skin.

Sixth, hand aerobics: Hands flat on the table, gently pressing down on them, lift a finger at a time, try to give high, stretch palms and fingers, can make the hands agile. Raise your hands above your head, clench your fist, then try to stretch each finger out as far as you can. For 5 minutes, you can reduce the amount of tension in the back of your hands, relieve the tension, and make your hands soft.

Seventh, perfect protection: Finally coated anti-wrinkle cream or hand cream, strengthen moist skin and lock absorbed nutrients, let the hands quickly reply and delicate skin smooth.

After basic understanding of the importance of hand care, it is time to look for a care product. Ordinary emulsion cannot fully meet the needs of the hands, so you need to specifically for hand care products, try to choose the sort of hydrated, non-greasy texture. hand care product is best to carry, which can daub at any time.

2. Nail care

(1) Use 75% of alcohol to disinfect yourself and your customers.

(2) Check your nails, wash your nail polish, and stain your nails.

(3) File a form with a fine file.

(4) Use warm water for 3 minutes to soften the dead skin.

(5) Coat your nails with softener and exfoliate skin and cut off excess dead skin.

(6) Use a sponge file to remove excess oil from the nail.

(7) Use a three-color polishing file to polish and polish the paraffin (gently, quickly).

(8) Surround your nails with nourishing oil and massage.

六、真甲保养护理

(一) 对话

(M=美甲师,C=顾客)

C:什么是真甲保养护理?

M:让自己的指甲光亮、干净、整齐。

C:好的,我知道了,谢谢。

(二) 课文

手部护理和指甲护理

1. 手部的护理

漂亮嫩滑的双手得益于精心的护理,选择合适的护手产品固然重要,但正确的护理方法更不可少。定期对双手进行全面护理,可使皮肤滋润滑腻,手指灵活柔软,建议最好每周做一次护理。

第一步,软化角质:在温水中滴入适量橄榄油,把双手完全浸入水中,保持 15 min,可改善皮肤干燥粗糙现象,使角质软化,中途若水变凉,请适当添加热水。

第二步,深层清洁:取少量去死皮霜或含有蛋白质的磨砂膏混合手部护理乳液均匀涂抹双手,然后轻轻按摩整个手掌及手腕,尤其是指甲周围容易产生硬皮和倒刺的部位以及虎口和指关节部位。深层清洁能够去除死皮和促进细胞新陈代谢,同时可趁机修剪指甲,此时指甲不易断裂。如果没有磨砂膏,可用洗面奶加盐进行清洁。做完后用温水洗掉。

第三步,舒缓修复:用有舒缓作用的手部修护乳涂抹于手部,注意选择含有维生素及蛋白质的产品,能帮助促进细胞新陈代谢及迅速改善皮肤弹性,令皮肤恢复柔软润泽。

第四步,手部按摩:均匀涂抹按摩霜,先从手背指尖开始按摩到手指根部,动作要从容柔和,然后螺旋状按摩手掌,并用指关节轻按手心上的穴位,最后用食指和中指夹住手指,从根部向指尖螺旋状旋转拉伸,每一根手指都按摩到。同样要注意虎口和关节部位,大约持续 10 min。同时涂抹较油性的滋润护肤膏于手背上的指节及粗糙位置,可软化粗糙皮肤及关节,还能充分滋润防护。

第五步,深层护理:涂上手膜后,用保鲜纸、热毛巾或棉手套包裹约 10 min,有助于巩固皮下组织及深层滋润肌肤。

第六步，手健美操：把双手平放在台面上，轻轻地向下压，每次举起一个手指，尽量举高，伸展手掌和手指，可使双手轻快敏捷；双手高举过头，握紧拳头，然后尽量向外伸展每根手指，做 5 min，可减少手背青筋暴露，解除紧张，使手部柔软。

第七步，完美保护：最后涂抹防皱润肤霜或护手霜，加强润泽肌肤及锁住已经吸收的养分，让双手皮肤迅速恢复娇嫩柔滑。

基本了解手部护理的重要性后，应该寻找合适的护理产品。普通的乳液不能完全满足手部护理的需要，所以需要专门的手部护理产品，尽量选择能够补充水分且不油腻的产品。手部护理产品最好随身携带，可以随时涂抹。

2. 甲部护理

(1) 美甲师用 75%的酒精消毒自己和顾客的双手。

(2) 检查指甲，清洗甲油，清洗甲缝的污垢。

(3) 用细锉修整甲形。

(4) 用温水泡手 3 min，软化死皮。

(5) 指甲周围涂上软化剂，并推移死皮，再将多余的死皮剪掉。

(6) 用海绵锉去掉甲面多余的油光。

(7) 用三色抛光锉依次抛光，并打磨甲蜡(力度要轻，摩擦的频率要快)。

(8) 在指甲的周围涂上营养油并按摩。

Ⅶ. The Type and Characteristics of the Nail

Part Ⅰ Dialogue

(M=Manicurist, C=Customer)

C: What is the characteristic of square nails?

M: Generally speaking, the square fingernails are individualized and lead the trend, not easy to break, liked by professional women.

C: What is the characteristic of fang yuan shaped nails?

M: Fang yuan shaped nails front and sides are straight, with a circular contour in edges and corners, this looks very strong shape and can give a person with soft feeling. For joints, fingers slender customers, fang yuan shape can make up for deficiencies.

C: What is the characteristic of oval nails?

M: Oval nails, starting from the free edge, the outline of the front end of the nail is oval, the traditional oriental nail.

C: What is the characteristic of the pointy nail?

M: Nails are easily broken because of small contact area, while the nail of the Asian is thin and not suitable for the tip shape.

Part Ⅱ Text

The Type and Characteristics of the Nail

In the life of the most commonly used shapes are square shape, fang yuan shape, oval shape, pointy shape, round shape, square, you can according to your own hand shape and preferences to choose the perfect shape.

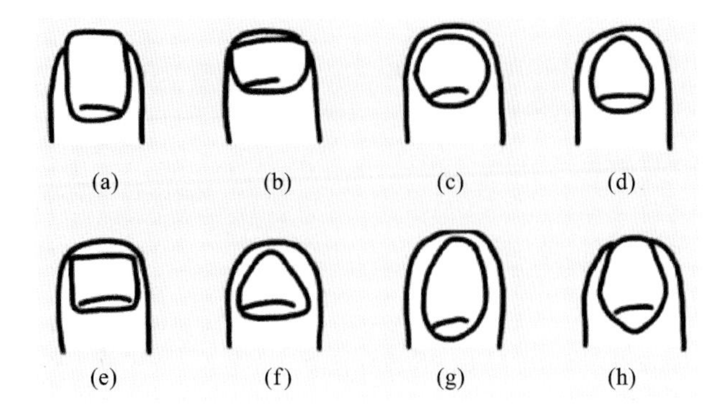

1. Square nails

Generally speaking, the square nails are individualized and lead the trend, not easy to break, liked by professional women.

2. Fang yuan shaped nails

Fang yuan shaped nails front and sides are straight, with a circular contour in edges and corners, this looks very strong shape and can give a person with soft feeling. For joints, fingers slender customers, fang yuan shape can make up for deficiencies.

3. Oval nails

Oval nails, from the beginning of the free margin, the outline of the front of the nail is oval, the traditional oriental nail.

4. Pointy nails

Pointy nails are easily broken because of small contact area, while the nails of the Asian are thin and not fit for the tip shape.

5. Circular nails

Suitable for long hands and good fingers.

6. squared nails

It is commonly found in a person with fragile nails, can be trimmed into square, or square circle.

七、甲形特点

(一)对话

(M=美甲师,C=顾客)

C:方形指甲的特点是什么?

M:一般来说,方形的指甲有个性,能够带领潮流,不易断裂,比较受职业女性喜欢。

C:方圆形指甲的特点是什么?

M:方圆形的指甲前端和侧面都是直的,棱角的地方为圆弧形,这种甲形看上去很结实,会给人以柔和的感觉,对于骨节明显,手指瘦长的顾客,方圆形可以弥补其手指不足之处。

C:椭圆形指甲的特点是什么?

M:椭圆形的指甲,从游离缘开始,到指甲前端的轮廓为椭圆形,属于传统的东方甲形。

C:尖形指甲的特点是什么?

M:尖形的指甲由于接触面积小,易断裂,而亚洲人的指甲较薄,不适合修成尖形。

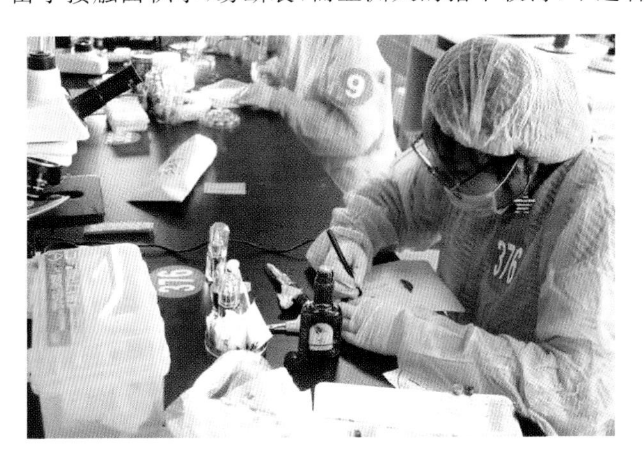

(二)课文

甲形特点

生活中常用的甲形有方形、方圆形、椭圆形、尖形、圆形和扇形六种,你可以根据自己的手形和喜好选择适合的甲形。

1. 方形指甲

一般来说,方形的指甲有个性,能够带领潮流,不易断裂,比较受职业女性喜欢。

2. 方圆形指甲

方圆形的指甲前端和侧面都是直的,棱角的地方为圆弧形,这种甲形看上去很结实,会给人以柔和的感觉,对于骨节明显,手指瘦长的顾客,方圆形可以弥补其手指不足之处。

3. 椭圆形指甲

椭圆形的指甲,从游离缘开始,到指甲前端的轮廓为椭圆形,属于传统的东方甲形。

4. 尖形指甲

尖形的指甲由于接触面积小,易断裂,而亚洲人的指甲较薄,不适合修成尖形。

5. 圆形

适合手修长,自身手指长得好看的人。

6. 扇形

常见于指甲易碎者,可修成方形或方圆形。

Section 2 Cosmetics

Ⅰ. The Meaning of Color Makeup and Facial Classification

Part Ⅰ Dialogue

A: Do you like makeup?

B: Yes.

A: When did you start to learn to make up for yourself?

B: When I was in high school, I saw my mother doing makeup, and I began to learn to dress like her.

A: So how can you create beautiful makeup?

B: Using the modern makeup technology with products, according to the physical characteristics of different individuals, the need of work and life, and reasonable design. Then we take a certain skin care cream makeup and a series of means and methods, so as to foster strengths and circumvent weaknesses to increase charm for the purpose of system theory and technology.

A: Yes, you can tell me concretely. How can I make myself more beautiful?

B: Ok, you have a long face because of higher forehead hair, bigger chin. We avoid showing all the face, so do a row of bangs, try to make both sides of the hair feel fluffy, do not suitable for straight hair.

Part Ⅱ Text

1. The concept and characteristics of beauty and makeup

What we call beauty makeup, is based on human medical science. On the basis of aesthetic psychology in human society, using the modern makeup technology with products, according to the physical characteristics of different individuals, the need of work and life, and reasonable design, we take a certain skin care, beauty, decoration, cosmetic and a series of means and methods to foster strengths and circumvent weaknesses to increase charm for the purpose of system theory and technology.

The mainstream of today's beauty makeup:

highlight the beauty of your personality.

The three states of makeup: basic makeup, correct makeup, style makeup.

Characteristics: we must suit our measures to different conditions in terms of locality, time, issue and people concerned.

2. Common face shapes

There are seven common face types: oval face, round face, long face, square face, equilateral triangle face, pour trigonometry face and lozenge face.

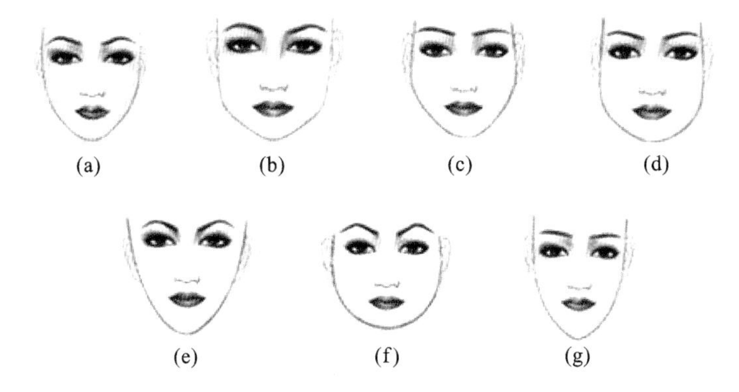

(1) Oval face: Oval face is shaped like goose egg, also called goose egg face. This is a more standard face, a lot of hairstyles can suit, and can achieve very harmonious effect.

(2) Round face: The round cheek is plump and the forehead and chin are round. Round face gives a person with gentle and lovely feeling. More hairstyles can suit it. We just need to modify a bit of hair on both sides to move forward.

(3) Long face: The long face has a higher forehead, a larger chin and a longer face. That not suitable for straight hair. Avoiding the face all reveal, the fringe should be made a row, try to make both sides of the hair feel fluffy.

(4) Square face: The square face lacks downy feeling, should pay attention to downy hairstyle when doing hairstyle, can grow a bit of hair. For example: long head hair, long hair, xiu zhi hairstyle.

(5) Equilateral triangle face: Equilateral triangle face like "pear", so it's also called the pear-shaped face. It has a narrower head and a wider lower jaw. Bang should be cut a thin layer, hanging down. It is better to be cut the length of the eyebrows, make it into your cheeks with more hair. Such as: students' hairstyle, shoulder-length hairstyle. This type of face is not suitable for long straight hair.

(6) Pour trigonometry face: Pour trigonometry face is a narrow, heart-shaped face, which is the opposite of the triangle face. The feature is opposite to the triangle face. When doing hairstyle, we should focus on forehead and chin. Bang can do together in a row. Hair length exceeds chin 2 centimeters to appropriate, and curl inside, increase chin width.

(7) Lozenge face: Lozenge face is a narrow frontal angle with prominent cheekbones and a pointed chin. When designing your hairstyle, we should focus on the prominent area of your cheekbones, use your hair to polish your front cheeks, and give your forehead a shaggy and

wide forehead.

3. Skin type

(1) Normal skin: The pH value is 5 to 5.6. The skin is red and delicate, rich in elasticity. The pore is small. The skin is neither dry nor oily.

(2) Dry skin: The pH value is 4.5 to 5, the pore is small, the skin is dry, easy to make small wrinkles.

(3) Oily skin: The pH value is 5.6 to 6.6. The pore is large. The sebum secretes a large amount. The skin is oily, bright and easy to have acne.

(4) Sensitive skin: The skin of sensitive skin is thinner. It is sensitive to external stimulation and swollen to wait for symptom. When applying makeup, we should reduce the stimulation of cosmetics to skin as far as possible, notice eye ministry and labial ministry, in order to reduce the occurrence of allergy.

4. The proportion of facial features

The length of the nose is one third of the face. The lower lip is a third of the lower margin of the nose to the lower margin of the nose. The widest part of the nose should not exceed the distance between the eyes. Looking up from the bottom, the nose is going to be a positive triangle. The height of the nose is 67% of the length of the nose. The width of the base of the nose is equal to the width of the nose which is half the width of the flanks. The angle between the nasal column and the philtrum is about 100 to 110 degrees for women, and men are about 90 to 95 degrees. Elongated by the nose and chin, the height of the nose is about 34 degrees for women and 36 degrees for men. From the side view, the tip of the nose, lower lip and chin are joined together in a straight line with the lower lip above the lower lip 1 to 2 mm, the lower lip is 1 to 2 mm above the chin. The width of both flanks and the width of a single eye are about a fifth of the distance between the two sides of the ear. The distance between the eyes and the pupil is 42% of the distance between the two ears.

一、彩妆的含义及面部分类

(一) 对话

A:你喜欢化妆吗?

B:喜欢呀!

A:你是什么时候开始学会给自己化妆的?

B:应该是上高中的时候吧,看见妈妈在化妆,我也开始学着她的样子给自己化妆。

A:那怎么样才能打造出漂亮的妆容呢?

B:运用现代化妆技术配合化妆产品,根据不同个体的身体特点、工作及生活的需要,进行合理的设计,并采取皮肤护理、美容、修饰、化妆等一系列手段和方法,以达到扬长避短、增加魅力的目的。

A:知道呀,那你能具体给我说说,我这种脸形怎么才能把自己打造得更漂亮?

B:好的,你是长脸。因为前额发际线较高,下巴较大且尖,脸庞较长,要避免把脸部全部露出，所以可以剪一排刘海，尽量让两边头发有蓬松感，不宜留长直发。

（二）课文

彩妆的含义及面部分类

1. 美容化妆的概念和特点

我们所说的美容化妆，是以人体医学科学为基础，以人类社会审美心理为标准，运用现代化妆技术配合化妆产品，根据不同个体的身体特点、工作及生活的需要，进行合理的设计，并采取皮肤护理、美容、修饰、化妆等一系列手段和方法，以扬长避短、增加魅力为目的的系统理论和技术。

当今美容化妆的主流：突出个性美。

化妆的三境界：基础化妆、矫正化妆、风格化妆。

特点：因人制宜、因时制宜、因地制宜。

2. 常见的脸形

常见脸形有七种：椭圆形脸、圆形脸、长方形脸、方形脸、正三角形脸、倒三角形脸及菱形脸。

（1）椭圆形脸：形似鹅蛋，故又称鹅蛋脸，是一种比较标准的脸形，适合多种发型，并能达到很和谐的效果。

（2）圆形脸：颊部比较丰满，额部及下巴偏圆。圆圆的脸给人以温柔可爱的感觉，较多的发型都能适合，只需稍向前修饰一下两侧头发就可以了。

（3）长方形脸：前额发际线较高，下巴较大且尖，脸庞较长。避免把脸部全部露出，可以剪一排刘海，尽量使两边头发有蓬松感，不宜留长直发。

（4）方形脸：较阔的前额与方形的腮部。方形脸缺乏柔和感，做发型时应注意选择柔和发型，可留长一点的发型，如：长穗发、长毛边或秀芝发型。

（5）正三角形脸：形似"梨"，又称梨形脸。头顶及额部较窄，下颚部较宽。刘海可削薄薄一层，垂下，最好剪成齐眉的长度，也可用较多的头发修饰腮部，如：学生发型、齐肩发型，不宜留长直发。

（6）倒三角形脸：上宽下窄，形似"心"，又称心形脸，特征与正三角形脸相反。设计发型时，应重点注意额部及下巴，可以做齐刘海，头发长度以超过下巴 2 cm 为宜，并向内卷曲，这样可以增加下巴的宽度。

（7）菱形脸：上额角较窄，颧骨突出，下巴较尖。设计发型时，重点考虑颧骨突出的地方，用头发修饰一下前脸颊，把额部头发做蓬松以拉宽额部的宽度。

3. 皮肤类型

（1）中性皮肤：pH 值为 5～5.6，皮肤红润细腻，富有弹性，毛孔较小，皮肤不干不油。

（2）干性皮肤：pH 值为 4.5～5，毛孔细小，皮脂分泌量少，皮肤较干燥，易产生细小皱纹。

（3）油性皮肤：pH 值为 5.6～6.6，毛孔粗大，皮脂分泌量多，皮肤油腻光亮，易产生粉刺、痤疮。

（4）敏感性皮肤：皮肤较薄，对外界刺激很敏感，易出现局部红肿、刺痒等症状。化妆时，应尽量减少化妆品对皮肤的刺激，注重眼部、唇部的护理，以减少过敏的发生。

4. 面部五官的比例关系

鼻子的长度是面长的 1/3。鼻子下缘至上下唇交接处占鼻子下缘至下巴长度的 1/3。鼻翼最宽部分不要超过两眼间距。由下往上看，鼻子要成正三角形。鼻头的高度为鼻子长度的 67%。鼻根部的宽度与鼻翼的宽度等长，即鼻根部的宽度为两侧鼻翼宽度的一半。鼻柱与人中之间的倾斜角度，女性为 100°～110°，男性为 90°～95°。从鼻根到下巴下拉一垂直线，鼻子隆起的高度与该垂直线形成的最佳角度，女性约为 34°，男性约为 36°。从侧面观，鼻尖、下唇、下巴三点共同连成一条直线。上唇比下唇突出 1～2 mm，下唇比下巴突出 1～2 mm。两侧鼻翼宽度与单眼的宽度各占两耳外侧距离的 1/5。双眼瞳孔间距离是两耳内侧距离的 42%。

Ⅱ. Makeup with different makeup

Part Ⅰ Dialogue

A: My sister is getting married. She brought beautiful wedding dress and jewelry.

B: That should also require beautiful bridal makeup to make her the most beautiful girl.

A: Yes, I want to make one of the most beautiful bridal makeup.

B: Do you know how to make the bridal makeup?

A: I'm not sure. Can you teach me?

B: Yes, the bridal makeup features: festive and elegant, makeup is round and gentle, gorgeous and not charming, generally with warm-tone.

(1) Foundation: The foundation should be suited to the characteristics of the bride's skin, such as defective shortcomings, not liquid and transparent powdery bottom, should choose hiding power stronger foundation, and put the naked together with colors to join.

(2) Blusher: Applying blusher should give all rosy blush bridal makeup shining and attitude. We start with the temples on your cheeks, eyes, cheekbones, and above the mandible, with a pale pink on top to show fullness and health of the face. Then we apply a bright red color around the outer corners of your eyes to connect with the surrounding colors.

(3) Powder: We should choose the transparent powder that is similar to skin color to fix makeup. It is to make the makeup face durable and not fall off. Powder must be uniform.

(4) Eye shadow: Eye shadow should be light brown. Warm-tone has a good sense of pleasure, don't paint complex multi-colored eye shadow.

Part Ⅱ Text

Makeup with Different Makeup

(1) Daily makeup is also called life makeup, light makeup. The daily makeup appears in the sunlight environment, for the ordinary person's daily life and work, the makeup must be done under the sunlight illuminant. Features: light elegance, natural coordination, no makeup.

(2) Evening makeup is also called the dinner makeup, heavy makeup. In the light

environment of a dance, party and other social occasions, the makeup must be made under similar lights. Characteristics: gorgeous and elegant, enchanting, display personality charm. The dress is gorgeous, the makeup color is bright and the contrast is slightly stronger, the facial painting can be moderately exaggerated.

(3) Bridal makeup characteristic is festival and elegance. Makeup is lubricious. Gorgeous makeup is not charming, generally with warm-tone. Bridal makeup should be lively, enchanting, natural, soft beauty. Vary according to season and dress.

(4) The dream makeup is also called body art painting, which originated from the ancient Indian tattoo. In the late 20th century, it rose makeup competitions and advertising design in Europe, the United States. Dream makeup is carrier with the body, use special makeup in the human body with bold and exaggerated gimmick to express a kind of the ideal that is like a dream. Features: originality.

(5) Makeup characteristics of different seasons

①Spring: The color of makeup can choose beige and ivory. Eye shadow can choose pale pink, pink, shallow green and so on. Lipstick can choose peach red, coral red and so on.

②Summer: Usually is given priority to with light or cold color. Base color should choose darker foundation, also can choose wet and dry powder; always use waterproof eyeliner, eyebrow, eye shadow and lipstick as shallow as possible.

③Fall: Generally warm color is given priority to. Choose the yellow, orange, wine red eye shadow and blush of the warm color department. Eyeliner and eyebrow can be sketched with deep coffee color, lipstick can use bright color such as orange, bright red, magenta.

④Winter: Makeup color is given priority to with warm color, the makeup of cool color department can also give a person a kind of noble and mysterious feeling. Moisturizing lotion and foundation are winter's preferred items. Powdery bottom can choose to be oily and heavy, but unfavorable too thick, want to have white tender transparent feeling. Eyebrow can choose deep coffee or dark grey, eye shadow should match with dress, lipstick can be bright red or orange red.

二、不同妆型的化妆

(一) 对话

A:我的姐姐要结婚了,她买了漂亮的婚纱、首饰。

B:那应该还需要美美的新娘妆,让她成为最美的女孩。

A:是的,我要给她化一个最漂亮的新娘妆。

B:那你知道,新娘妆怎么化吗?

A:我不太清楚,你能教我吗?

B:可以,新娘妆的特点是喜庆典雅,妆面圆润柔利,艳而不媚,一般以暖色调为主。

(1) 粉底:根据新娘本身皮肤深浅特点涂上合适的粉底,如有瑕疵缺点,不宜涂液体和透明粉底,应选择遮盖力较强的粉底,并将裸露的地方涂上粉底,颜色要衔接。

(2) 腮红:新娘妆应脸色红润,表现出神采飞扬的姿态,在面颊上从太阳穴开始,眼部、颧骨到下颌角以上，淡淡地涂上一层浅红，显示面部的丰满与健康；然后在外眼角周围涂一层较艳的红色，要与周围色衔接。

(3) 定妆粉：选用与肤色相近的透明粉定妆，是为了使妆面持久不脱落，粉要扑得均匀。

(4) 眼影：眼影要用浅棕色，偏暖的色调有喜悦感，不要涂复杂的多色眼影。

(二) 课文

不同妆型的化妆

(1) 日妆：也称生活妆、淡妆，日妆出现在日光环境下，用于一般人的日常生活和工作。化妆必须在日光光源下进行。其特点是清淡典雅、自然协调，不露化妆痕迹。

(2) 晚妆：也称晚宴妆、浓妆，晚妆一般出现在舞会、宴会等社交场合的灯光环境下，化妆时必须在类似的灯光下进行。其特点是艳丽高雅，妩媚，能够显示个性魅力。服装华丽鲜艳，妆色明暗对比略强，五官描画可适度夸张。

(3) 新娘妆：新娘妆的妆面应该明快、妩媚、自然、柔美。根据季节、服饰的不同而有所变化。其特点是喜庆典雅，妆面圆润柔利，艳而不媚，一般以暖色调为主。

(4) 梦幻妆：梦幻妆又称人体艺术彩绘，起源于古印第安人的纹身术。20 世纪末，兴起于欧美的化妆比赛和广告设计。梦幻化妆是以人体为载体，用特殊的化妆品在人体上以大胆夸张的手法表达一种如梦如幻的理想境界的化妆方法。其特点是创意。

(5) 不同季节的化妆特点

①春季：妆色偏亮丽，基础底色可选择米色、象牙色。眼影可选择浅桃红、粉红、浅绿色等。口红可选择桃红、珊瑚红等。

②夏季：一般以清淡或冷色为主。基础色应选用较深色粉底，也可选用干湿两用粉。画眼线一定要选用防水眼线液，眉笔、眼影、口红尽量浅淡。

③秋季：一般以暖色调为主，宜选偏暖色系的黄色、橙色、酒红色的眼影和腮红。眼线和眉毛可以用深咖啡色来勾画，口红可以用橙红、大红、玫红等艳丽的色彩。

④冬季：妆色以暖色系为主，冷色系的妆也会给人高贵神秘的感觉。保湿性强的化妆水、乳液和粉底是冬季首选用品。粉底可选用偏油性的，但不宜太厚，要有白嫩透明感。眉毛可选用深咖啡色或深灰色，眼影应与服装搭配，口红可用大红色、橙红色。

Section 3 Tattoo Makeup

Part Ⅰ Dialogue

(B=Beautician, C=Customer)

B: Good morning. Sit down, please. What can I do for you?

C: I'd like to know about Korean style semi-permanent.

B: Have you ever seen that before?

C: Yes, but I know little about it.

B: Sit up straight now, let me see. Do you often draw eyebrows?

C:Painting,I just think it's too much trouble to paint every day,so I want to make a semi-permanent.

B:Let me design it for you first.

C:Ok.

B:Ok,now I'm going to help you design it. You look at the shape of your face,so we should raise the eyebrows a little bit. The brow peaks a little bit,so that you can modify your face shape. Can you look at this eyebrow shape?

C:I don't think two eyebrows are the same. One is tall and one is low.

B:Beauty,each of us has two eyebrows that are different,not exactly the same. We are more rigid about the symmetry of the eyebrow.

C:Will it make me feel very painful?

B:No. It's all a stabilizer in advance. The whole process is painless.

C:What should be noticed after the operation?

B:Don't dip water in a week. Don't catch it with your hands. You should eat less spicy food and use a restorer sooner or later.

C:I see.

Part Ⅱ Text

Tattoo makeup

Tattoo beauty is the name of the lip embroidered eyebrow. Tattoo beauty is like embroidery. Its method is the method of the embroidery eyebrow shape,and then put the pigment into the subcutaneous tissue around 0. 2 to 0. 3 cm,make the pigment on the skin, and do not fade for a long time to achieve the goal of beauty. Tattoo beauty is a kind of beauty for the purpose of beauty, which will reproduce special exogenous pigment in the human body skin by professionals with the help of a special professional equipment,draw text into specific patterns that can exist for a long time,reaching the cosmetic effect.

1. Basic introduction

Tattoo is actually a traumatic skin color that creates a stable color in the skin tissue. Because the epidermis is thin and translucent,the pigment passes through the epidermis, showing the effect of color and lustre to conceal the defects and beautify. The pigment that penetrates the skin is smaller than one micron,and soon it is surrounded by collagen,unable to be swallowed by phagocytes,which forms the symbol and is also called tattoo. In ancient times,embroidery on silk was called "tattoo". In contrast to the literature,when the Han Dynasty embroidered on the cloth,it became known as "embroidery". These mainly refer to the continuation of the ancient "embroidery",and then to develop into the body. Now tattoo has three lines:eyebrow cut,eye line cut,lip line cut.

2. Tattoo common sense

Tattoo is a traumatic skin color that has certain risks. The tattoo is only allowed to succeed,so the operation of the tattoo should be done in safe condition. There must be specialized instrument tools, sterilized sanitation. It also requires the tattoo operators to have certain medical, aesthetic knowledge, and training skills. Environment, atmosphere, light and so on are the important conditions that relate to the tattoo.

The tattoo is applied to the dermis and the epidermis, the protection function of the skin after puncturing the epidermis is decreased, the bacteria are easy to invade, so the operation room of the tattoo beauty is to keep the air clean.

Calm and gentle atmosphere, good atmosphere can mediate a person's nerve to make a person calm and stability. In this environment, the tattoo artist can concentrate on the operation. The tattooed person should actively cooperate in the good environment to ensure the quality of the tattoo.

Lighting is an important condition for the cosmetic operation. Soft and comfortable light can reduce the visual fatigue of the tattoo operators, ensuring the quality of the tattoo.

3. Tattoo disinfection

The process of tattoo is a direct puncture of the skin, so as a qualified beautician, there should be strict disinfection and hygiene.

(1) Beauty operation instrument disinfection: All appliances exposed to the customer's skin should be sterilized. Tattoo needle should guarantee one person one needle, the item USES disposable disinfectant to prevent cross infection.

(2) Beauty environmental health: With a variety of pathogenic and harmful microorganisms in external environment, be sure to keep the operating environment clean and health, the air circulation, for sterile processing on a regular basis.

(3) Sterilize the skin of the tattoo area for the customer.

(4) The aseptic operation of the beautician: The beautician should clean the hands and then put on the sterile gloves. Once sterilized, the area is treated as a sterile area and only in the sterile area during operation.

(5) If the effect of medical cosmetology is not good, it may be related to sterilization.

4. Tattoo methodology

(1) Block processing.

When doing the labial surface coloring, the entire lip surface can be divided into four regions. Each region can be divided into 2-3 pieces according to the number of eyebrow tattoo machine. When you do it, you should carefully fill each piece of grain, and then make another piece. Note: Before you make the next piece, you must suck the liquid out of the pin cap, re-stain the color, and ensure the color effect.

When making the lip line your should take the paragraph treatment, it is to divide the lip line into a number of paragraphs, a paragraph of ground to carry a needle. When making the lip line it also should be stained with color and ensure the flow color is smooth. If you meet with a blood point or a stacking condition, you should avoid this area and do the color next time.

(2) Layer by layer of color method.

Operate labial surface coloring when using the length of the needle cap to adjust the length of the needle, for the first time coloring clockwise shows 1 mm long, the second time coloring clockwise with 1.5 mm long, third time coloring clockwise above 2 mm long. Note: The needle must be lowered when the needle is switched on, and the needle will be placed on the skin to ensure that all the needles are penetrated into the skin and the needle angle is 90 degrees.

(3) Reduce wipes but more permeation.

Operation of the lip color can't wipe edge profile thorn to let every time after coloring pigment on the lips keep 2-3 minutes, then wipe gently. Besmearing again after finish is wrong.

(4) Five words definitely.

①Light: The operator should be light. The left hand cannot pull the skin vigorously. The needle in the right hand is not too heavy.

②Soft: The needle movement is gentle. The momentum is the same. The force is consistent to maintain vertically into the needle.

③Quick: The movement of the needle is fast, and the speed is stable, not suddenly quick or slow.

④Post: The needle cap is on the skin. We let the needle all penetrate the skin, make sure the color is even.

⑤Tight: The needle line should be thick, no matter how method, the color of the needle must be uniform.

5. Applicable crowd (simple tattoo technique)

Tattoo is different from person to person. Apply the gentle technique, the natural color, then achieve the beautification eyebrow, eye, lip with leaving no obvious embellishment mark.

(1) Embroider eyebrow.

Objective: Shape the eyebrows, improve the shape of the eyebrows, improve the shape of the face and beautify the forehead.

Applicable crowd: Eyebrows are sparse, scattered, pendulous or a long thrush.

(2) Embroider eye line.

Objective: Beautify the shape of the eyes and make them appear to be smooth.

Applicable crowd: Eyebrows are sparse, long-term eyeliner, not natural after double eyelid plasty.

(3) Embroider lip.

Objective: Soften the lip shape, improve the lip color, the visual effect to accentuate the face, beautify the face quality.

Applicable crowd: The lip is too thin or too thick. The edge of lip is not clear. Upper lip proportion is maladjusted. Lip color is not healthy.

6. Tattoo notes

(1) The color of a tattoo: The color of a tattoo is based on hair color, skin color, eye

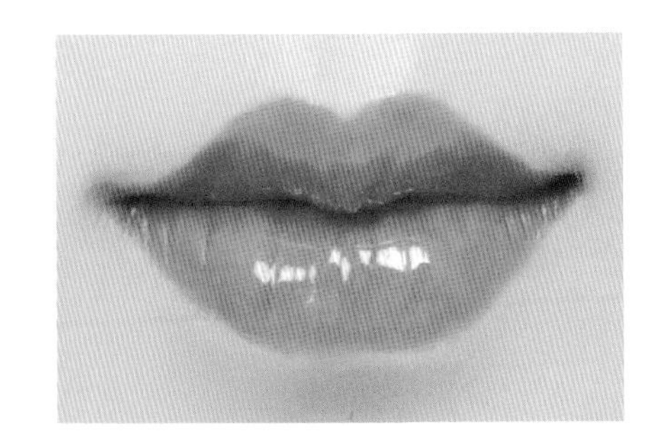

color, age, and occupation.

(2) The design of a tattoo: A tattoo is based on the face, occupation, character and natural condition.

(3) The tattoo artist professional degrees: The technical requirements of tattoo artists are strict due to the combination of the density and the strength of the wrist. With the growth direction of the natural eyebrow, the tattoo can be used to achieve a three-dimensional effect, with a sense of virtual reality and a sense of space.

(4) Tattoos are related to skin tissue: The human skin is divided into three layers, the dermis, the epidermis and the subcutaneous tissue. The epidermis is the outer layer of skin, has a protective effect. The epidermis physiological organization outside-in is divided into five layers, namely the corneous layer, the transparent layer, the granular layer, the stratum spinosum, and the base layer. The place of the tattoo is in the granular layer of the epidermis, the skin has no blood vessels, but there are many nerve endings, so there is no bleeding in the tattoo and no red or swelling, no need for any protective measures.

（一）对话

(B＝美容师，C＝顾客)

B：早上好，请坐。你有什么需要我帮助的？

C：我想了解一下韩式半永久。

B：你以前了解过吗？

C：是的，但是没有深入了解。

B：请坐，我看一下。你经常画眉毛吗？

C：画，就是觉得每天画眉毛太麻烦，想做一个半永久。

B：那我先帮你设计一下。

C：行。

B：好的，现在我开始帮你设计。你看你的脸形为由字形，所以眉毛应该稍微有点弧度，眉峰稍微靠后一些，这样会修饰一下您的脸形。你看这个眉形可以吗？

C：我怎么觉得两个眉毛不一样，一个高、一个低。

B：美女，我们每个人的两个眉毛都是不一样的，没有完全一样的，太要求对称就会比较死板。

C：这个会不会很痛？

B：不会呀，现在纹眉都是提前敷稳定剂的，整个过程是无痛的。

C:这种术后要注意什么?

B:一周内不宜沾水,少吃辛辣刺激的食物,不要抠痂皮,早晚涂修复膏。

C:我明白了。

(二)课文

纹　　绣

纹绣美容的方法犹如绣花,是以刺青的方法绣出眉毛形状,然后把色素注入皮下组织0.2～0.3 cm,使色素附于皮肤,长期不褪色,以达到美容的目的。纹绣美容是一种以美容为目的,由专业人员实施,将专门的外源性色素借助特制的专业器具转载到人体皮肤内,绘制成特定的图案、文字等并能长期存在,以达到美容效果的美容方法。

1. 基本简介

纹绣美容实际上是创伤性皮肤着色,将色素植入皮肤组织内形成稳定的色块,由于表皮很薄,呈半透明状,色素通过表皮层,呈现出色泽以起到掩盖瑕疵、扬长避短、修饰美化的作用。刺入皮肤的色素均为直径小于1 μm的小颗粒,能够很快被胶原蛋白包围,无法被吞噬细胞吞噬,从而形成标记。古代在丝帛上刺绣,称为“纹绣”,以区别于文锦。至汉代在布帛上绣花,才统称为“刺绣”。这些延续主要参见古代的“纹绣”“刺绣”,而后才得以发展成为人体美。现在纹绣有三种,即纹眉、纹眼线、纹唇。

2. 纹绣常识

纹绣美容是创伤性皮肤着色,具有一定的风险性。纹绣美容只许成功不许失败,因此纹绣美容的操作应保证在安全的条件下进行。必须有专门的器械工具并且消毒卫生,还要求纹绣操作者有一定的医学、美学基础知识及训练有素的操作技巧。环境、气氛、光线等都是与纹绣美容有关的重要条件。

清洁舒适的环境。纹绣美容实施于真皮和表皮,刺破表皮皮肤后保护功能减退,细菌容易侵入。因此纹绣美容的操作间要保持清洁、空气清新。

安静柔和的气氛。良好的气氛可以调解人的神经,使人心情宁静、情绪稳定。在这种环境中纹绣,操作者可以集中精力进行操作。被纹绣者在良好的气氛中应积极配合,确保操作者的工作质量。

光线适度的照明。照明是纹绣美容操作的重要条件。柔和舒适的光线可以减少操作者的视觉疲劳,保证纹绣质量。

3. 纹绣消毒

纹绣是直接刺破皮肤注入色素的,所以作为一名合格的美容师,应该具备严格的消毒与卫生观念。

(1) 美容操作器具的消毒:所有接触顾客皮肤的器具都要进行消毒处理,纹绣针片应保证一人一针,采用一次性的消毒灭菌用品以防止交叉感染。

(2) 美容环境的卫生:外界环境生长着各种各样致病和有害的微生物,所以一定要保持操作环境的洁净与卫生、空气流通,定期进行无菌消毒处理。

(3) 对所要纹绣部位的皮肤进行消毒。

(4) 无菌操作:美容师应对手进行清洁,然后戴上无菌手套。无菌物品和手为无菌区域,操作只能在无菌区进行，并保持至美容操作全部完成。

(5) 医疗美容的效果不佳，可能与消毒灭菌未做好有关，应弄清原因后设法予以补救。

4. 技巧方法

(1) 分块处理：在做整个唇面上色时将上下唇面分为四个区域，每块区域根据纹眉机针的数量又可分为2～3块，操作时要认真地、细细地、满满地把每块纹好，纹好一块后再进行下一块。注意：在纹下一块前一定要将针帽里的液体吸出，重新蘸色料，确保上色效果。

在做唇线时应采取分段处理，就是把唇线分为若干个段落，一段一段地运针，在做唇线时也要勤蘸色料，确保流色顺畅，遇有出血点或色料堆积情况，要避开此处，做下一遍上色时再补针。

(2) 层层入色法：操作唇面上色时利用针帽的伸缩调节针的长度，做第一遍上色时针露出长度为1 mm，第二遍上色时针露出长度为1.5 mm，第三遍上色时针露出长度为2 mm。注意：一定要在开机的情况下调针，运针时针帽贴在皮肤上确保所需针长全部刺入皮肤，入针角度为90°。

(3) 少擦拭多渗透：操作唇部上色时不能边纹刺边擦拭，每次上色后要让色料在唇部上保留2～3 min，然后再轻轻擦拭，整个唇部做完后再涂上色料的做法是错误的。

(4) 五字决。

①轻：操作者动作要轻，左手不可大力拉动皮肤，右手下针不可太重。

②柔：运针动作轻柔，顺势走，力度一致，保持垂直角度入针。

③快：运针动作要快，并且速度稳定，不可忽快忽慢。

④帖：针帽贴在皮肤上，让所露针长全部刺入皮肤，确保上色均匀。

⑤密：运针路线要密，不论用何种运针针法，上色效果必须均匀。

5. 适用人群(单纯纹绣技术)

纹绣因人而异，突出个性设计，并且应用轻柔的手法、自然的色泽，进而做到既美化眉、眼、唇，又不留明显的修饰痕迹。

(1) 绣眉。

目的：塑造眉形，改善稀疏或散乱眉形，可以从视觉上改善脸形，美化眉、眼和额头。

适用人群：眉毛稀疏、散乱、下垂或长期画眉者。

(2) 绣眼线。

目的：美化眼睛形状，使之修长流畅，从视觉上使睫毛浓密，瞳仁黑亮。

适用人群：眉毛稀疏或重睑术后不自然或长期化眼线者。

(3) 绣唇。

目的：柔化唇形，改善唇色，从视觉上衬托面色，美化脸形。

适用人群：唇形过薄或过厚，唇口边缘不清，上下唇比例失调或唇色不健康者。

6. 注意事项

(1) 色的调配：纹绣应根据发色、肤色、眼球色、年龄、职业配制相应的颜色。

(2) 型的设计：纹绣应根据脸形、职业、性格、天然条件进行设计。

(3) 纹绣师专业度：纹绣师的技术要求严格，源于纹绣的疏密度和力度、腕力的配合，随着天然眉的生长方向一根根进行纹绣，讲究立体效果，具有虚实感和空间感。

（4）纹绣与皮肤组织的关系：人的皮肤分为三层，由外向内为表皮、真皮和皮下组织。表皮是皮肤最外一层，有保护作用，表皮的生理组织由外向内分为五层，即角质层、透明层、颗粒层、棘层和基底层。纹绣是在表皮中的颗粒层进行的，表皮没有血管，但有许多神经末梢，所以纹绣中不会出血、不红不肿，无需任何防护措施。

Unit 7 Traditional Chinese Medicine Cosmetology

Ⅰ. Acne

Part Ⅰ Dialogue

(C=Customer,B=Beautician)

B:Hello,can I help you?

C:Hi,look at my face,especially my forehead. I've been having acne lately?

B:I'm going to show you,well,you have acne,look at your situation,it's the beginning of acne,and there's only pimples and red papules.

C:Acne?

B:Yeah,do you know about acne?

C:I don't know much about it.

B:Then I'll just tell you about acne. Acne is a chronic inflammatory disease of hair follicle and sebaceous glands common in adolescence. It is one of the most common lesions of the beauty,and the general skin lesions are the pimples. The severity of the disease is divided into three phases. Namely,the initial stage:pimple,papules. Metaphase:tubercle cyst. Later: acne.

C:Oh,is that a good treatment for me?

B:You're in the early stages of acne. You will soon be good,as long as you live regular and eat light,guaranteed adequate sleep,do not squeeze with your hands,cooperate with our treatment.

Part Ⅱ Text

Acne

Acne is a form of facial skin disease. More young people suffer from acne symptoms,so acne will bring skin parts of health hazards. You should know more about acne,and you should be alert to the complications of acne. What are some of the complications associated with acne?

1. Cysts

On the basis of nodules,the accumulation of a large number of pus cells in the sebaceous glands of the hair follicle has the presence of the suppurated bacterial residues,sebum,and keratinocytes, as well as the inflammatory infiltration. The structure of the hair follicle and sebaceous gland is completely damaged and the feeling of the cyst is felt, and the effect of the extrusion can be seen in the fact that it is very harmful to the acne.

2. Pimples

Pimples are the most basic form of acne. Under the condition of the hair follicle and sebaceous glands mouth blockage, hair follicle and sebaceous glands anoxic environment is formed, anaerobic acne propionic acid bacillus massive reproduction, decomposition of sebum, produce chemical chemokines, leukocytes aggregation and inflammatory papules, so this kind of papule belongs to the inflammatory damage. The pustules are further aggravating the inflammatory papules. A large number of neutrophils are gathered in the sebaceous glands and the hair follicle, and the inflammation of the propionic acid bacillus is swallowed, and a large number of pus cells accumulate into pustules. This condition is easy to form scar, mainly depression scar.

3. Scar

Scar is the most serious harm of acne. Severe damage above the inflammatory papules is damaged by dermal tissue, which can then be repaired by the connective tissue. Different people, different ages, the degree of acne scar has a lot of changes, there are atrophic scar (concave scar) and hypertrophic scar and so on, once formed, not easy to self-heal. Nodules, on the basis of the pustules, a large number of keratinocytes, the sebum, the pus cells in the sebaceous glands and the hair follicle, resulting in the destruction of the sebaceous glands and the hair follicle and forming the red nodules above the surface of the skin. The base has obvious infiltration, hot flashes and tenderness.

一、痤疮

(一)对话

(C=顾客,B=美容师)

B:你好,有什么可以帮助你?

C:你好,你看看我脸上,特别是额头,最近一直长痘。

B:我给你看看,你这是痤疮,看你的情况,是属于痤疮的初期,只有一些粉刺和红色的丘疹。

C:痤疮?

B:是的,你对痤疮了解吗?

C:我不是很了解。

B:那我简单地给你说一下痤疮。痤疮是青春期常见的一种毛囊、皮脂腺的慢性炎症性病变,是常见的损美性疾病之一,一般皮损为粉刺、丘疹、脓疱、囊肿、暗疮、结节。按病情的轻重分为三期,分别为初期(粉刺、丘疹)、中期(结节、囊肿)、后期(暗疮)。

C:哦,那我这种情况好治吗?

B:你这种情况属于痤疮的初期,好治,只要你配合我们的治疗,生活规律、饮食清淡,保证充足的睡眠,不要用手挤压痤疮,你脸上的痤疮会逐渐好转。

（二）课文

痤　疮

痤疮是面部皮肤疾病中的一种，较多的年轻人会出现痤疮，因此会给皮肤的健康带来危害，大家应当多学习了解痤疮的知识，还应当警惕痤疮的并发症。让我们来认识一下痤疮的相关并发症。

1. 囊肿

囊肿在结节的基础上发展而来，毛囊、皮脂腺结构内有大量脓细胞聚集，既有脓液、细菌残体、皮脂和角质细胞，又有炎症浸润将毛囊、皮脂腺结构完全破坏。触摸起来有囊肿样感，挤压可有脓、血溢出。

2. 丘疹

丘疹是最基本的痤疮损害。在毛囊、皮脂腺口堵塞的情况下，形成毛囊、皮脂腺内缺氧的环境，厌氧性痤疮丙酸杆菌大量繁殖，分解皮脂，产生化学趋化因子，白细胞聚集而发生炎性丘疹，所以这类丘疹属于炎性损害。脓疱是炎性丘疹的进一步进展、加重。毛囊、皮脂腺内大量中性粒细胞聚集，痤疮丙酸杆菌发生炎性反应，大量脓细胞堆积形成脓疱。这种情况预后易形成瘢痕，主要为凹陷性瘢痕。

3. 瘢痕

瘢痕是痤疮损害中最严重的。严重程度在炎性丘疹以上，因真皮组织遭到破坏，愈后结缔组织修补从而形成瘢痕。不同的人，不同的年龄，痤疮瘢痕的程度有很大的变化，有萎缩性瘢痕(凹洞)、增生性瘢痕等等。瘢痕一旦形成，不易自愈。结节在脓疱的基础上发展而来的，毛囊、皮脂腺内有大量的角质细胞、皮脂、脓细胞储存，使毛囊、皮脂腺结构遭到破坏而形成高出皮肤表面的红色结节。基底有明显的浸润，潮红，触之有压痛。

Ⅱ. Black Mole

Part Ⅰ　Dialogue

(C=Customer, B=Beautician)

B: Hello, can I help you?

C: Hello, what are these black spots on my face?

B: Let me show you. How long have you had these black moles?

C: It's been a few years ago. Some people say this is freckle. Do you think it's freckle?

B: This one on your face is black mole, also known as nevoid lentigo.

C: Nevoid lentigo? Isn't that freckle?

B: Although they have freckles, they are not freckles.

C: Oh, I thought it was the same. What's the difference between freckle and black mole?

B: Freckle and black mole are pigmented skin diseases, but there are obvious differences between them. Black moles can occur anywhere in the skin, but freckles mainly occur in the face, neck, and the back of hands. Black moles are slightly raised on the skin. Freckles are flat with the skin. The distribution of black moles on the face are relatively scattered, and the suntan relationship is not big, and freckles are more densely distributed in the face, which is more affected by the sunlight.

C: Can you get rid of that?

B: If you want to get rid of it, it doesn't work very well by cosmetics. But you can remove it by surgery.

Part Ⅱ Text

Black Mole

Black moles are also known as freckles, but they are two different types of pigmented skin diseases. In ancient Chinese medicine, the black mole is called a simple lentigo. It can be distributed in all parts of the skin, skin and mucosa junction or eye combined with membrane, characterized by brown or dark brown spots, some slightly higher, assumes the circular, generally about tip to sesame seed size. The surface of the spot may have slight desquamation, but its fine skin texture has not changed. The pigmentation is evenly distributed, the edges gradually fade away and close to normal skin color.

Black moles are born in childhood, but they have been growing in an average annual rate. There are also sudden and sudden mass emergence, which has been gradually reduced over the years. There are other common types of black moles. The difference between black moles and freckles is that the spots of the former are darker and the distribution is sparse and dispersed. The color is not deepened after sunburn and the number is not increased. Freckles had close relation with the sun. Its distribution is limited to the sun, and increased due to the sun in summer. The number of freckles increased, the color deepened, the damage increased, while in winter the number decreased, the color became lighter, and the damage decreased.

Black mole is a pigmentation disorder, sometimes one of the clinical manifestations of some hereditary syndrome. There is not any special or effective way to treat it. If the patient requires treatment, surgical removal can be performed. Freckles, though linked to genetics, are closely associated with sun exposure, reducing sunlight exposure, using sunscreen cream, and reducing freckles.

二、黑子

(一) 对话

(C＝顾客,B＝美容师)

B:你好,有什么可以帮助你?

C:你好,我脸上这些黑色的点点是什么?

B:我给你看看,你这些黑色的点点,多长时间了?

C:这个已经有好几年了,之前有人说是雀斑,你说这个是不是雀斑?

B：你脸上的这个是黑子，也叫雀斑样痣。

C：雀斑样痣？那不是雀斑吗？

B：虽然说都有雀斑两个字，但不是雀斑。

C：哦，我还以为是一样的呢，那雀斑和黑子有什么不同呢？

B：雀斑和黑子都属于色素性皮肤病，但两者有明显的不同。黑子在皮肤的任意部位都可以发生，但雀斑主要在面部、颈部、手背日晒部位发生。黑子稍微高出皮肤，而雀斑与皮肤相平。黑子在面部分布得比较分散，和日晒关系不大，而雀斑在面部分布较密集，受日晒影响会加重。

C：那这个可以去除掉吗？

B：你要去掉的话是可以的，用产品、药物效果不大，可以通过外科手术切除。

（二）课文

黑　子

黑子又叫雀斑样痣，但是黑子与雀斑是截然不同的两种色素性皮肤病。黑子在我国古代医书中称为黑子痣。它可以分布在皮肤的任何部位，如皮肤黏膜交界处、眼结合膜，其表现为褐色或黑褐色的斑点，有的略微突起，呈圆形，一般为针尖至芝麻大小。斑点表面可有轻微的脱屑，但其皮肤纹理没有变化。色素沉着均匀一致，边缘逐渐变淡且接近正常皮肤颜色。

黑子多发生于幼年，但一直到成年均可以逐渐增多，亦有突然弥散性大量出现者，也有经过多年逐渐减少而消失的。另有泛发性黑子病及面正中黑子病，属于黑子的特殊类型。

黑子与雀斑的不同之处在于，黑子较雀斑颜色深，分布比较稀疏和分散，日晒后颜色不加深，数目亦不增多。而雀斑与日晒关系极为密切，其分布仅限于日晒部位，夏天由于日晒增多，雀斑数目增多，颜色加深，损害变大；而冬季相反，数目减少，颜色变淡，损害缩小。

黑子是一种色素沉着性疾病，也是某些遗传性综合征的临床表现之一。治疗上没有什么特殊有效的方法，如果患者要求治疗，可以通过外科切除。而雀斑虽与遗传有关，但与日晒关系密切，减少日光照射，使用遮光剂及防晒霜，则会减少雀斑的发生。

Ⅲ. Face Telangiectasis

Part Ⅰ　Dialogue

(C=Customer, B=Beautician)

B: Hello! What can I do for you?

C: Hello, can you help me to see what's wrong with my face? It always looks red.

B: How long have you been in this situation?

C: It's been years. It's getting worse and worse.

B: Have you ever been treated?

C: No.

B: Is there any other disease?

C: No.

B: You belong to face telangiectasis, also known as red blood silk. It is a common kind of loss of beauty.

C:What shall we do?

B:We have a special program for red blood silk,you can be assured,and we will teach you how to care daily.

Part Ⅱ Text

Face Telangiectasis

In everyday life we often see some facial skin redness,characterized by filamentous star point or line,part of the red or purple,look closely to see many red blood vessels,skin like a little bit red head,this is the face telangiectasis,commonly known as the red blood silk.

Red blood silk is frequently found in women,divided into primary and secondary. There is a relationship between primary and family heredity. The resistance of the capillaries exceeds the normal range, resulting in the rupture of the capillary dilated due to the stimulation of the wind and the temperature changes. Excessive sun exposure can also cause chronic light-linear dermatitis, resulting in dry skin. People live for a long time in the relatively poor living environment as thin air,oxygen to the skin,compensatory enlargement, leading to increased number of red blood cells and gradually vasoconstrictor function disorder,causing permanent capillary dilated. Those who have long exposed to the wind and cold can also cause the expansion of capillaries. Hormone dilatation is the sequelae of inappropriate treatment,such as the use of external drugs in the face,improper cosmetics and long period of skin care. The acidity of the skin care product seriously destroys the protective effect of cuticle and the elasticity of capillary,so that the capillaries expand or rupture.

To get rid of red blood, we can eat a lot of milk, soy products and fresh fruits and vegetables to enhance our skin resistance. Note to enhance the strength and elasticity of blood vessel walls, improve the microcirculation, and improve the problem of skin redness and red blood. Fish,shrimp,crab and so on can easily cause skin allergy,need to avoid edible as far as possible.

三、面部毛细血管扩张

(一)对话

(C=顾客,B=美容师)

B:您好!请问有什么可以帮助您?

C:您好,能帮我看一下我的脸是怎么了吗,总是看着红红的?

B:您出现这种情况多长时间了?

C:好几年了,越来越严重了。

B:请问您有做过治疗吗?

C:没有。

B:您有没有其他方面的疾病?

C:没有。

B:您这是属于面部毛细血管扩张,也就是俗称的红血丝,是比较常见的损美性疾病。

C:怎么办呢?

B:您放心吧,我们有专门针对红血丝的治疗方案,同时也会教您日常的护理方法。

(二)课文

面部毛细血管扩张

在日常生活中我们常常看到一部分人面部皮肤泛红,出现丝状、点状、星芒状或线状红斑,部分呈红色或紫红色,仔细看能见到皮肤上许多红色血管,就像红线头,这就是面部毛细血管扩张,俗称红血丝。

红血丝多发于女性,分为原发性和继发性两种。原发性红血丝和家族遗传有一定关系,继发性红血丝是由于风吹日晒、温度变化的刺激,导致毛细血管的耐受性超过了正常范围,引起毛细血管扩张破裂。过度的日晒还会引起慢性光线性皮炎,造成皮肤干燥等。长期生活在较为恶劣的环境中,如空气稀薄的高原,会导致皮肤缺氧,红细胞数量增多及血管代偿性扩张,久而久之血管收缩功能障碍,引起永久性毛细血管扩张。长期接触风、冷、热的水手、厨师、农民和运动员也会出现毛细血管扩张。激素性扩张是不恰当治疗的后遗症,如面部滥用外用药物、化妆品或长期护肤不当引起的后遗症。护肤产品的酸性成分会严重破坏皮肤角质层的保护作用和毛细血管的弹性,使毛细血管扩张或破裂。

要祛除红血丝,饮食上可多摄入一些牛奶、豆制品及新鲜的蔬菜、水果,以增强皮肤的抵抗力。注意增强血管壁韧性和弹性,改善微循环,从根本上缓解和改善皮肤泛红和红血丝问题。鱼、虾、蟹等极易引起皮肤过敏的食物,要尽量避免食用。

Ⅳ. Black Eye

Part Ⅰ Dialogue

(C=Customer,B=Beautician)

B:Hello! What can I do for you?

C:Hello,the color is dark under my eyes. I don't know what's wrong.

B:How long have you had the problem?

C:Several days. It's getting worse and worse.

B:Have you got a good sleep recently? Do you stay up late?

C:Sleep is not so good. I've been working overtime every day.

B:What's wrong with your body?

C:No.

B:Don't worry too much. You have black eyes,commonly known as panda eyes.

C:What shall I do?

B:We have a special program for black eyes. You can rest assured that you should take a rest after you go back,and try to avoid overtime.

Part Ⅱ Text

Black Eye

The eye is the window of the heart, also it is the top priority of the overall image, so, avoid the invasion of black eye, the maintenance of eye skin is urgent. Black eye, that is, we often say that the panda eye, due to stay up late, mood swings, eye fatigue, aging, which lead to the eye skin vascular blood flow speed too slow and the formation of viscous flow, organization oxygen deficiency, the metabolic waste accumulation in blood vessel is overmuch, cause eye pigmentation. Generally older people, the subcutaneous fat around the eyes become thinner, so the dark circles are more obvious.

Black eye can usually be divided into two types: One is a pelious black eye, which is caused by a vein of blood trapped in the microvessel, which appears to be a dark blue. It is common for teenagers to avoid normal life. One is dark circles of tea, produced by melanin generation and metabolism. The causes of black eye in tea are related to the growth of the age, and the long-term sun exposure can cause pigmentation to linger in the eyes. Life is not normal, sleep is not enough, eye fatigue, pressure, anaemia and so on, and the skin is too dry, can cause black eye.

We should keep a regular routine, get enough sleep, have a good spirit, limit tobacco and alcohol, strengthen eye hygiene, avoid eye strain, balance diet to ensure adequate intake of vitamin A and C, regular physical examination. If other organs are found to be problematic, we should find timely diagnosis and treatment.

四、黑眼圈

(一) 对话

(C＝顾客，B＝美容师)

B：您好！请问有什么可以帮助您？

C：您好，我的眼睛下面颜色好深，不知道是怎么了。

B：您出现这种情况多长时间了？

C：好几天了，越来越严重了。

B：请问您最近睡眠好吗？有没有熬夜？

C：睡眠一直不怎么好，最近工作又忙，天天加班。

B：身体有什么其他的不舒服吗？

C：没有。

B：不用太担心，您这是黑眼圈，俗称熊猫眼，很常见的。

C：怎么办呢？

B：您放心，我们有专门针对黑眼圈的治疗方案，回去之后要注意休息，尽量避免加班熬夜。

（二）课文

黑 眼 圈

眼睛是心灵的窗户，也是影响一个人整体形象的重中之重，所以，避免黑眼圈的入侵，眼部皮肤的保养刻不容缓！黑眼圈也就是我们常说的熊猫眼，由于熬夜、情绪波动大、眼疲劳、衰老导致眼部皮肤血管血流速度过于缓慢，从而形成滞流，组织供氧不足，血管中的废物代谢减慢、积累过多，造成眼部色素沉着。一般年纪越大的人，眼睛周围的皮下脂肪变得越薄，所以黑眼圈就越明显。

黑眼圈通常可分为两种：一种是青黑色黑眼圈，这是因为微血管的静脉血液滞留，从外表上看，皮肤出现暗蓝色，常见于青少年，生活作息不正常的人尤难避免。一种是茶色黑眼圈，因黑色素生成与代谢不全而产生。茶色黑眼圈的成因与年龄增长有关，长期日晒会造成色素沉淀在眼周，挥之不去。

生活作息不正常、睡眠不足、眼睛疲劳、压力、贫血等因素，以及皮肤过度干燥，都会导致黑眼圈形成。

保持作息规律，睡眠充足，精神愉快，节制烟酒；加强用眼卫生，避免眼疲劳；均衡膳食，保证充足维生素 A 和 C 的摄入；定期体检，有其他脏器问题尽早发现、及时诊治。

Ⅴ. Verruca Vulgaris

Part Ⅰ Dialogue

B: Hello, may I help you?

A: Well, hello, I recently encountered some problems in a beauty salon, and there is a skin disease called warts. Do you know?

B: Of course, there was a patient who had a flat wart.

A: Yes, yes, yes, I just want to ask, flat warts are flat uplift, and there are other, more prominent, and contagious ones, too, called flat warts?

B: Oh, no, the one you're talking about is a common wart.

A: Well... Common wart? What's a common wart? What's the difference between a flat wart and common wart?

B: Simple from the shape, the flat wart is flat and uplifted, the protrusion is not obvious, and the surface is smooth, and the common wart is obviously protruding from the surface of the skin, and the surface is coarser.

A: Well, I understand some things, thank you very much.

B: Well, you're welcome.

A: Flat wart and common wart should belong to the same kind of disease, just divide the type, besides these two kinds of types, still have other types?

B: Oh, you're so smart. You're right. Besides these two types, there are other ones.

A: Can you tell me more about the other types?

B: Sure, please come to the office with me. We'll discuss it in detail.

Part Ⅱ Text

Verruca Vulgaris

Verruca is a common verruca of the epidermis caused by human papilloma virus (HPV). Modern medicine thinks, verruca vulgaris is human papilloma virus (HPV) infection, which can be spread by direct or indirect contact. Trauma or damaged skin is also an important factor for HPV infection. Wart course has important relationship with the body's immune. This is similar to the etiology of flat wart, which is a type of wart. Clinical manifestations as the beginning of as needlepoint big pimple, large or larger gradually extend to the peas, assumes the circular or polygonal, rough surface, the angle is changed obviously, qualitative hard, a further the unclean yellow or brown, continue to develop a proliferation of papilloma samples, friction or collision is easy to bleed. In addition, it is common on the fingers, the back of the hand and foot, the number of the number of different, the beginning of more than one, later can be developed to dozens, general lack of self-conscious symptoms, occasional tenderness, often born in adolescents. Chronic disease, partial self-healing. Note that there is a wart called the plantar wart, which is a verruca vulgaris occurred in a foot. Because local oppression friction, surface formed yellow callosity, using a knife to cut to this layer, the visible white soft thorn wart body, surface often scattered in a small black spots. It occurs in the case of a simple keratosis, and the typical verrucous damage occurs when the skin is invaded and treated. If it spreads to the nail, causes the nail to rise, destroys the growth of the nail, causes the tear, pain and secondary infection. There are some symptoms before the wart fading, and the itching can occur in the basal part of the wart. The lesion can suddenly become large and unstable, or the individual wart may subside or have small new warts. There are other special types, such as filamentous wart occurs in the place such as the eyelid, neck, and chin, mostly single soft filament is outstanding. Finger warts: on the basis of the same soft occurred more than a bunch of jagged finger bumps, tips for cutin thorns, amount, on the scalp, often also can send to the face and between the toes.

Treatment of local or systemic immune responses, regulation of the local skin growth and destruction of verruca are the main methods, including systemic treatment and topical treatment. The following two points should be noted when treating the whole body. ①Traditional Chinese medicine (TCM), such as cure wart soup, purslane mixture, radix isatidis injection, chai hu injection, etc. ② Interferon, for multiple and stubborn refractory warts, can be combined with systemic or diseased local injection interferon, and alone interferon effect is not certain. In the local drug treatment, because most patients with warts can regress in the onset of 1 or 2 years, a lot of patients even with destructive depth treatment, a third warts will relapse, so the curative effect of various local treatment for warts valuation should be particularly cautious, for some way to cause permanent scarring unfavorable use.

There is also photodynamic therapy, system or partial use of photosensitizer amino acetyl propionic acid or aminolevulinic acid (ALA), which causes local cell death after

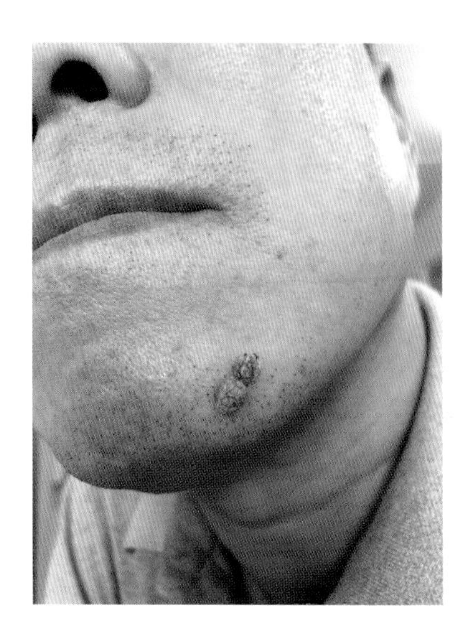

exposure to light, can cure some common warts. Physical therapy is also common, such as cryotherapy, the treatment of cauterizing, laser treatment, infrared coagulation therapy for an unusually small number of warts. Surgical resection can also be selected, but it is easy to relapse after surgery.

五、寻常疣

(一) 对话

B:您好,请问有什么可以帮助您的吗?

A:嗯,您好,我最近在美容院实习遇到一些问题,有个皮肤病叫作疣,您知道吗?

B:当然,之前有一个患者是扁平疣,上次你还碰到过呢……

A:对对对,是的,我就是想问一下,扁平疣是扁平隆起的,那还有其他的比较突出的,也有传染性的那种,也是叫扁平疣吗?

B:哦,不是的,你说的那个应该是寻常疣。

A:嗯……寻常疣?什么是寻常疣呢?与扁平疣有什么区别呢?

B:简单从形状上区分,扁平疣是扁平隆起的,突起不明显,并且表面较光滑,而寻常疣明显突起于皮肤表面,并且表面比较粗糙。

A:嗯,大概明白一些了,非常感谢。

B:嗯嗯,不客气。

A:那扁平疣和寻常疣应该是属于同一种疾病了,只是分型不同,那除了这两种分型,还有没有其他的分型呢?

B:哦,你太聪明了,说得很对,除了这两种分型,还有其他的分型。

A:那能再给我说说其他的分型吗?

B:当然可以,请跟我一起到办公室来,我们详细讨论。

（二）课文

寻 常 疣

疣是由人乳头状瘤病毒（HPV）引起的表皮良性赘生物。临床上常见的有寻常疣、跖疣、扁平疣等。

现代医学认为，寻常疣是人乳头状瘤病毒（HPV）感染引起的，可通过直接或间接接触传播，外伤或皮肤破损对 HPV 感染也是一个重要的因素，疣的病程与机体免疫有重要的关系。这跟之前学习的扁平疣的病因一样，扁平疣属于疣。临床表现为初起为针尖大的丘疹，渐渐扩大到豌豆大或更大，呈圆形或多角形，表面粗糙，角化明显，质地坚硬，呈灰黄、污黄或污褐色，继续发育呈乳头瘤样增殖，摩擦或撞击易出血。另外，寻常疣好发于手指、手背、足缘等处，产生时数目不等，初起多为一个，以后可发展为数个到数十个。一般无自觉症状，偶有压痛，多发生于青少年。病程缓慢，部分可自愈。要注意的是，有一种疣称为跖疣，是发生于足底的寻常疣。由于局部压迫、摩擦，表面形成黄色胼胝状，如用小刀削去此层，可见白色软刺状疣体，表面常有散在小黑点。发生在甲缘者，表现为单纯性角化，待侵及皮肤时才出现典型赘疣状损害。若向甲下蔓延，使甲掀起，破坏甲的生长，会导致裂口、疼痛及继发感染。疣在消退前会有一些征兆：突然瘙痒，疣基底部发生红肿，损害突然变大，趋于不稳定状态，个别疣可消退或有细小的新疣发生。还有其他的特殊类型，比如，丝状疣：好发于眼睑、颈、颏部等处，多为单个细软的丝状突出。指状疣：在同一个基础上发生一簇参差不齐的多个指状突起，尖端为角质样棘刺，数目多少不等，常发于头皮，也可发于面部、趾间。

治疗以破坏疣体、调节局部皮肤生长、刺激局部或全身免疫反应为主要手段，包括全身治疗和局部治疗。全身治疗可用中药方剂和干扰素治疗。①中药方剂：平肝活血方、治疣汤、马齿苋合剂、板蓝根注射液、柴胡注射液等。②干扰素：对多发性且顽固难治的疣，可配合全身或病损局部注射干扰素，单独用干扰素疗效不肯定。在局部药物治疗时，由于多数疣患者在发病 1～2 年内能自行消退，不少患者即使采用深度破坏性治疗方法，有 1/3 的疣仍会复发，因此对疣的各种局部治疗的疗效估计应特别慎重，对一些能造成永久性瘢痕的方法不宜使用。

还有光动力治疗，系统或局部使用光敏剂氨基乙酰丙酸或氨基酮戊酸（ALA），经光照射后引起局部细胞死亡，可治疗部分寻常疣。物理治疗也比较常见，如冷冻疗法、电灼疗法、激光疗法、红外凝固疗法等适用于数目少的寻常疣。也可以选择外科手术切除，但是术后易复发。

Ⅵ. Eczema

Part Ⅰ Dialogue

(C＝Customer，B＝Beautician)

B：Hello，may I help you?

C：I had these red rashes a few days ago. What's the special itch?

B：According to my diagnosis，you have eczema.

C：What is eczema?

B:Eczema is a common allergic skin inflammation intense itching to rash diversity, symmetric distribution, recurrent, easily turn into chronic. Any part in any season can happen at any age. But often recurrence or worse after winter.

C:What should I pay attention to in life?

B:Classification of Chinese medicine eczema: hot and humid, blood deficiency and wind dryness. Hot and humid disease: general itching, papule type to stop itching and inhibition of bacteria, eczema patient should avoid drinking, coffee, spicy stimulation and fried food, food should be light, eat fruits and vegetables. The patient can eat green beans, white gourd, lotus seeds, bitter gourd and other clearing hot and wet food. Blood deficiency and wind dryness symptom: blood deficiency and wind dryness symptom should pay attention to the liver fire is exuberant, the food is as light as possible, eat green vegetables, celery, carrot more. It is important to avoid alcohol and tobacco, and can be used to relieve itching, dry and dry heat, and repair skin damage. Stay in a good mood, do not constipation. Most people who are allergic to do not contact allergen.

C:How about the treatment?

B:We can use both drugs and non-drug treatments simultaneously.

C:Ok, then help me make a treatment plan.

Part Ⅱ Text

Eczema

Eczema is an inflammation of the skin caused by various internal and external factors. Its access stage can be divided into the acute, sub-acute and chronic stages clinically. The acute stage has the tendency of exudate, and the slow period is infiltrated with hypertrophic and some patients have the characteristics of chronic eczema skin lesion with pleomorphic, symmetry, pruritus and recurrent attacks.

The etiology of eczema is complex and often results from interaction. Internal cause such as chronic digestive system diseases, mental tension, insomnia, fatigue, mood changes, endocrine disorders, infections, metabolic disorders, and so on, external causes, such as living environment, climate change, food and so on all can influence the occurrence of eczema. External stimuli such as cold, dry, hot sun, hot water scald, and all kinds of animal fur, synthetic fibers, plant, cosmetics, soap, all can induce eczema. Eczema is a late onset of allergy caused by complex internal and external factors.

1. Diagnosis

It is mainly based on the history, the form of the rash and the course of the disease. Generally eczema lesion is given priority to polymorphism, erythema, papules, mound herpes, rash in central obviously, gradually to spread out around, boundary is not clear, diffuse, has a tendency to seep, more chronic hypertrophy. The course is irregular, repeated, itchy. The etiology of eczema is complicated, it is easy to relapse after treatment, but it is difficult to cure. Because the clinical form and the position have their own characteristics, the medicine is different for different person.

(1) General control principles: Seek out possible triggers, such as work environment, lifestyle, diet, hobbies, thoughts and emotions, as well as chronic lesions and visceral organ diseases.

(2) Internal use therapy: Use antihistamines to stop itchy, and use them when necessary. Common eczema can be taken orally or injected with glucocorticoid, but not in long-term use.

(3) External therapy: Appropriate dosage forms and medications should be selected according to the skin lesions. Acute eczema can use local saline and 3% boric acid.

2. Eczema prevention

(1) Avoid your own potential triggers.

(2) Avoid various external stimuli, such as hot water washing, excessive scratching, cleaning and contact of sensitive substances such as fur preparations. Less contact with chemical components, such as soap, washing powder, detergent, etc.

(3) Avoid possible allergens and irritating foods such as chili, strong tea, coffee, and alcohol.

六、湿疹

(一) 对话

(C＝顾客,B＝美容师)

B:你好,请问有什么可以帮助你的吗?

C:我前几天身上出现了这些红疹子,特别痒。请问是什么?

B:根据我的诊断,你所患的是湿疹。

C:什么是湿疹呢?

B:湿疹是一种常见的过敏性皮肤炎症反应,以皮疹多样性、对称分布、剧烈瘙痒、反复发作、易演变成慢性疾病为特征,可发生于任何年龄、任何部位、任何季节,但常在冬季以后复发或加剧。

C:那生活上我应该注意什么呢?

B:中医在临床上将湿疹分为湿热证、血虚风燥症。湿热证:治疗以止痒抑菌为主,湿疹患者应避免食用含有酒精的食物,以及咖啡、辛辣刺激与油炸的食物,饮食应清淡,多吃水果蔬菜。患者可多吃绿豆、冬瓜、莲子、苦瓜等具有清热利湿功效的食物。血虚风燥证:血虚风燥证应注意肝火旺盛,饮食尽量以清淡为主,多吃青菜。一定要注意避免烟酒,同时可以用一些调理气血的药物止痒润燥、清热利湿、修复皮损。保持心情舒畅、防止便秘等。最重要的是过敏的患者尽量不要接触过敏原。

C:那这个怎么治疗呢?

B:我们可以通过中医药物和非药物两种方法同时治疗。

C:好的,那帮我制订一个治疗方案吧。

（二）课文

湿 疹

湿疹是由多种内外因素共同作用引起的瘙痒剧烈的一种皮肤炎症反应，分急性、亚急性、慢性三期。急性期有渗出倾向，慢性期则有浸润、肥厚的表现。有些患者直接表现为慢性湿疹。皮损具有多形性、对称性、瘙痒和易反复发作等特点。

湿疹病因复杂，常为内、外因相互作用的结果。内因有慢性消化系统疾病、精神紧张、失眠、过度疲劳、情绪变化、内分泌失调、感染、新陈代谢障碍等。外因如生活环境、气候变化、食物等均可导致湿疹的发生。外界刺激如日光、寒冷、干燥、炎热、热水烫洗以及各种动物皮毛、植物、化妆品、肥皂、人造纤维等均可诱发湿疹。湿疹是复杂的内、外因素引起的一种迟发型超敏反应。

1. 湿疹诊断和治疗

湿疹主要根据病史、皮疹形态及病程来进行诊断。一般湿疹的皮损为多形性，以红斑、丘疹、丘疱疹为主，皮疹中央明显，逐渐向周围散开，边界不清，弥漫性。急性者有渗出倾向，慢性者则有浸润、肥厚的表现。病程不规则，反复发作，瘙痒剧烈。

湿疹病因复杂，治疗好转后仍易反复发作，难以根治。因临床形态和部位各有特点，故用药因人而异。

(1) 一般防治原则：寻找可能诱因，如工作环境、生活习惯、饮食、嗜好、思想情绪等，以及有无慢性病灶和内脏器官疾病等。

(2) 内用疗法：选用抗组胺药止痒，必要时两种配合或交替使用。泛发性湿疹可口服或注射糖皮质激素，但不宜长期使用。

(3) 外用疗法：根据皮损情况选用适当的剂型和药物。急性湿疹可局部使用生理盐水、3%硼酸溶液。

2. 湿疹预防

(1) 避免自身可能的诱发因素。

(2) 避免各种外界刺激，如热水烫洗，过度搔抓、清洗及接触可能敏感的物质如皮毛制剂等。少接触化学成分用品，如肥皂、洗衣粉、洗涤精等。

(3) 避免可能致敏和刺激性食物，如辣椒、浓茶、咖啡、酒类。

Ⅶ. Fat Granule

Part Ⅰ Dialogue

(C=Customer, B=Beautician)

B: Hello, may I help you?

C: What are these white spots around my eyes?

B: Oh, let me see. Through my diagnosis, there is fat granule.

C: What is a fat granule?

B: Fat granule is a benign organism caused by fat metabolism disorder.

C: Why did I grow fat granule?

B: There are two main causes of fat granule. One is endocrine imbalance, which causes the secretion of facial oil, which can clog pores. Another is the cosmetics, the cosmetic ingredients in a high, skin nutrition surplus.

C: Oh, that's true. What should I pay more attention to in my life?

B: To light some daily diet, avoid greasy food and oily cosmetics. The fat granule can be extruded with a needle prick, but be careful not to infect. Eat more vegetables, do not eat greasy things.

C: How can I treat this?

B: Fat granules can be treated by acne needles, which can be picked out by the content objects or by the method of acupuncture point therapy to eliminate the fat granules.

C: That's good. Please help me to make a treatment plan.

Part Ⅱ Text

Fat Granule

The fat granules are divided into two types. One is the mature type of grease and the other is the immature type of grease. The former is very easy to identify, white, round (very regular dot), protruding from the skin, feel hard by hand. The immature fat granule is a small white dot that feels less hard. You should pay attention to the cleaning of the eyes, so as to ensure the normal excretion and absorption of the skin.

There are several methods to dispel fat granule.

(1) Mature grease grain is very difficult to metabolize, if grease grain grows on the cheek and two buccal, to use the method of pick, gently pick out can not damage skin.

(2) If the grease grain around the eye, if choose the above way, the risk is bigger, easy to bleeding, scar, stimulate blood capillary, it is recommended to use fat elimination class to protect skin to taste, eliminate pouch of eye cream, for instance, or dilute oil grain eye cream and so on, sometimes will be slower, sometimes at 2-3 months. If the time is too long, you can needle a little bit, don't squeeze out to prevent bleed, then apply some professional gels to eye, also will fall off.

(3) The formation of fat granule is a long-term process, but don't be afraid, if older, depends on the fat granule formation age, state of fat granule, and according to the different parts to regulate, The younger one should pay attention to nursing care, then problems will not be too big.

In short, the girl who often has fat granule, should take prevention first. Professional correct preventive measure is indispensable. Don't wait for fat granule mature to worry again. Fat granules appear mostly due to the imbalance of the body's internal secretion, resulting in excess facial oil secretion. The skin is not thoroughly cleaned, resulting in the formation of clogged pores. To light some daily diet, avoid greasy food intake, avoid using oily cosmetics, the fat granule can be extruded with a needle prick. You can squeeze out the fat with a needle, but be careful not to get infected. You should also eat more vegetables, not greasy.

七、脂肪粒

(一)对话

(C=顾客,B=美容师)

B:你好,请问有什么可以帮助你?

C:请问我眼周出现的这些白色点点是什么啊?

B:哦,我看看,通过我的诊断,你眼周出现的是脂肪粒。

C:什么是脂肪粒啊?

B:脂肪粒是由脂肪代谢障碍引起的良性赘生物。

C:那为什么会长脂肪粒呀?

B:引起脂肪粒的原因主要有以下两个。

①内分泌失调,导致面部油脂分泌过盛,从而堵塞毛孔。

②使用化妆品不当,使用的化妆品营养成分高,皮肤营养过剩。

C:哦,原来是这样啊,那我在生活中应该注意什么呢?

B:平日饮食要清淡一些,避免油腻类食物的摄入,避免使用油性化妆品。脂肪粒可以用针挑破挤出,但要注意防止感染。挤出后可以用眼部精华霜。

C:那请问脂肪粒怎么治疗呢?

B:脂肪粒可以通过粉刺针挑治,将内容物挑出,也可以通过火针点疗的方法,将脂肪粒消除。

C:那好啊,请帮我制订一个治疗方案吧。

(二)课文

脂 肪 粒

脂肪粒分为两种,一种是成熟型的脂肪粒,一种是未成熟型的脂肪粒。前者很好辨认,白色,圆形(非常有规律的圆点),突出于皮肤表皮,用手摸感觉硬硬的为成熟型脂肪粒。未成熟型的脂肪粒是一个小白点,摸起来没那么硬。平时应注意眼部的清洁,以保证皮肤正常的排泄和吸收。

祛除脂肪粒的方法有以下几种。

(1) 已经成熟的脂肪粒很难自己代谢,如果脂肪粒长在腮部和两颊,一般用挑的方法,轻轻挑掉不会损害表皮。

(2) 如果脂肪粒长在眼部,眼部的脂肪粒如果也用挑的方式,危险性比较大,容易出血,形成瘢痕,刺激毛细血管,建议使用消油脂类护肤品,如消除眼袋的眼霜,或者淡化脂肪粒的眼霜等,时间会慢一些,有时会长达2~3个月。如果嫌时间太长,可以挑破一点点,不要挤出来,防止出血,然后敷上一些专业护眼的啫喱,脂肪粒也会自己脱落。

(3) 脂肪粒的形成是个长期的过程,不过别害怕,如果年龄偏大,就要根据脂肪粒形成的时间、脂肪粒的状态进行调理,要根据部位的不同进行针对性的调理,对于年龄偏小的患者,注意护理,问题就不会太大。

总之,经常长脂肪粒的女孩,要以预防为主。专业正确的预防措施必不可少,不要等到脂肪粒成熟再发愁。出现脂肪粒大多是由于身体内分泌失调,面部油脂分泌过剩,皮肤没有得到彻底清洁,导致毛孔阻塞。平日饮食要清淡一些,避免油腻类食物的摄入,避免使用油性化妆品,对于脂肪粒可以用针挑破挤出,但是要注意防止感染。挤出后可以用眼部精华霜。

Unit 8 Physical Cosmetology

Section 1 Laser Cosmetology

Part Ⅰ Dialogue

(C=Customer,B=Beautician)

B:Welcome. What can I do for you?

C:Hello,recently I found my skin was a little flabby and I am afraid of affecting my looks.

B:Well,the optical instruments work very well on this kind of situation. After the skin recovery,the gloss and firming of the skin will be improved obviously.

C:Is it as good as you say?

B:Of course. Let me introduce in detail. Under the action of light,the various fibers of dermis will shrink after stimulation heat,and the dermis layer of the skin becomes thicker and appendages of the skin functions better. Thus,it recovery our skin elasticity and promote the lineament and gloss of skin.

C:How many times need I take?

B:You will see the obvious effect of this therapy after only one time. You can choose to do this beauty project again according to your skin situation.

C:Oh,thank you for your explanation.

B:You're welcome.

Part Ⅱ Text

Basic Knowledge of Cosmetic Laser Equipment

The application of modern laser technology in cosmetic dermatology treatment is one of the breakthrough developments in Department of Dermatology in China in recent years. In about ten years' time,laser technology,which has formed a relatively complete theoretical system and clinical practice,has become one of the main treatments of the Department of Cosmetic Dermatology.

As more attention has been paid to ultraviolet protection,skin sunburn,skin tanning and skin photo aging have become a hot research topic. A new scientific field is taking shape,and the concept of "light cosmetic medicine" is vividly portrayed. The development of cosmetic laser equipment has gone through three stages.

Fundamental research stages (1960s): The first significant laser device was introduced in 1960 by Maiman, which includes a ruby and emits lasers at 694 nm wavelengths. In 1961, the ruby laser was used to solder the stripped retina. In 1963, Goldman L began using the ruby laser in the treatment of benign skin lesions and tattoos, and achieved success. In 1965, Goldman L reported that using a ruby laser to remove tattoos was very relaxed and free of scarring. After that, ND: YAG laser was used to remove tattoos and treat superficial vascular malformations. Nd: YAG laser was developed in Shanghai in 1968.

Clinical application stage (1970s): In 1970, Goldman L and others used continuous carbon dioxide laser in the treatment of basal cell carcinoma and cutaneous hemangioma for the first time. Due to providing effective laser power and energy density continuously, the device overcame the early pulse laser of low power and low efficiency, which set off the first wave of lasers used in medical treatment both at home and abroad.

Mature stage (1990s): In early 1990s, application of Q switch laser treatment of pigmented diseases (such as nevus of Ota, tattoos) had achieved nearly perfect treatment effect; in the middle and late 1990s, variable pulse width 532 nm laser in the treatment of vascular diseases also achieved great result. At this time, complete sets of advanced laser cosmetic instruments were introduced quickly in the United States, Israel, Germany and other countries, and tended to be popular. Some domestic laser cosmetic instruments have been widely used at home and gradually replaced the leading position of foreign products.

Three effects of light on tissue.

1. Photothermal effect

Photothermal effects contain thermal effect, thermal coagulation and thermal stripping. Thermal effect: when the temperature rises to a certain level of about 60-65 ℃, collagen fibers can shrink to 1/3 of the original length and diameter of collagen increases, but the integrity of the corresponding tissue structure is not affected. Thermal coagulation: when the temperature is higher than 75 ℃, and the coagulation time is long enough, which will cause thermal tissue, a certain amount of dispersed solidification will bring lasting obvious skin reconstruction effect. Thermal stripping: when the temperature is higher than 300 ℃ and last long enough, tissue vaporization, combustion, and molecular lysis and plasma formation will lead to tissue trauma.

2. Photodynamic reaction

Irradiating tissue can cause a series of chemical reactions which result in the effective treatment or improvement of the lesion site (target tissue).

3. Photostimulation

Photostimulation is a series of stimuli that are caused by light. Among them, infrared, ultraviolet, visible, X-ray and laser can be used to stimulate acupoint skin, which can play a therapeutic role.

(一) 对话

(C=顾客,B=美容师)

B:您好,欢迎光临。有什么可以帮助你?

C:您好,最近我发现我的皮肤有些松弛下垂,我害怕影响我的容貌。

B:好的,这类情况用光学类仪器治疗效果很好。治疗完,皮肤恢复后,皮肤的光泽度和紧致感会得到明显改善。

C:能有你说的这样好吗?

B:是的。我给您详细讲一下,我们的皮肤在光的作用下,真皮层的各种纤维受到热的刺激后,会收缩,这样皮肤的真皮层就会变厚,皮肤里面的附属器官功能增强,可以恢复我们皮肤的弹性、提升轮廓、明显改善光泽度。

C:那需要做几次?

B:这样的治疗做一次就会看到明显的效果,后期根据您皮肤的情况,可以再次选择做这样的美容项目。

C:哦,谢谢你的解释。

B:您客气了。

(二) 课文

激光的基础知识

现代激光技术应用于美容皮肤科学的治疗领域,是近年来我国皮肤科专业内突破性的发展之一,在短短的十年左右的时间里,激光技术已经形成了一套较完整的理论体系和临床实践,成为美容皮肤科主要的治疗手段之一。

紫外线防护受到全面重视,皮肤晒伤、皮肤晒黑及皮肤光老化等成为热门研究课题,一种新的科学领域正在形成,“光美容医学”的概念呼之欲出。激光美容仪器的发展大致经历了以下三个阶段。

基础研究阶段(20 世纪 60 年代):第一台有意义的激光器是 1960 年由 Maiman 引入临床的,它包括一根红宝石激光管并能发射波长为 694 nm 的激光。1961 年,将红宝石激光用于剥离视网膜的焊接;1963 年,Goldman L 开始将红宝石激光应用于良性皮肤损害和纹身治疗并取得成功;1965 年 Goldman L 使用红宝石激光有效祛除纹身而治疗后非常轻松没有瘢痕,之后应用 ND:YAG 激光来消除纹身及治疗表浅血管畸形。1968 年上海研制出 ND:YAG 激光。

临床使用阶段(20 世纪 70 年代):1970 年 Goldman L 等人首次用连续二氧化碳激光治疗基底细胞癌和皮肤血管瘤,由于连续地提供有效的激光功率和能量密度,克服了早期脉冲激光功率低、效率低的缺点,从而掀起了国内外激光医疗的热潮。

发展成熟阶段(20 世纪 90 年代):90 年代初期应用 Q 开关激光治疗色素性疾病如太田痣、纹身等已取得了近乎完美的治疗效果;90 年代中、后期可变脉宽 532 nm 激光治疗血管性疾病也取得了较好的疗效。此时,美国、以色列、德国等国生产的先进成套的激光美容仪迅速引进国内,并趋向普及,一些国产的激光美容仪在国内也得到了越来越多的应用并逐步取代国外产品的领先地位。

光对组织有以下三个作用。

1. 光热作用

光热作用分为热效应、热凝固、热剥脱。热效应是指温度升高到一定程度(60～65 ℃)时,胶原纤维即可收缩到原来长度的1/3且胶原的直径增加,相应组织结构的完整性未受到影响。热凝固是指当温度高于75 ℃,且时间足够长时会引起组织的热凝固,一定量分散的热凝固会带来持久的明显的皮肤重建效果。热剥脱是指当温度大于300 ℃且时间足够时,热可导致组织气化、燃烧,分子裂解和等离子形成,进而导致组织创伤。

2. 光动力反应

光照射组织,引起组织内产生一种或一系列化学反应而导致病变部位(靶组织)被有效治疗或改善的一种方法。

3. 光刺激

光刺激就是由于光受到的一系列刺激。其中有利用红外线、紫外线、可见光、X线和激光刺激穴位皮肤,可以起治疗作用。

Section 2 RF Beauty

Part Ⅰ Dialogue

(C=Customer,B=Beautician)

B:Welcome. Is there anything I can do for you?

C:Hello,recently I found my skin was a little flabby and I am afraid of affecting my appearance.

B:Well, the RF beauty works very well on this kind of situation. After the skin recovery,the gloss and firming of the skin will be improved obviously.

C:Is it as good as you say?

B:Of course. Let me introduce in detail. After the preparation work,under the action of the instrument, the various fibers of dermis will shrink after stimulation heat. Thus, it recovers our skin elasticity and promotes the lineament and gloss of skin.

C:How many times need I take?

B:You will see the obvious effect after only one time. You can choose to do this beauty project once a week because it is considered to be the first choice of daily skin care.

C:Oh,thank you for your explanation.

B:You're welcome.

Part Ⅱ Text

RF Beauty

RF beauty is a new non-surgical, non-invasive, high security anti-aging cosmetic technique.

An electric wave will resonate through molecules, and friction will produce heat. When RF acts on the skin, it produces high frequency oscillations of ions (the number of vibrations can reach millions of times per second). The physical thermal field is formed under the double action of the natural stress reaction and the Brown motion of the target tissue on the high frequency vibration of the ion. Under the thermal field effect, the structure of collagen hydrogen bonds is destroyed, and the three-helix structure changes in collagen molecules, which leads to the immediate contraction and rearrangement of collagen. If the target tissue is regularly treated with RF, the collagen chain will be continuously tightened, and then the three-dimensional contraction of the target tissue can be achieved to the most extent. Finally the achievements of reconstruction and remodeling of the skin have great curative effect on firming, promoting facial contour, whitening, removing wrinkles and fading scars.

（一）对话

（C＝顾客，B＝美容师）

B：您好，欢迎光临。有什么可以帮助你？

C：您好，最近我发现我的皮肤有些松弛下垂，我害怕影响我的容貌。

B：好的，这类情况用具有射频技术的仪器治疗效果很好。治疗后，皮肤的光泽度和提升度会得到明显改善。

C：能有你说的这样好吗？

B：是的。我给您详细讲一下，准备工作做完后，仪器作用在皮肤上，真皮层的各种纤维会受热，然后纤维收缩，这样皮肤的弹性、轮廓、光泽度都得到了改善。

C：那需要做几次？

B：这样的治疗做一次就会看到明显的效果，每周可以做一次，是日常保养的首选美容项目。

C：哦，谢谢你的解释。

B：您客气了。

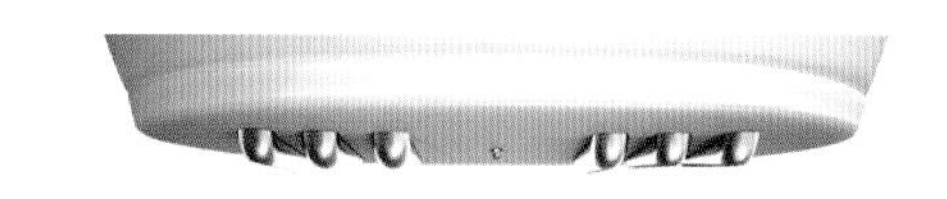

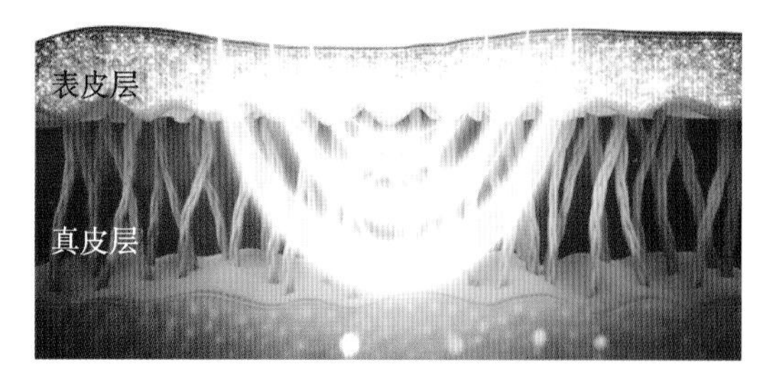

（二）课文

射频美容

射频美容是一种全新的非手术、非介入、无创伤、高安全性的抗衰老美容技术。

电磁波通过分子会发生共振，摩擦产生热量。当射频作用于皮肤时，会产生离子高频振动现象(其振动次数可达每秒数百万次),在靶组织对离子高频振动的自然应激反应和布朗运动的双重作用下,形成了物理性热场,在热场效应作用下胶原氢键结构被破坏,胶原分子中的三螺旋结构发生变化,导致胶原收缩并重新排列。如对靶组织进行适当定期的射频治疗,可使胶原链不断收紧,继而达到最大程度的靶组织三维立体收缩,实现皮肤的重建和重塑,以达到紧肤、提升面部轮廓、美白、祛除皱纹、淡化瘢痕等疗效。

Part Two

Cosmetic English Conversations

Section 1 Skin Physiology and Pathology

1. Standards of Healthy Skin

A:Hello,could you tell me what the healthy skin is like?

B:According to the norms of beauty,healthy skin contains five points:

Skin is neat.

Skin looks ruddy gloss and fair.

Skin feels smooth and flexible.

Skin is not withered or yellow,without wrinkles,spots,etc.

Skin can resist aging.

A:Ok,thanks.

2. Ten Factors Affecting Skin

A:How many factors will affect our skin? Tell me specifically?

B:There are ten factors.

A:What are they?

B:①Age. ②Sleep. ③Heredity. ④Exercise. ⑤Gender. ⑥Health. ⑦Solarization. ⑧Drug taking habit. ⑨Diet. ⑩Proper skin care.

3. Features of Dry Skin

A:Could you tell me which kind of skin I have? Dry skin or oily skin?

B:Let me see. It is dry skin.

A:What does dry skin look like?

B:①Susceptible to wrinkles and freckles. ②Lack of water. ③ Fragile and sensitive.

4. The Causes,Symptoms and Care of Wrinkles

A:Could you please help me to check my skin? It is flabby and full of wrinkles. What caused it?

B:It can be divided into internal and external factors. The external factors:①Dry and lack of water. ② Ultraviolet rays—the main cause to skin aging. ③Unbalanced work and rest,smoking,drinking,excessive fatigue,etc.

The internal factors:①Subcutaneous tissue damage. ②Aging. ③Elastin changing,etc.

A:What can be called flabby skin?

B:①Skin is flabby without flexibility. ②Wrinkle.

A:How to care it? Could you give me some advice?

B:①Do more face massage and care in your spare time. ② Have proper diet and have more antioxidant food (vitamin C and E), like oranges. ③ Have proper work and rest. ④Avoid high temperature.

Section 2 Fundamental Beauty Nursing

1. Cleanser

A:There are all kinds of cleansers,could you give me some advice?

B:Sure. There are foam cleanser, solvent cleanser, foam free cleanser and collagen cleanser. The solvent cleanser is mainly applied to the oily skin. The foam free cleanser combines the right amount of oil with some surfactant. The collagen cleanser is warm and comfortable,which can effectively remove melanin,including the face cutin and dirt,making the skin restore the transparent relaxed state.

A:What kind of cleanser should I choose?

B:Different cleansers have different functions,and you should apply them according to your skin type.

A:Skin type? Could you introduce it in detail?

B:There are oily skin,mixed skin,normal skin,dry skin and sensitive skin. The sebum secretion of the oily skin is more than that of the average person,so it needs to choose some stronger cleaning products. Normal skin should choose foam type and dry skin should choose foam free type. For sensitive skin,you'd better choose antiallergic type.

A:Thanks.

2. Day Cream

A:Hello,I want to know something about the day cream.

B:Hello,madam. It is my pleasure to help you. Please sit down.

A:What is the effect of the day cream?

B:It has the effects of moisture,repair and anti-aging.

A:What is the proper time to use the day cream?

B:After shower or face washing,you can apply it to the skin.

A:Thanks for your introduction.

3. Night Cream

A:Is the night cream applied to care our skin before sleeping?

B:Yes,according to the scientific research,the skin is easier to lose water at night than in the day time.

A:What is the effect of the night cream?

B:The night cream has the similar effect to the day cream.

A:How to apply the night cream?

B:Both the day and night cream should be applied after thorough face cleaning.

4. Eye Gel

A:How many types of eye gel can be divided into?

B: According to the effect, it can be divided into moist eye gel, anti-aging eye gel, anti-allergic eye gel and firming eye gel.

A: How to choose a proper eye gel?

B: When you choose a proper eye gel, the following factors should be considered:

①With efficient hydrating and water-lock up function. ②Refreshing and easy to absorb. ③Safe ingredients without stimulus.

5. Sunscreen

A: There are all kinds of sunscreen creams, how to choose?

B: My advice: ①Do a precise skin check before choosing. ②Work out the SPF. Generally speaking, the higher the SPF is, the longer time it can protect the skin. ③Know something about different sunscreen creams for different people. ④ Choose refreshing sunscreen products.

A: Should it be chosen according to skin, for example, what kind of cream suits oily skin?

B: For the oily skin, choose water dosage form without oil formula. It is refreshing and won't clog pores. For the dry skin, choose a mild and hydrating effect and enhance skin immunity. For the sensitive skin, choose products which are specifically formulated for it or which are marked with "allergy-proofed".

6. Moisturizing Cream

A: Now there are all kinds of moisturizing products, how to choose?

B: It should be suitable for your skin. We have normal skin, oily skin, mixed skin and dry skin. For the normal skin, firstly, apply the moist lotion, then the moist essence, last the moist night cream. For the dry skin, do the same as normal skin. For the oily skin, first apply the whiten lotion, then the moist lotion. For the mixed skin, moist lotion, moist essence, and cream.

7. Lotion

A: What is the effect of the lotion?

B: The lotion has three effects: cleaning, moisturizing, nourishing.

A: How to apply lotion?

B: According to different kinds of lotions, you can choose tap and apply.

8. Essence

A: How to apply essence for the dry skin, normal skin and oily skin?

B: ①Dry skin should choose moist essence. ②Normal skin should choose whitening or wrinkle-removing essence. ③Oily skin chooses pure focus essence. ④Sensitive skin chooses comforting, antiallergic essence. ⑤Mixed skin should apply pure focus essence on T zone, and moist on cheeks.

A: What products are applied in the routine care steps?

B: Routine care steps:

Common facial care products: ① Cleanser. ② Hydrabio exfoliating cream. ③ Massage cream. ④Mask gel. ⑤Purifying & balancing toner, firming toner. ⑥Essence. ⑦Eye gel. ⑧Whitening repair and insulation milk.

A: How to choose a proper essence?

B: Firstly, choose essence which is suitable for your skin, and then choose it according to the seasons. In summer, choose moist ones. In winter, choose thick type lotion.

9. Make-up Water

A: How to choose the make-up water according to different skins?

B: ① For the oily skin, choose firming lotion. ②The normal skin, choose a toner. ③The dry skin, choose a smoothing toner. ④The mixed skin, apply a firming toner on T zone, both toner and smoothing toner can be applied on other parts. ⑤ The sensitive skin, choose a repair water or you can choose whitening water.

A: I have bought several kinds of make-up water, how to store them?

B: You can either store them in normal temperature or in special circumstance, like in the refrigerator.

10. Foundation

A: What are the effects of a foundation?

B: The foundation is a base of a make-up look, such as, adjust the skin color, and conceal flaws. Some foundations can block the UV and care the skin.

A: How to choose a foundation?

B: It depends on the types of skin. For the mixed skin, we can choose a liquid foundation. Dry powder is also preferred. For the pale skin, we should choose farinaceous powdery bottom. If you are of black skin, choose the liquid fluid foundation, which can make the skin appear more vigorous. As long as you choose the right products according to your skin characteristics, you will be a beautiful girl.

Section 3 Types and Functions of Cosmetics

1. Freckle Removing

B: Welcome to Haitang International, what can I do for you?

A: Hello, my skin is not good recently, I want to know the reasons.

B: This way please. Could you please fill in your information here? Then I will do a test for your skin.

A: I am wondering whether the spots on my skin can be dispelled.

B: What do you feel about your skin?

A: I feel my skin is dry and lack of burnish. There are more and more spots.

B: Face is a mirror of our body. The spots on your face show that there are toxins in your body which cannot be let out.

A: What should I do?

B: Do a detoxification for your skin, and it can return to moist and tender.

A: How long will the effect last?

B: Like Miss Wang here, she has made a difference in just one month. With our products, you can make it quickly.

A: Really good. Thank you very much.

2. Acne Removing

A: What are the effects of Haitang Wang?

B: Improve the oily skin, remove acne, and whiten the skin.

A: What kind of skin does it suit for?

B: It suits for acne, brandy nose skin and oily skin.

A: Why can the Haitang Wang cleanser prevent skin allergy? What kind of skin does it suit for?

B: It can remove acne, mites and toxin, as well as improve skin metabolism. It can cure acne, brandy nose and oily skin.

3. Desensitization

A: What kind of skin does Haitang Mei cleanser suit?

B: It suits the black and yellow complexion. Besides, it does well to the wind chill type of allergy, blood deficiency type allergy, allergic constitution, dry sensitive skin, senile itchy skin and blood stasis syndrome.

A: What are the effects of Haitang Mei cleanser?

B: It can activate blood circulation, eliminate stasis, contract pore, whiten skin and make the skin from inner to outer obtain perfect care.

4. Flat Wart Removing

A: What are the effects of flat wart cleanser?

B: It contains natural herbal essence, and can soften and dissolve hard masses, improve skin resistance, help to eliminate the hyperplasia in the skin, leave skin healthy and natural.

A: What kind of skin does it suit for?

B: It suits for flat wart, calculus, rough and tuberous acne skin.

5. General Questions Concerning Haitang Wang

A: Can Haitang Wang be taken for treating weak spleen and stomach acne?

B: Haitang Wang belongs to cool quality medicine, so weak spleen and stomach sufferers should not take Haitang Wang capsule.

A: Can Haitang Wang and Haitang Mei be taken at the same time?

B: They can be taken according to the symptoms of the syndrome. The treatment is based on syndrome differentiation.

A: What is the difference between Haitang Wang cleanser and other products?

B: Haitang Wang cleanser is made of pure Chinese herbal medicine, and pure herbal medicine accounts for 80 percent. It has the effect of eliminating bacteria and mites, controlling animalcule growth and improving skin oil secretion. Common acne removing cleanser is alkalescent, although after washing with it, one may feel refreshing. It can make skin dehydrated, and decrease skin resistance.

A: Can people with high blood pressure and heart disease take Haitang Wang?

B: Yes, they can. Since Haitang Wang has the effect of eliminating toxins and heat, regulating fat metabolism, it can also improve high blood pressure.

A: What is the effect of Haitang Wang oral products on toxin?

B: Haitang Wang oral products are all made from pure Chinese herbs, which can also be taken as nutrients, without any side effect.

6. General Questions Concerning Haitang Mei

A: Taking Haitang Mei capsule makes me feel nausea. Can I still take it?

B: Yes, you can. Nausea is caused by your wrong taking method. Usually we suggest you take it before dinner.

A: How long does one treatment of using Haitang Mei series products take?

B: According to the theory of TCM, one hundred days is called one big treatment, one month called one small treatment. So the balance of zang-organs can prevent from developing disease.

A: What is the difference between Haitang Mei series products and Taitai oral liquid and Baixiao Dan?

B: Taitai oral liquid and Baixiao Dan belong to single-ingredient medicine, it is general medicine instead of being specific treatment. While Haitang Mei products are specially composed to treat chloasma, so it has very good pertinence.

A: Can children take Haitang Mei series products?

B: Yes, they can. But the dosage must be reduced. Anorectic children can take Haitang Mei No. 3.

A: Does Haitang Mei have the effect for menopausal syndrome?

B: Haitang Mei can regulate zang-fu viscera, and incretion. If it is taken with Shugan Quban Dan, it can relieve irritation of menopause, and with Yang Yin Dan to cool the heat resulting from yin deficiency.

Vocabulary

1. symptom['simptəm] n.[临床]症状,征兆
2. essence['esns] n.本质,实质,精华,香精
3. foundation[faun'deiʃən] n.基础,地基,基金会,根据,创立
4. desensitization[ˌdisənsətə'zeʃən] n.[免疫]脱敏,[感光]减感
5. pathology[pə'θɔləʤi] n.病理学(复数 pathologies)
6. norm[nɔːm] n.规范,基准,定额,分配的工作量
7. ruddy['rʌdi] adj.红的,红润的
   adv.极度,非常
   vt.使变红

vi. 变红

8. gloss[glɔs] n. 光彩，注释，假象
vt. 使光彩，掩盖，注释

9. withered['wiðəd] adj. 枯萎的，憔悴的，凋谢了的，尽是皱纹的
v. 干枯，减弱，羞愧(wither 的过去分词)

10. heredity[hi'rediti] n. 遗传，遗传性（复数 heredities）

11. solarization[ˌsəulərai'zeʃən，-ri'z-] n. 日晒，[摄] 负感作用

12. fragile['frædʒail] adj. 脆的，易碎的

13. internal[in'tə:nəl] adj. 内部的，内在的，国内的

14. external[ik'stə:nəl] adj. 外部的，表面的，[药] 外用的，外国的，外面的

15. excessive[ik'sesiv] adj. 过多的，极度的，过分的

16. fatigue[fə'ti:g] n. 疲劳，疲乏，杂役
vt. 使疲劳，使心智衰弱

17. antioxidant[ˌænti'ɔksidənt] n. [助剂] 抗氧化剂，硬化防止剂，防老化剂

18. foam[fəum] n. 泡沫，水沫，灭火泡沫

19. solvent['sɔlvənt] adj. 有偿付能力的，有溶解力的

20. surfactant[sə'fæktənt] n. 表面活性剂
adj. 表面活性剂的

21. melanin['melənin] n. 黑色素

22. cutin['kju:tin] n. [生化] 角质，蜡状质，表皮素

23. transparent[træn'spærənt，-'peə-，trænz-，trɑ:n-] adj. 透明的，显然的，易懂的

24. antiallergic[ˌæntiə'lə:dʒik] n. 抗过敏药
adj. 抗变应性的，抗过敏症的

25. moist[mɔist] adj. 潮湿的，多雨的，含泪的
n. 潮湿

26. hydrating adj. 保湿的，吸水的

27. stimulus['stimjuləs] n. 刺激，激励，刺激物

28. dosage['dəusidʒ] n. 剂量，用量

29. formula['fɔ:mjulə] n. [数] 公式，准则，配方，婴儿食品

30. clog[klɔg] v. 阻塞，障碍
n. 障碍，木底鞋

31. pore[pɔ:] n. 气孔，小孔

32. immunity[i'mjunəti] n. 免疫力，豁免权，免除

33. nourishment['nʌriʃmənt] n. 有营养的，滋养的

34. hydrabio 水活保湿

35. insulation[insju'leiʃən，insə-] n. 绝缘，隔离，孤立

36. flat wart 扁平疣

37. hyperplasia[ˌhaipə'pleʒə] n. [医]增殖，增生物

38. calculus['kælkjuləs] n. 微积分学，[病理] 结石

39. tuberous['tju:bərəs] adj. 块茎状的，有结节的，隆凸的

40. viscera['visərə] n. 内脏，内容(viscus 的复数)
41. meridian[mə'ridiən] adj. 子午线的，最高点的
    n. [天] 子午线，[天] 经线，顶点
42. relapse[riː'leæps, ri'læps, 'riːlæps] n. 复发，再发，故态复萌，恢复原状
43. incretion[in'kriːʃən] n. 内分泌，内分泌物，激素
44. menopause['menəpɔːz] n. [医]停经，绝经(期)，活动终止期
45. anorectic[ˌænə'rektik] adj. 食欲缺乏的，厌食的
46. pertinence[pəːti'nəns] n. 有关性，相关性，针对性
47. nausea['nɔːziə, -siːə, -ʃə] n. 作呕，恶心，反胃，极度厌恶
48. dehydrated[ˌdiː'haidreitid] adj. [医]脱水的
49. alkalescent['ælkə'lesənt] adj. 弱碱性的，碱性的
50. animalcule[ˌæni'mælkjuːl] n. 微生物
51. herbal['həːbəl, həː-] adj. 药草的，草本的
52. syndrome['sindrəum] n. 综合征
53. spleen[spliːn] n. 坏脾气，怒气，怨气，脾脏

Part Three

Cosmetology of TCM Conversations

Section 1 Traditional Chinese Medicine

Dialogue One

Foreigner: More and more people have recognized TCM's noticeable treatment effect, haven't they?

Chinese: That is true. And you know what? TCM's contribution to the world is not only an original medical system but also a part of Chinese traditional culture.

Foreigner: Yeah! Yesterday I read one book on TCM. Some of the basic theories are so difficult to understand, like the concept of yin and yang. I still can't understand it until now.

Chinese: It is difficult for a foreigner to understand Chinese traditional philosophy. But the key point is to keep balance and harmony, whether inside or outside your body.

Foreigner: That is a wonderful idea!

Chinese: Right! TCM is such a broad and deep system. It deserves more attention and study.

Foreigner: I'm trying to learn more about TCM. Would you help me?

Chinese: Sure! Let's study together.

Dialogue Two

Foreigner: Hello, Mr. Wang! Can you generally describe the basic theories of TCM?

Chinese: Well, like other branches of traditional Chinese medicine, TCM cosmetology is also based on the basic theories of TCM, such as the theory of yin and yang, the theory of five phases, the theory of zang-organs and fu-organs and the theory of qi, blood and body fluid.

Foreigner: Really? That sounds interesting!

Chinese: The theory of yin and yang believes that the world is material, and that the material world generates, develops and changes with the mutual actions of yin-qi and yang-qi.

Foreigner: Do you mean that it is a concept of unity of opposites?

Chinese: Right. There exists not only the opposition and struggle between yin and yang, but also the interdependence and mutual promotion, the waxing and waning of yin and yang, the transformation between yin and yang.

Foreigner: Yes.

Chinese: Well, the theory of five phases holds that wood, fire, earth, metal and water are the indispensable and basic materials to form the material world. The mutual generation and restriction and movement and changes of the five phases constitute the material world.

Foreigner: That is a good lesson! Thanks a lot! See you next time!

Chinese: See you!

Dialogue Three

Foreigner: Hello, Nice to see you again!

Chinese: Nice to see you! How are you going on with your TCM?

Foreigner: I found qi, blood and body fluid was much easier to gain.

Chinese: Really? That's not an easy job for you.

Foreigner: They are the components of the human body to maintain the life activities of human body.

Chinese: Yes. Qi is the vital substances which move endlessly with very powerful activities.

Foreigner: Yes, I think so. See you later.

Chinese: See you!

Dialogue Four

Foreigner: Good morning, Mr. Wang. I am very glad to see you again.

Chinese: Good morning, Jone. I am very glad to see you again, too.

Foreigner: Yesterday, I read the part of five zang-organs.

Chinese: Really? That must be interesting.

Foreigner: Yeah! From the book, I know TCM divides all the functions of the human organism into five types, each being put under the name of one of five zang-organs.

Chinese: Right! You got it!

Dialogue Five

(P=Patient, D=Doctor)

D: Hello, Mr. Smith. How are you feeling today?

P: Much better, thank you! I can sit straight today. The acupuncture is really great.

D: Yes. Traditional Chinese medicine has a history of more than 5,000 years. It has a complete theory about the occurrence, development and treatment of diseases. Acupuncture is only one of the most effective ways to treat diseases such as your pleurapophysis.

P: All the same, many of my colleagues feel much puzzled about TCM. Would you please tell me more about it?

D: Sure. According to TCM theory, the occurrence of diseases is the incoordination between yin and yang and the treatment of diseases is the reestablishment of the equilibrium between them.

P: Oh, what's yin and yang?

D: They are the two concepts from ancient Chinese philosophy and they represent the two contradictories in everything. In TCM theory, yin and yang are used to explain physiological and pathological phenomena of the body. They are also the principles of diagnosing and treating diseases.

P:Then how do you treat your patients by using this theory?

D:Roughly speaking,there are two common ways of TCM curing diseases:drug therapy and non-drug therapy.

P:That's very interesting. What drugs do you often apply?

D:As for drug therapy,traditional medicines are used such as herbs,mineral,animals.

Dialogue Six

(P=Patient,D=Doctor)

D:Good morning. Sit down please. What seems to be bothering you?

P:Oh,my neck is stiff since I got up this morning. I feel pain when I turn my head.

D:Have you had a similar problem before?

P:Yes,but not as painful as this time.

D:Now please sit straight,put your arms down and relax your shoulders. I'd like to examine you. Now I'll press your neck muscles with my thumbs. Does it hurt here?

P:A little bit. Further up and to the right.

D:Now please try to turn your head to both sides.

P:Ouch. It hurts.

D:Well,try to raise and lower your head.

P:That hurts too.

D:Your movement is limited. I'd like to have a look at your tongue. Stick it out,please. Well,it's dark purplish. Now let me take your pulse. Lay your wrist on the mattress with your palm facing upward. OK,give me another wrist. Your pulse is wiry and tense.

P:Doctor,what's the matter with me?

D:Your neck is stiff. We also call it torticollis.

P:How did I get it?

D:From the point of view of TCM,your problem is usually due to improper positioning of the neck during sleep,a pillow of improper height or long term stress of the jugular condyle and muscles. These factors can cause localized channel qi and blood blockage. You're suffering from a typical qi stagnation blood stasis torticollis. Acupuncture and moxibustion are very effective in treating it.

P:Oh,I see. I'll try it. But this is my first acupuncture treatment. Does it hurt a lot?

D:Don't worry. You'll have soreness,numbness,pain and distended feeling after insertion. This is called needling sensation,also known as the arrival of qi. It's perfectly tolerable. Only if the qi is drawn to the area can you get the full benefit of the treatment.

P:I've heard something about fainting. What's that all about?

D:That is due to nervousness,weakness,overwork or improper positioning. You won't faint if you are relaxed.

P:OK,I'll follow your advice.

(The doctor starts the treatment.)

P:Doctor,what acupoints are you using?

D:I have chosen stiff neck point,Ashi point and Houxi for you.

P:What's the use of these acupoints?

D:Stiff neck point is an extra point which can cure torticollis. Ashi point can open up the channel qi in painful areas. And Houxi is related to the small intestine meridian,whose meridian and muscle region are distributed in the base of the neck. According to the channel theory,acupuncture and moxibustion can be used to treat disease in the area governed by its pertaining meridian. They help to open up the channel qi from the far end. How is your needling sensation now?

P:A bit painful. I also feel numbness in my elbow.

D:Good. Now I'll start needling manipulation. Try to gently rotate your neck when I turn the needle. OK,rotate gently. How are you feeling?

P:I can move my neck and the pain have been eased off.

D:Good. I'll keep these needles in place for 20 minutes. You'd better keep still. I'm going to give you a circling moxibustion treatment on Fengchi and Dazhui. Both of these points activate the meridians and collaterals.

P:What is the moxa stick for?

D:We use moxa to affect the flow of qi and blood. In this way moxibustion works to stimulate the points to warm the channel,expel the cold,and induce the smooth flow of qi and blood. Do you feel better now?

P:Great! I don't feel any more pain and my neck doesn't feel so stiff. Thanks a lot.

D:You are welcome.

P:Is there anything else I need to do?

D:Use a medium high pillow,keep your neck warm,and massage your neck if you have the time.

P:All right. Thank you very much.

Section 2 Common TCM Cosmetic Methods

Dialogue One

(P=Patient,D=Doctor)

P:Good morning,doctor.

D:Good morning,please sit down.

P:I have been feeling tired recently. I have no appetite. Sometimes I even feel short of breath.

D:OK. Let me feel your pulse. Open your mouth and say "Ah".

P:What's wrong with me,doctor?

D: Nothing serious. You are just sub-healthy. I suggest you try cosmetic health preservation.

P:Could you tell me something about that,doctor?

D:No problem. TCM believes that cosmetic and health preservation are closely related. Health preservation for prevention of diseases is an important guiding ideology of TCM. TCM believes that health preservation not only prevents diseases and aging to prolong life span,but also makes people remain young and healthy. The TCM theory of cosmetic health preservation can be summarized as the following six aspects: adaptation to seasonal variations,normal daily life,proper adjustment of work and rest,movement of the limbs and body,stable emotional activities and reasonable diet.

P:Sounds great! I will follow your instructions. Hope it will work on me.

D:Certainly it will.

P:Thanks a lot. Goodbye.

D:You are welcome. Goodbye.

Dialogue Two

(S=Student,P=Professor)

S:Good morning,Professor Li.

P:Good morning,Peter.

S:I am reading some books about TCM cosmetic health preservation. In the book,it refers to adaptation of seasonal variations. I can't quite understand it. Can you explain it for me?

P:Ok. TCM holds that seasonal variations are closely related to human health,diseases and complexion. Yang is nourished in spring and summer, when yang-qi is sufficient, everything is full of life and human metabolism is relatively vigorous. Therefore,if life style is not changed in accordance with seasonal variations,diseases will occur and health will be harmed.

S:I understand now,professor. Thank you!

P:You are welcome!

Dialogue Three

(P=Patient,D=Doctor)

P:Good morning,doctor.

D:Good morning,madam. What's the matter with you?

P:I don't know,but I feel weak. Look at the color of my face.

D:Let me see. Your skin is sallow and a little wrinkled.

P:Can you give me some suggestions about TCM cosmetic methods?

D:Of course. First of all,you should know aging starts from skin. This shows that the most obvious region of aging lies in skin,especially the facial skin which reflects the degree of aging. In order to prevent aging and facial wrinkles,some diseases affecting looks should be treated on the basis of the patient's individual condition with the choice of cosmetic methods of traditional Chinese drugs, medicinal diet, acupuncture and moxibustion, massage and qigong in addition to normal daily life.

P: Oh, I see. So please tell me some practical TCM cosmetic methods.

D: You can achieve aesthetic effect with traditional Chinese drugs. On the basis of the medication and mechanism of action, the drugs are classified into two groups: cosmetic drugs for oral medication and cosmetic drugs for external use. On the basis of the preparation, cosmetic traditional Chinese drugs for external use are classified into cosmetic powder, cosmetic liquid, cosmetic soft extract, cosmetic facial film, etc.

P: It is so amazing! I will have a try.

Dialogue Four

(M=Mary, L=Lily)

M: Hi, Lily! I haven't seen you for a long time.

L: Hi, Mary! How are you?

M: I'm fine. You look great, especially your skin color. What have you done to yourself?

L: I have been trying cosmetology with acupuncture and moxibustion. It really works.

M: Really? I have never heard about that. Tell me about it.

L: No problem. Cosmetic with acupuncture and moxibustion means that by means of acupuncture and moxibustion, the meridians and acupoints are stimulated to bring internal organic factors into play, regulate the functions of all zang-fu organs and tissues, promote blood circulation and qi flow, and activate meridians and collaterals. By doing so, the purpose is achieved of resisting exogenous pathogenic invasion, delaying aging and making complexion beautiful. Besides, acupuncture increases muscular elasticity, eliminates canthal and frontal wrinkles, and eradicates pigmented spots and comedo.

M: Oh, it has so many functions.

L: Yeah, There are mainly three kinds of acupuncture, which are cosmesis with acupuncture, cosmesis with moxibustion and cosmesis with otopuncture.

M: That's wonderful. I will try all of them.

Dialogue Five

(T=Teacher, S=Student)

S: Excuse me, Miss Shi. Can you spare some time for me? I want to consult you something.

T: OK. What's up?

S: It is said that cosmetic with massage is very effective. Can you explain it to me in detail?

T: OK. Actually, cosmetic with massage concentrates on the head and face. TCM holds the head and face which are the convergences of yang meridians of foot starting from the head and face. It has been clinically proven that regular and constant massage of the skin of the head and face or certain acupoints of the two regions treats premature grey hair and pathological baldness, regulates qi and blood, radiates people's vigor, delays aging, prevents wrinkles and increases skin elasticity and luster.

S: Fabulous! Miss Shi, can you introduce me some cosmetic massage manipulations?

T: Well, the cosmetic massage manipulations are various. They should be selected on the basis of individual conditions. In case of scalp massage, hold the two side of the scalp with two hands and rub forward to the top of the head. Or with the hands comb or push and knead along the frontal hairline, posterior neck hairline, the lateral hairline of ears and the top of head. The principle of facial skin massage is massage of the facial skin and muscles along the course of the veins from the upper to the lower, from the inner to the outer, with soft manipulation and moderate force. Finally, it should be pointed out that cosmetic massage does not show the effects in a short time. It should be kept constantly.

S: I see. Thank you so much.

T: You are welcome.

Dialogue Six

(D=Doctor, P=Patient)

P: Good morning, doctor!

D: Good morning, madam. What's the matter with you?

P: Look at my face, please. There appear some freckles and several fine wrinkles.

D: I see. Here I suggest you try cosmetic with medicinal diet. As early as in the Zhou Dynasty, there were professional medical dietitians in China, working on medicinal diets. Cosmesis with the medicinal diet is in fact a cosmetic prescription which is processed with proper drugs and food on the basis of personal requirements. It regulates qi and blood, nourishes zang-fu organs to achieve the cosmetic purpose.

P: It sounds attractive. I will have a try.

Vocabulary

1. dietitian[ˌdaiə'tiʃən] n. 饮食学家，营养学家
2. prescription[pri'skripʃən] n. [医]药方，处方
3. freckle['frekəl] n. 雀斑，斑点
4. moderate['mɔdərit] adj. 温和的，适度的，中等的
5. vein[vein] n. 静脉
6. lateral['lætərəl] adj. 侧面的，横向的
7. knead[niːd] vt. 揉捏(面团、湿黏土等)，按摩，揉捏(肌肉等)
8. scalp[skælp] n. 头皮
9. manipulation[məˌnipjʊ'leiʃn] n. (熟练的)操作，操纵，控制
10. radiate['reidieit] vt. 辐射，发射，使向周围扩展
11. baldness['bɔːldnis] n. 光秃，枯燥
12. convergence[kən'vəːdʒəns] n. 集合，会聚
13. otopuncture n. [医]耳针

14. cosmesis[kɔz'mesis] 美容术
15. comedo['kɔmidəu] n. 粉刺
16. pathogenic['pæθə'dʒenik] adj. 引起疾病的
17. exogenous[ek'sɔdʒinəs] adj. 外生的，外成的，外因的
18. collateral[kə'lætərəl] n. 担保物，旁系亲属
19. acupoint['ækjuˌpɔint] n. 穴位
20. meridian[mə'ridiən] n. 子午圈，子午线，顶点
21. extract[eks'trækt] n. 精，汁，榨出物，摘录
22. medication[ˌmedi'keiʃən] n. 药物，药剂，药物治疗，药物处理
23. moxibustion[ˌmɔksi'bʌstʃən] n. 灸术，艾灼
24. massage['mæsɑːʒ] n. 按摩，推拿
    vt. 按摩，推拿
25. sallow['sæləʊ] adj.（尤指面色）蜡黄色的
26. complexion[kəm'plekʃən] n. 肤色，面色，气色
27. limb[lim] n. 肢，翼，大树枝
28. ideology[ˌaidi'ɔlədʒi] n. 思想（体系），思想意识，意识形态
29. pathological[ˌpæθə'lɔdʒikəl] adj.［医］病理学的，由疾病引起的
30. physiological[ˌfiziə'lɔdʒikəl] adj. 生理学的，生理的
31. contradictory[ˌkɔntrə'diktəri] adj. 矛盾的，反驳的，抗辩的
32. reestablishment[ˌriːi'stæbliʃmənt] n. 重建，恢复
33. incoordination[ˌinkəʊɔːdi'neiʃən] n. 不同等，不配合
34. pleurapophysis[ˌpluərə'pɔfisis] n. 椎骨侧突，椎肋
35. acupuncture['ækjʊˌpʌŋktʃə] n. 针灸（疗法）
36. indispensable[ˌindi'spensəbl] adj. 不可缺少的，绝对必要的
37. transformation[ˌtrænsfə'meiʃən] n. 变化，转换
38. interdependence[ˌintədi'pendəns] n. 互相依赖
39. mutual['mjuːtʃuəl] adj. 共有的，共同的，相互的，彼此的
40. fluid['fluːid] n. 液体，流体
41. cosmetology[ˌkɔzmə'tɔlədʒi] n. 美容术，整容术
42. noticeable['nəutisəbl] adj. 显而易见的，明显的，引人注目的
43. torticollis[ˌtɔːti'kɔlis] n. 斜颈，歪头（因颈部肌肉收缩导致的）
44. condyle['kɔndil] n. 髁状突，（节肢类动物）硬壳间的接触点
45. moxa['mɔksə] n. 艾，灸料

Appendix

Appendix A The Commonly-Used Traditional Chinese Drugs of Cosmetic Purpose

Chinese Names	English Names
丁香	Caryophylli Flos
大黄	Radix et Rhizoma Rhei
山药	Rhizoma Dioscoreae
山茱萸	Fructus Corni
川芎	Rhizoma Ligustici Chuanxiong
女贞子	Fructus Ligustri Lucidi
女贞叶	Folium Ligustici Lucidi
马齿苋	Herba Portulacae
天冬	Radix Asparagi
木香	Radix Aucklandiae
五味子	Fructus Schisandrae
升麻	Rhizoma Cimicifugae
丹参	Radix Salviae Mitiorrhizae
乌梅	Fructus Mume
玉竹	Rhizoma Polygonati Odorati
石膏	Gypsum Fibrosum
白及	Rhizoma Bletillae
白术	Rhizoma Atractylodis Macrocephalae
白芷	Radix Angelicae Dahuricae
白附子	Rhizoma Typhonii
白蒺藜	Fructus Tribuli
白鲜皮	Cortex Dictamni Radicis
白僵蚕	Bombyx Batryticatus

豆蔻	Fructus Ammomi Rotundus
冬瓜	Wax Grourd
冬瓜子	Semen Benincasae
冬葵子	Fructus Malvae
半夏	Rhizoma Pinelliae
甘草	Radix Glycyrrhizae
地黄	Radix Rehmanniae
地肤子	Fructus Kochiae
当归	Radix Angelicae Sinensis
红花	Flos Carthami
麦冬	Radix Ophiopogonis
远志	Radix Polygalale
苍术	Rhizoma Atractylodis
芦荟	Aloe
杏仁	Semen Armeniacae Amarum
鸡内金	Endotheliun Corneum Gigeriae Galli
细辛	Herba Asari
珍珠	Margarita
栀子	Fructus Gardeniae
枸杞子	Fructus Lycii
柿子	Fructus Kaki
胡桃仁	Semen Juglandis
荆芥	Herba Schizonepetae
茯苓	Poria
茺蔚子	Fructus Leonuri
牵牛子	Semen Pharbitidis
桃仁	Semen Persicae
莲子	Semen Nelumbinis
莲花	Flos Nelumbinis
莲须	Stamen Nelumbinis
益母草	Herba Leonuri
浮萍	Herba Spirodelae
桑叶	Folium Mori
桑椹	Fructus Mori
桑白皮	Cortex Mori Radicis
菟丝子	Semen Cuscutae

菊花	Flos Chrysanthemi
黄芩	Radix Scutellariae
黄连	Rhizoma Coptidis
黄精	Rhizoma Polygonati
蛇床子	Fructus Cnidii
猪胰	Pancreas Suis
猪蹄	Pettitoes
鹿角胶	Colla Cornus Cervi
紫草	Radix Arnebiae Seu Lithospermi
蒲公英	Herba Taraxacum
蜂蜜	Mel
滑石	Talcum
蔓荆子	Fructus Viticis
墨旱莲	Herba Ecliptae
薏苡仁	Semen Coicis
薄荷	Herba Menthae

Appendix B A Comparison between the Nomenclature of Traditional Chinese Medicine and that Modern Medicine

Traditional Chinese Medicine	Modern Medicine
热疮、热气疮	herpes labialis
黄水疮	running sore
羊胡子疮、燕窝疮	sycosis
面游风毒	angioedema
日晒疮	solar dermatitis
鬼脸疮	lupus erythematosus discoides
桃花癣	pityriasis simplex
面游风	seborrheic dermatitis
唇风	exfoliative cheilitis
油风脱发	alopecia areata
发蛀脱发	seborrheic alopecia
头屑风	dry seborrheic dermatitis

白驳风	vitiligo
青记、蓝记	nevus of Ota
黧黑斑、面尘	chloasma
黧黑斑、面尘	melanosis
黑痣、黑子	pigmented spots
肺风粉刺、面疱	acne
酒糟鼻	rosacea
瘾疹、风疹块、鬼纹疙瘩	urticaria
水疥、水风疮、水疱湿疡	papular urticaria
扁瘊	flat wart

Appendix C The Commonly-Used Terms and Expressions in Traditional Chinese Medicine

整体观念	holism
辨证论治	treatment based on syndrome differentiation
标本兼治	treating both manifestation and root cause of disease
草药	medicinal herb
方剂	prescription
汤剂	decoction
阴阳	yin yang
五行	five phases
针刺法	acupuncture technique
针灸	acupuncture
经络	meridian
经穴	channel point/meridian point
精气	essential qi
津液	fluid
五脏	five zang viscera
六腑	six fu viscera
脏腑	zang and fu viscera
七窍	seven orifices
七情	seven emotions
元气	primordial qi
命门	vital gate
人中	philtrum

炮制	processing drugs
配伍禁忌	incompatibility of drugs in a prescription
气功	qigong
养生	health preservation
气虚	qi deficiency
气血失调	disorder of qi and blood
气滞	qi stagnation
瘀血	static blood
风邪	wind pathogen
寒邪	cold pathogen
湿邪	dampness pathogen
暑邪	summer-heat pathogen
痰饮	phlegmatic fluid
益气	benefiting qi
补血	replenishing blood
补阳	tonifying yang
扶正固本	strengthening body resistance
固表止汗	consolidating exterior for arresting sweating
活血调经	promoting blood flow for regulating menstruation
活血化瘀	promoting blood circulation for removing blood stasis
清热解毒	clearing heat-toxin
清热生津	clearing heat and promoting fluid production
清咽	clearing heat from throat
祛暑解表	dispelling summer-heat to relieve exterior syndrome
去毒生肌	detoxication and promoting granulation
润肺化痰	moistening lung for removing phlegm
疏肝解郁	dispersing stagnated liver qi for relieving qi stagnation
舒筋活络	relieving rigidity of muscles and activating collaterals
通经活络	activating the meridians and collaterals
安神	tranquillization
消食化滞	resolving food stagnation
滋阴	nourish yin
推法	pushing manipulation
推拿	tuina/massage
拔罐疗法	cupping therapy
四诊	four diagnostic methods
望诊	inspection
闻诊	listening and smelling/auscultation and olfaction
问诊	inquiry
切诊	palpation and pulse taking

脉象	pulse manifestation
舌诊	tongue inspection
舌苔	fur/tongue coating
舌质	tongue quality
虚火上炎	deficient fire flaring up
阴阳失调	yin-yang disharmony
正邪相争	struggle between vital qi and pathogen
证候禁忌	incompatibility of drugs in pattern
治病求本	treatment aiming at its pathogenesis
中成药	traditional Chinese patent medicines and simple preparations
中西医结合	integration of traditional and western medicine
中药	Chinese materia medica/traditional Chinese medicine
中药师	traditional Chinese pharmacist
中医师	traditional Chinese physician

Appendix D　Cosmetic Terms

Ⅰ. Cosmetics, Skin-care Products and Cleansers

洗面奶	facial cleanser/face wash (foaming, milky, cream, gel)
爽肤水	toner/astringent
紧肤水	firming lotion
柔肤水	toner/smoothing toner (facial mist/facial spray/complexion mist)
护肤霜	moisturizers and creams
隔离霜，防晒	sunscreen/sun block
露	lotion
霜	cream
日霜	day cream
晚霜	night cream
眼部啫喱	eye gel
眼膜	eye mask
面膜	facial mask/masque
剥撕式面膜	pack
剥落式面膜	peeling
磨砂膏	facial scrub
润肤露(香体乳)	lotion/moisturizer
护手霜	hand lotion/moisturizer

沐浴露	shower lotion
彩妆	cosmetics
遮瑕膏	concealer
修容饼	shading powder
粉底	foundation (compact,stick)
粉饼	pressed powder
散粉	loose powder
闪粉	shimmering powder/glitter
眉粉	brow powder
眉笔	brow pencil
眼线液(眼线笔)	liquid eye liner,eye liner
眼影	eye shadow
多色眼影	multi-color eye shadow
睫毛膏	mascara
口红护膜	lip coat
唇线笔	lip liner
唇膏	lip color/lipstick
腮红	blush
卸妆水	makeup remover
卸妆乳	makeup removing lotion
指甲油	nail polish
去甲油	nail polish remover
护甲液	nail saver
洗发水	shampoo
护发素	hair conditioner
焗油膏	hairdressing gel
摩丝	mousse
发胶	styling gel
冷烫水	perming formula

Ⅱ. Cosmetic Applicators

粉刷	cosmetic brush,face brush
粉扑	powder puff
海绵扑	sponge puff
眉刷	brow brush
睫毛夹	lash curler
眼影刷	eye shadow brush/shadow applicator
口红刷	lip brush
胭脂扫	blush brush
转笔刀	pencil sharpener

电动剃毛器	electric shaver
电动睫毛卷	electric lash curler
描眉卡	brow template
纸巾	facial tissue
吸油纸	oil-absorbing sheets
化妆棉	cotton pads
棉签	cotton swab
卷发器	rollers/ perm rollers
美容仪器	beauty apparatus
皮肤测试仪	skin analysis apparatus
健胸仪	breast strengthening apparatus
减肥仪	weight reducing apparatus
扫斑仪	fleck removal apparatus
纹眉机	eyebrow-tattooing apparatus

Ⅲ. Famous Cosmetic Brands

AVON	雅芳
Avène	雅漾
Biotherm	碧欧泉
Borghese	贝佳斯
Chanel	香奈尔
Christian Dior(CD)	迪奥
Clarins	娇韵诗
Clinique	倩碧
DECLEOR	思妍丽
Estée Lauder	雅诗兰黛
Guerlain	娇兰
Helena Rubinstein	赫莲娜
H_2O^+	水芝澳
JUVENA	柔美娜
Kanebo	嘉娜宝
KOSE	高丝
Lancome	兰蔻
L'Oréal	欧莱雅
DeBON	蝶妆
Max Factor	蜜丝佛陀
MAYBELLINE	美宝莲
Nina Ricci	莲娜丽姿
Olay	玉兰油
Revlon	露华浓

Shiseido	资生堂
Sisley	希思黎
SK Ⅱ	SK Ⅱ
VICHY	薇姿
Yves Saint Laurent(YSL)	依夫·圣罗兰
ZA	姬芮
SunVana	姗拉娜
PRETTIEAN	雅姿丽
LA DEFONSE	黎得芳
ANELIN	颜婷
Elizabeth Arden	伊丽莎白·雅顿
Givenchy	纪梵希
Evian	依云

参考文献

References

[1] 杨顶权.中医美容之路——传承创新,整合发展[J].皮肤科学通报,2017,34(6):629-630.

[2] 吴恒.中医美容史略[J].中医文献杂志,2016,34(3):61-65.

[3] 王宏瑾.皮肤组织基本结构在皮肤美容护理中的作用[J].齐鲁护理杂志,2012,18(32):102-104.

[4] 诗锦.2018年,成熟的中药化妆品成分[J].中国化妆品,2018(Z1):88-91.

[5] 王爱娟,寇然,陈颖,等.中医药美容的现状及对策[J].皮肤科学通报,2017,34(6):631-635.

[6] 李金金,罗长浩,袁秀平.中草药在化妆品行业中的研究现状及发展策略[J].杨凌职业技术学院学报,2018,17(3):8-10.

[7] 张红梅.皮肤激光美容治疗中应用综合性护理干预的效果[J].中外女性健康研究,2018(22):124-125.

[8] 金瑶,袁辉春.射频用于面部美容和痤疮治疗的可行性研究[J].中国实用医药,2016,11(2):105-106.

[9] 孔诗曼琦.中国当代女性美甲艺术研究[D].上海:东华大学,2016.

[10] 许秀枝.探析基础皮肤护理与角质层保护的问题[J].世界最新医学信息文摘,2018,18(53):200-202.